VOLUME ONE HUNDRED AND THIRTEEN

VITAMINS AND HORMONES

Vasopressin

VOLUME ONE HUNDRED AND THIRTEEN

VITAMINS AND HORMONES

Vasopressin

Series Editor

GERALD LITWACK, PhD
Toluca Lake, North Hollywood, California

ELSEVIER

ACADEMIC PRESS
An imprint of Elsevier

Cover photo credit:
The illustration on the book cover is taken from two figures in the book. On the left is a ribbon structure in a 3D model of the vasopressin receptor from Fig. 4A of chapter 8 entitled "Vasopressin actions in the kidney renin angiotensin system and its role in hypertension and renal disease" by A.A. Gonzalez, N. Salinas-Parra, F. Cifuentes-Araneda, and C. Reyes-Martinez. On the right is a surface representation of neurophysin II (blue) with bound vasopressin (green) from Fig. 3B of chapter 3 entitled "Amyloid-like aggregation of provasopressin" by M. Spiess, N. Beuret, C.P. Baschong, and J. Rutishauser.

Academic Press is an imprint of Elsevier
50 Hampshire Street, 5th Floor, Cambridge, MA 02139, United States
525 B Street, Suite 1650, San Diego, CA 92101, United States
The Boulevard, Langford Lane, Kidlington, Oxford OX5 1GB, United Kingdom
125 London Wall, London, EC2Y 5AS, United Kingdom

First edition 2020

ISBN: 978-0-12-820816-8
ISSN: 0083-6729

For information on all Academic Press publications
visit our website at https://www.elsevier.com/books-and-journals

Publisher: Zoe Kruze
Acquisitions Editor: Ashlie M. Jackman
Editorial Project Manager: Joanna Collett
Production Project Manager: Denny Mansingh
Cover Designer: Victoria Pearson

Typeset by SPi Global, India

Contents

Contributors

Daniel F. Alonso
Laboratory of Molecular Oncology, Science and Technology Department, National University of Quilmes, Buenos Aires, Argentina

Nicole Beuret
Biozentrum, University of Basel, Basel, Switzerland

Jonathan S. Bogan
Department of Internal Medicine, Section of Endocrinology and Metabolism; Department of Cell Biology, Yale University School of Medicine, New Haven, CT, United States

Flavia Cifuentes-Araneda
Instituto de Química Pontificia Universidad Católica de Valparaíso, Valparaíso, Chile

Benjamin L. Clarke
Department of Biomedical Sciences, School of Medicine, University of Minnesota-Duluth, Duluth, MN, United States

Darron A. Cullen
Molecular Developmental Physiology and Signal Transduction Group, Zoological Institute, KU Leuven, Leuven, Belgium

Leopold Duerrauer
Center for Physiology and Pharmacology, Medical University of Vienna, Vienna, Austria

J. Paul Elliott
Colorado Brain and Spine Institute, Denver, CO, United States

Juan Garona
Laboratory of Molecular Oncology, Science and Technology Department, National University of Quilmes, Buenos Aires, Argentina

D.V. Golosova
Sechenov Institute of Evolutionary Physiology and Biochemistry of the Russian Academy of Sciences, St. Petersburg, Russia

Alexis A. Gonzalez
Instituto de Química Pontificia Universidad Católica de Valparaíso, Valparaíso, Chile

Christian W. Gruber
Center for Physiology and Pharmacology, Medical University of Vienna, Vienna, Austria

Estifanos N. Habtemichael
Department of Internal Medicine, Section of Endocrinology and Metabolism, Yale University School of Medicine, New Haven, CT, United States

Taroh Iiri
Department of Endocrinology and Nephrology, The University of Tokyo, Tokyo; Department of Pharmacology, St. Marianna University School of Medicine, Kawasaki, Japan

Susan M. Jones
Swedish Medical Center, Englewood, CO, United States

Alexander Jurkevich
Molecular Cytology Research Core Facility, University of Missouri, Columbia, MO, United States

Seong W. Kang
Poultry Science Center, University of Arkansas, Fayetteville, AR, United States

Wayne J. Kuenzel
Poultry Science Center, University of Arkansas, Fayetteville, AR, United States

Chunling Li
Department of Pathophysiology, Zhongshan School of Medicine, Sun Yat-sen University, Guangzhou, China

Don T. Li
Department of Internal Medicine, Section of Endocrinology and Metabolism; Department of Cell Biology, Yale University School of Medicine, New Haven, CT, United States

Suchun Li
Department of Pathophysiology, Zhongshan School of Medicine, Sun Yat-sen University, Guangzhou, China

Noriko Makita
Department of Endocrinology and Nephrology, The University of Tokyo, Tokyo, Japan

Katsunori Manaka
Department of Endocrinology and Nephrology, The University of Tokyo, Tokyo, Japan

Sally A. Mayasich
Department of Biomedical Sciences, School of Medicine, University of Minnesota-Duluth, Duluth, MN, United States

Emilie Monjon
Molecular Developmental Physiology and Signal Transduction Group, Zoological Institute, KU Leuven, Leuven, Belgium

Edin Muratspahić
Center for Physiology and Pharmacology, Medical University of Vienna, Vienna, Austria

Yu.V. Natochin
Sechenov Institute of Evolutionary Physiology and Biochemistry of the Russian Academy of Sciences, St. Petersburg, Russia

Marina Pifano
Laboratory of Molecular Oncology, Science and Technology Department, National University of Quilmes, Buenos Aires, Argentina

Cristina Prescianotto Baschong
Biozentrum, University of Basel, Basel, Switzerland

Cristian Reyes-Martinez
Instituto de Química Pontificia Universidad Católica de Valparaíso, Valparaíso, Chile

Giselle Ripoll
Laboratory of Molecular Oncology, Science and Technology Department, National University of Quilmes, Buenos Aires, Argentina

Stephen M. Rogers
Department of Zoology, University of Cambridge, Cambridge, United Kingdom

Jonas Rutishauser
Biozentrum, University of Basel, Basel, Switzerland

Nicolas Salinas-Parra
Instituto de Química Pontificia Universidad Católica de Valparaíso, Valparaíso, Chile

Junichiro Sato
Department of Endocrinology and Nephrology, The University of Tokyo, Tokyo, Japan

Martin Spiess
Biozentrum, University of Basel, Basel, Switzerland

Jozef Vanden Broeck
Molecular Developmental Physiology and Signal Transduction Group, Zoological Institute, KU Leuven, Leuven, Belgium

Weidong Wang
Department of Pathophysiology, Zhongshan School of Medicine; Department of Nephrology, The Seventh Affiliated Hospital, Sun Yat-sen University, Shenzhen, China

Emil Zeynalov
Swedish Medical Center, Englewood, CO, United States

Preface

Vasopressin and oxytocin are two octapeptide hormones released from the posterior pituitary. They are identical in structure with the exception of two amino acid residues that allow specific binding of each one to its cognate receptor resulting in completely different actions. A central activity of vasopressin is involved in water regulation. In this case, its action in the distal kidney is to induce the reabsorption of water into the bloodstream. This occurs in the following scenario. Arginine vasopressin (AVP or antidiuretic hormone, ADH) is stored in vasopressinergic nerve endings in the posterior pituitary. In response to signaling of osmoregulators or a decrease in blood volume, AVP is released from this site. AVP also can be released by nicotine (smoking). The gene encoding mRNA for AVP contains responsive elements at its N-terminus: osmolar response element, glucocorticoid response element, estrogen response element, and *fos/jun* response element. The polypeptide precursor liberates AVP, neurophysin II, and copeptin. Neurophysin II is a carrier protein that is complexed with AVP through its transport in the bloodstream presumably to protect AVP from inactivation. At some point before associating with a distal kidney cell, AVP dissociates from the AVP–neurophysin II complex. The liberated AVP binds to its cognate receptor in the distal kidney cell membrane, and the activated receptor complex activates cellular protein kinase C. Protein kinase C then phosphorylates aquaporin 2 subunits, four of which form a water channel in the apical membrane. The increase in aquaporin channels in the apical membrane facilitates the uptake of water that is transported across the cytoplasm in a vesicle from the apical side to the basolateral side of the cell. There, water is expelled into the extracellular space through different aquaporin channels, fixed within the basolateral membrane, and eventually reaches the bloodstream to expand its water volume. This is one aspect of the activities of AVP. In this book, there are reviews of many aspects of this fascinating hormone.

In the opening chapter, S.A. Mayasich and B.L. Clarke report on "Vasotocin and the origins of the vasopressin/oxytocin receptor gene family." "Oxytocin/vasopressin-like neuropeptide signaling in insects" is recounted by E. Muratspahić, E. Monjon, L. Dürrauer, S.M. Rogers, D.A. Cullen, J. Vanden Broeck, and C.W. Gruber. M. Spiess, N. Beuret, C.P. Baschong, and J. Rutishauser describe "Amyloid-like aggregation of provasopressin."

Next, N. Makita, K. Manaka, and T. Iiri review "V2 vasopressin receptor mutations." Opposite to its activity, vasopressin can be inactivated as related in a chapter by D.T. Li, E.N. Habtemichael, and T.S. Bogan entitled "Vasopressin inactivation: Role of insulin-regulated aminopeptidase." Related to AVP action on the distal kidney, S. Li, C. Li, and W. Wang discuss "Molecular aspects of aquaporins." Following this, a review of "The vasotocinergic system and its role in the regulation of stress in birds" is authored by W.J. Kuenzel, S.W. Kang, and A. Jurkevich. In the relationship of vasopressin to kidney disease, A.A. Gonzalez, N. Salinas-Parra, F. Cifuentes-Araneda, and C. Reyes-Martinez offer "Vasopressin actions in the kidney renin angiotensin system and its role in hypertension and renal disease." "Vasopressin receptor subtypes and renal sodium transport" is next owing to the contributions of Y.V. Natochin and D.V. Golosova. Relative to clinical problems, J. Garona, M. Pifano, G. Ripoll, and D.F. Alonso describe "Development and therapeutic potential of vasopressin synthetic analog [V^4Q^5]dDAVP as a novel anticancer agent." Following in this vein, a chapter on "Vasopressin and vasopressin receptors in brain edema" by E. Zeynalov, S.M. Jones, and J.P. Elliott completes the presentation.

The illustration on the book cover is taken from two figures in the book. On the left is a ribbon structure in a 3D model of the vasopressin receptor from Fig. 4A of chapter 8 entitled "Vasopressin actions in the kidney renin angiotensin system and its role in hypertension and renal disease" by A.A. Gonzalez, N. Salinas-Parra, F. Cifuentes-Araneda, and C. Reyes-Martinez. On the right is a surface representation of neurophysin II (blue) with bound vasopressin (green) from Fig. 3B of chapter 3 entitled "Amyloid-like aggregation of provasopressin" by M. Spiess, N. Beuret, C.P. Baschong, and J. Rutishauser.

I am grateful for the efforts of Joanna Collett of Elsevier (Oxford, UK) and of the production group in Chennai, India.

Gerald Litwack
North Hollywood, California, United States
September 16, 2019

CHAPTER ONE

Vasotocin and the origins of the vasopressin/oxytocin receptor gene family

Sally A. Mayasich*,†, Benjamin L. Clarke
Department of Biomedical Sciences, School of Medicine, University of Minnesota-Duluth, Duluth, MN, United States
*Corresponding author: e-mail address: maya0010@d.umn.edu

Contents

Abstract

We previously characterized the arginine vasotocin receptor sequences in the jawless vertebrate sea lamprey. These gene and protein sequences provide clues to the origins of the various arginine vasopressin and oxytocin receptor family members in jawed vertebrates. However, orthological relationships between the jawless and jawed receptors

† Current address: US Environmental Protection Agency Mid-Continent Ecology Division, 6201 Congdon Blvd, Duluth, MN 55804, United States.

Vitamins and Hormones, Volume 113
ISSN 0083-6729
https://doi.org/10.1016/bs.vh.2019.08.018

is unclear. The current work is a closer examination and comparison between these G protein-coupled receptor sequences of the lamprey, the early jawed vertebrate elephant shark, and the boned fish and tetrapods. Our objective was to gain more insight into the differentiation of key signaling domains, which may then aid in discerning the pattern and timing of whole genome duplication early in the vertebrate lineage. The lamprey receptors remain less differentiated than shark receptors, due in part to the single vasotocin ligand in the lamprey and the selection pressure of a second ligand, oxytocin, in the shark. However, variation in G proteins utilized among the V1A, V1B and oxytocin receptor types has also contributed to differentiation, as well as leading to a change in second-messenger signaling pathway in the V2-type receptors. Conservation of gene regulatory elements may provide additional evidence of receptor gene orthology. These molecular evolution studies can ultimately be informative in applications such as drug discovery and environmental toxicology to determine cross-species sensitivity to chemicals.

1. Introduction

The major physiological roles of the arginine vasopressin (AVP)/oxytocin (OXT) hormone-receptor system in osmoregulation, metabolism, stress response, reproductive function and cardiovascular responses as well as behavior including courtship, bonding and vocalization, have been reviewed extensively in mammals. Compared with mammals, other jawed (gnathostome) vertebrates including fish (Balment, Lu, Weybourne, & Warne, 2006), birds (Kuenzel, Kang, & Jurkevich, 2013), amphibians and reptiles (Wilczynski, Quispe, Muñoz, & Penna, 2017) exhibit basic similarities of these same functions, although utilizing slightly different ligands (e.g., arginine vasotocin (AVT) rather than AVP). The similarity of function and tissue expression patterns of these G protein-coupled receptors (GPCRs) are the logical consequence of conserved protein and DNA sequences. These include conserved protein domains, or structural/functional elements that have remained relatively unchanged within the protein sequences during evolution, and conserved DNA motifs in gene regulatory regions, across vertebrate species.

During evolution, gene duplication events produced paralogs, or a set of genes derived from duplication within the genome of a species. The paralogs within these "gene families" generally evolve new, potentially related functions. When two species diverge, a gene derived from the same ancestral gene that retains the same function in both species is called an ortholog. Crucial to understanding the expansion of functional diversity through the creation of gene families are theories on the number and timing of whole-genome duplication (WGD) events in ancestral vertebrates, and other scenarios involving chromosome-scale duplication events (Onimaru & Kuraku, 2018; Smith & Keinath, 2015; Smith et al., 2018). Because at least one round (1R) of WGD is generally believed to have occurred shortly before the ancestral jawless (agnathan) vertebrates diverged from the vertebrate lineage, extant agnathans such as the sea lamprey have been the focus of efforts to trace the origins of orthologous and paralogous relationships within gene families. Studying the AVP/AVT/OXT receptor (AVPR/AVTR/OXTR) family (Mayasich & Clarke, 2016) and other gene families (Ellens, Kittilson, Hall, Sower, & Sheridan, 2013; Haitina et al., 2007; Sower, Decatur, Joseph, & Freamat, 2012) in lampreys has been providing a new understanding of how ancestral genes may undergo sub- or neo-functionalization after gene duplication. In turn, deciphering the patterns of paralogy and orthology through the determination of specific evolutionary amino acid changes has current applications in drug discovery (Pifano et al., 2018) and environmental toxicology (LaLone et al., 2016).

As many as six AVPR/AVTR/OXTR paralogs are found in different classes of jawed vertebrates, including two V1 types (V1AR and V1BR), the OXTR, and three V2 types (V2AR, V2BR and V2CR) (Lagman et al., 2013; Mayasich & Clarke, 2016). The phylogeny of the receptor family is illustrated in Fig. 1. Lamprey genomes have been sequenced and constructed to the level of contiguous lengths of DNA called scaffolds (Mehta et al., 2013; Smith et al., 2013) or longer, near-chromosome level sequences (Smith & Keinath, 2015; Smith et al., 2018). Our analyses of neighboring genes (gene synteny) along the lengths of these genome constructs show that two AVTR types are present on each of two different scaffolds in lampreys (Mayasich & Clarke, 2016). This points to a tandem duplication of a single receptor, creating a V1-type and a V2-type on the same chromosome prior to 1R WGD. The 1R WGD then provided a 2–4 expansion in agnathans, and fodder for diversification of function. A fifth receptor found in sea (*Petromyzon marinus*) and arctic or Japanese (*Lethenteron camtschaticum*: syn. *Lethenteron japonicum*) lampreys may be a local, lamprey-specific duplication based on synteny and phylogeny results (Mayasich & Clarke, 2016). However, this is speculative owing to the incomplete state of genome construction. In many birds, amphibians, reptiles and fish studied, a V1AR gene shares a chromosome with the V2CR type. In fish, another chromosome carries both an OXT-type (isotocin) receptor and a V2BR type. Birds have no V2AR or V2BR-type; amphibians and reptiles have no V2BR; fish have

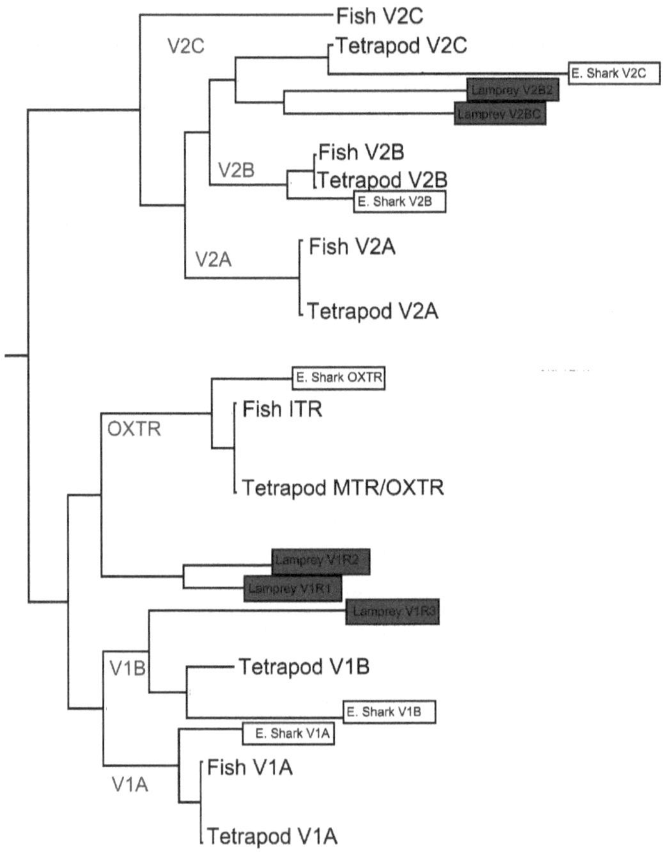

Fig. 1 Phylogenetic tree depicting the vasopressin/vasotocin receptor (V1AR and V1BR) and oxytocin/mesotocin/isotocin receptor (OXTR/MTR/ITR) family of G protein-coupled receptors. The tree is simplified by grouping the jawed vertebrate taxa into bony fish and tetrapods (including mammals, birds, amphibians and reptiles) to highlight clustering of the jawless sea lamprey (*Petromyzon marinus*) vasotocin receptors and jawed cartilaginous elephant shark (*Callorhinchus milii*) oxytocin and vasotocin receptors. The method and basis for tree topology is described in Mayasich and Clarke (2016).

no V1BR. The AVPR/OXTR gene family in humans and other mammals is comprised of four-receptor genes on four separate chromosomes: AVP receptor genes AVPR1A, AVPR1B and AVPR2A; and the oxytocin receptor OXTR.

The existence of six receptor-gene types and their chromosomal configurations in gnathostomes has been postulated to reflect a second round (2R) of WGD. However, the V1BR gene and V2AR genes, in all species studied thus far that have retained these types, exist singly on separate chromosomes.

Because these chromosomes contain neighboring genes with shared orthology (Lagman et al., 2013), the fact that no seventh or eighth receptor type has been found could reflect gene losses. The most parsimonious explanation (i.e., requiring the fewest evolutionary changes) of the existing orthology groups, however, is based on the proposal of Smith et al. (2018). They propose that one round of WGD took place in the ancestral lineage before the divergence of the jawless from the jawed vertebrate lineage along with potential smaller scale duplications and translocations, sustaining fewer subsequent gene losses.

The sea lamprey genes were sequenced and deposited in the GenBank database, and for lack of clear orthology with gnathostome receptors, the lamprey receptors have been termed V1R1 (Accession no. KC731437), V1R2 (KJ813004), V1R3 (KJ813003), V2BR2 (KF031008) and V2BC-like (KM015305) (Mayasich & Clarke, 2016). These genes are present within the first, somatic sea lamprey genome build (Smith et al., 2013). The arctic lamprey genome has been sequenced from reproductive (testis) tissue (Mehta et al., 2013). Highly similar orthologs are present in the arctic lamprey genome based on basic local alignment search tool (BLAST) queries (Mayasich & Clarke, 2016). It is important to note that the lamprey goes through a process known as programmed gene rearrangement (PGR) during development in which approximately 20% of its genome is eliminated by 3 days post fertilization (Smith, Antonacci, Eichler, & Amemiya, 2009; Smith et al., 2018). However, all of the AVTR genes and no additional AVTR-family genes were present in a search of the most recent, germline sea lamprey genome (Smith et al., 2018; https://genomes.stowers.org/organism/Petromyzon/marinus). This indicates that there are no additional genes in this family in lampreys that were unaccounted for due to PGR, and these genes likely function in the lamprey throughout its life, as is true for jawed vertebrates.

In this chapter we pose theories on three key elements in the evolution of the AVPR/AVTR/OXTR family based on the state of these genes in the agnathan sea lamprey compared with those of gnathostomes: (1) Emergence of the second ligand after lamprey divergence may have driven amino acid changes in the ligand binding domains. The pattern of differentiation may help identify orthological and paralogical relationships as well as potential for different responses to agonists and antagonists. (2) The V2A receptor may have evolved cAMP signaling capability and new functionality through amino acid changes in the intracellular domains. Tracing these changes may reveal the history of genetic duplication. (3) Conservation of specific

DNA motifs that function as gene regulatory elements may contribute to conserved protein function. Identification of these elements may therefore provide evidence of orthology.

2. Emergence of the second ligand and derivation of the oxytocin receptor

2.1 Neuropeptide hormones

There are several reviews that discuss the origins of mammalian nine-amino-acid neuropeptide hormones AVP and OXT (Donaldson & Young, 2008; Goodson, 2005; Goodson & Kabelik, 2009). The AVP hormone gene ortholog in all non-mammalian vertebrates is AVT; the OXT hormone gene ortholog in fish is isotocin (IT), and that in birds, reptiles and amphibians is called mesotocin (MT) (Stafflinger et al., 2008) (Table 1).

Gwee, Tay, Brenner, and Venkatesh (2009) provide evidence that the AVP- and OXT-family of neurohypophysial hormones evolved in a common ancestor of jawed vertebrates through tandem duplication of the ancestral AVT gene, as occurred to produce two-receptor genes on the same chromosome. They sequenced the region of this tandem duplication in both the jawed vertebrate elephant shark *Callorhinchus milii*, where both AVT and OXT genes were found, and the Japanese lamprey, where only the AVT gene was present. In our previous work (Mayasich & Clarke, 2016) we searched both the Japanese and sea lamprey (*P. marinus*) genomes and found no evidence of a paralog of the AVT hormone gene. These results support the contention that the duplication took place post-WGD, and after the jawless lampreys diverged from the vertebrate lineage.

Table 1 Arginine vasopressin- and oxytocin-like hormone amino acid sequences in vertebrate animals.

Arginine vasopressin (AVP)	CYFQNCPRG	Mammals
Lys-vasopressin	CYFQNCPKG	Pig, some marsupials
Phenypressin	CFFQNCPRG	Some marsupials
Arginine vasotocin (AVT)	CYIQNCPRG	Non-mammalian vertebrates
Oxytocin (OXT)	CYIQNCPLG	Mammals
Isotocin (IT)	CYISNCPIG	Fish
Mesotocin (MT)	CYIQNCPIG	Birds, reptiles, amphibians

The AVT nonapeptide differs by only one amino acid from mammalian AVP with isoleucine instead of phenylalanine in the third position and differs from mammalian OXT only in the eighth position with an arginine instead of leucine (Table 1). This confers a combination position between the AVP and OXT hormones. The AVT hormone in the lamprey is therefore postulated to bind with all of its AVTR-family proteins, which are produced by gene paralogs that were likely minimally differentiated within the relatively short time period since the genomic duplication events. The additional OXT-type ligand would drive the fixation of specific sequence differences between receptor paralogs in the dual-ligand species, especially selecting those ligand-interacting amino acids having greater affinity for the second (OXT/IT/MT) ligand. The amino acids with AVT affinity would remain in the other receptor paralogs. Although some receptor mutations have likely occurred to enhance OXT-type ligand affinity, analysis of the lamprey AVTR sequences suggests that many amino acids specific to the OXTR type in gnathostomes already existed in certain lamprey AVTRs. To illustrate this concept, we used phylogeny, sequence identity and similarity alignments, and comparisons of amino acids in the specific regions of known ligand interaction.

2.2 Orthology and paralogy of the AVP/OXT receptor family

As shown in the phylogenetic tree (Fig. 1), the lamprey V1R1 and V1R2 cluster most closely with the OXTR type in jawed vertebrates, and the lamprey V1R2 clusters with the V1BR. In both cases, however, these branches are not well supported by the statistical (bootstrap) analysis (see Mayasich & Clarke, 2016). We have here conducted a new analysis using bit scores from BLAST searches of the GenBank database to gain more insight into possible orthology. Each of the three lamprey V1-type receptors, the elephant shark (*C. milii*) V1AR, V1BR and OXTR, and the human OXTR protein sequences were used to individually query the entire non-redundant protein database using the default blastp (protein-protein BLAST) parameters. The bit score presents a constant statistical indicator of the alignment significance between the query sequence and each protein sequence found in the database, taking into account length, coverage, number of identical amino acids and number of similar amino acids. Similarity accounts for the probability that mutation commonly results in substitution between certain amino acids, e.g., leucine, isoleucine and valine. The highest bit score was selected for alignments between each of the query genes and V1AR, V1BR and

OXTR proteins of a representative of each of three taxonomic groupings (fish, mammals and either a bird, amphibian or reptile). The query genes were also compared to V2-type receptors and gonadotropin releasing hormone receptors (GnRHR) within their own species as outgroups (more distantly-related proteins) to serve as references. Because bony fish have no V1BR, the coelacanth (*Latimeria chalumnae*) was often the highest-scoring fishlike V1BR. Bit scores vary by protein length, so were normalized by taking the percentage of the maximum possible bit score (i.e., 100% identity covering the entire length of the protein) for each query sequence. Results are presented in Fig. 2.

The lamprey V1-type receptors exhibit very little difference in percentage of maximum bit score among the V1AR, V1BR and OXTR protein sequences in jawed vertebrates. The highest score was 54% between the lamprey V1R2 and the bonytongue fish (*Scleropages formosus*) V1AR and the lowest score was 42% between the lamprey V1R3 and the garter snake (*Thamnophis sirtalis*) OXTR type (mesotocin receptor). For the jawed, cartilaginous elephant shark, however, each receptor exhibited distinct similarity to its ortholog in the other jawed species. Comparatively, the human OXTR was highly similar to the white rhinoceros with a score of 91% (primates were excluded from the analysis). The human OXTR also showed a higher alignment score with the spotted gar OXTR (isotocin receptor) ortholog (65%) than any of the lamprey receptor scores, all of which were in the same range as the scores for the shark and human query proteins when compared with their non-orthologs in other species. Scores for all query sequence alignments with V2-type receptors showed a distinct drop from the V1AR, V1BR and OXTR alignment scores. The scores dropped further for the GnRHR, a member of the same subfamily (A6) of rhodopsin-like GPCRs. These comparisons highlight the level of differentiation that has occurred among these receptors between the divergence of the agnathans and the cartilaginous jawed vertebrates, and on to the bony fish and tetrapods. In terms of amino acid sequence, the alignment scores are evidence that the lamprey receptors and the gnathostome V1AR, V1BR or OXTR types are paralogons from the V1-type receptor ancestor, but does not support assignment of orthology between individual lamprey and gnathostome receptor types. This being said, however, orthology by ancestral origin may exist but not show itself through overall sequence similarity. The key to ortholog and paralog differentiation as evolution progressed may be in finding vestigial signature amino acids among the conserved functional domains.

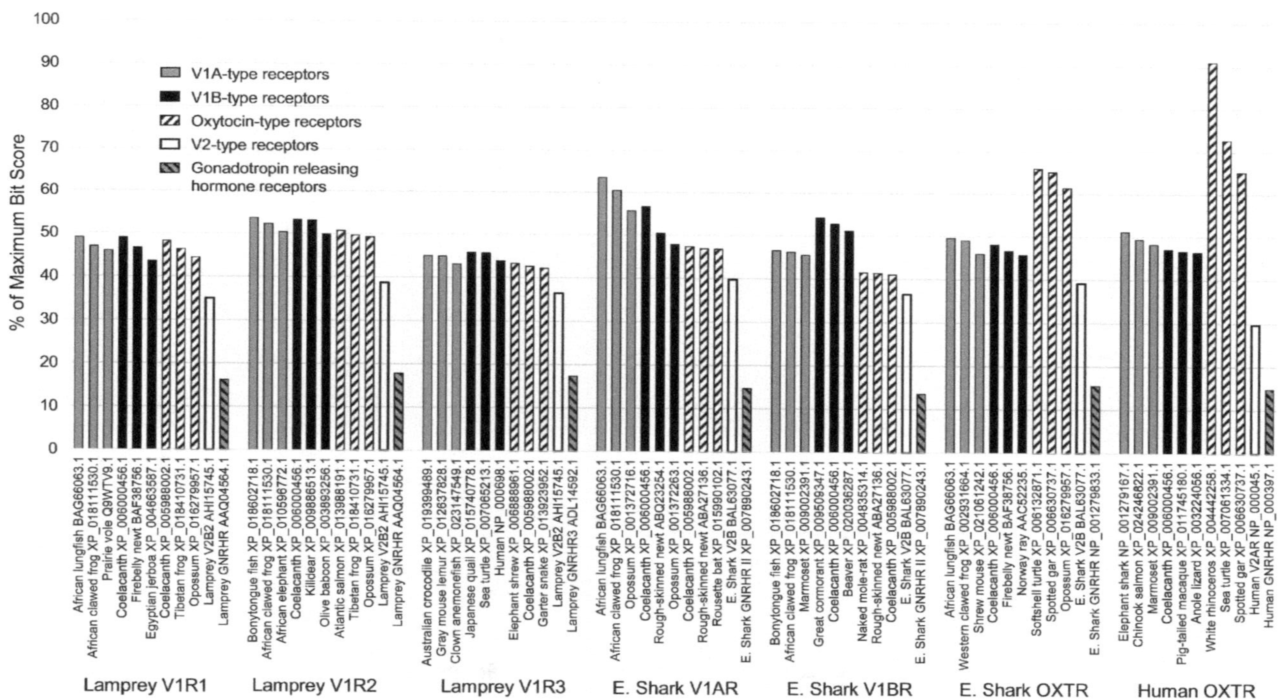

Fig. 2 Percent of maximum bit scores from basic local alignment search tool (BLAST) queries of GenBank, used to indicate protein sequence similarity. Each of the three lamprey V1-type receptors, the elephant shark (*Callorhinchus milii*) V1AR, V1BR and OXTR, and the human OXTR protein sequences were used to individually query the entire non-redundant protein database using the default blastp (protein-protein BLAST) parameters. The common name and accession number are shown for high-scoring representative species. The query genes were also compared to V2-type receptors and gonadotropin releasing hormone receptors (GNRHR) within their own species as outgroups.

2.3 Differentiation of receptor types

Because only the AVT peptide ligand is present in lampreys whereas both the OXT and AVT ligands are found in the sharks, we sought to determine those specific amino acids involved in differentiation. GPCRs can be divided into structural domain regions including the extracellular N-terminus, seven transmembrane (TM) domains, three extracellular loops (ECL), three intracellular loops (ICL), and the intracellular C-terminus. Among these domains, regions of known ligand or G protein interaction that have been established through experimentation (e.g., site-directed mutagenesis and X-ray crystallography) were determined from the literature. We compared these functional domain regions in the lamprey V1-type AVTR pre-differentiated protein sequences with the elephant shark and other gnathostome OXTR, V1AR and V1BR proteins. The human ortholog for each receptor was used in a protein BLAST (blastp) query of each taxonomic class individually for mammals, and the chicken ortholog was used to query birds, amphibians, reptiles and fish. Only six complete sequences were available for many of the amphibian and reptile receptor types. Therefore, the six highest bit scores were selected from each class, totaling 30 sequences for OXTR and V1AR, and 24 sequences for V1BR, which is lacking in fish. Sequence logos were then created for the functional domain regions in all receptors using WebLogo Version 2.8.2 (Crooks, Hon, Chandonia, & Brenner, 2004; https://weblogo.berkeley.edu/). The sequence comparisons are shown in Fig. 3.

Amino acids known to interact with the OXT, AVP or AVT ligands are noted based on studies by Koehbach et al. (2013), Acharjee, Do-Rego, Oh, Ahn, et al. (2004) and Chini and Fanelli (2000), and are numbered according to the human OXTR sequence (Accession no. NP_001341583.1). Fig. 3 shows that of the 32 known or putative ligand or G protein-interacting positions, 16 are 100% conserved for all three receptor types in gnathostomes as well as the elephant shark and the three lamprey receptors. The amino acids arginine (R), aspartate (D), valine (V), proline (P) and methionine (M) each appear at one conserved position, while the other 12 are dominated by phenylalanine (F), glutamine (Q), tryptophan (W) and lysine (K). Ten positions have substitutions of functionally similar amino acids, with consensus amino acids of threonine (T), isoleucine (I) or V at six positions, F or tyrosine (Y) at two, serine (S) at one, M or alanine (A) at one, and one position with no consensus (310). These 27 positions are therefore not considered to provide evidence of orthology. Five positions, however, have non-similar, low probability exchanges between receptor types, four of which involve F or Y in some receptors and S, T, V or I at others.

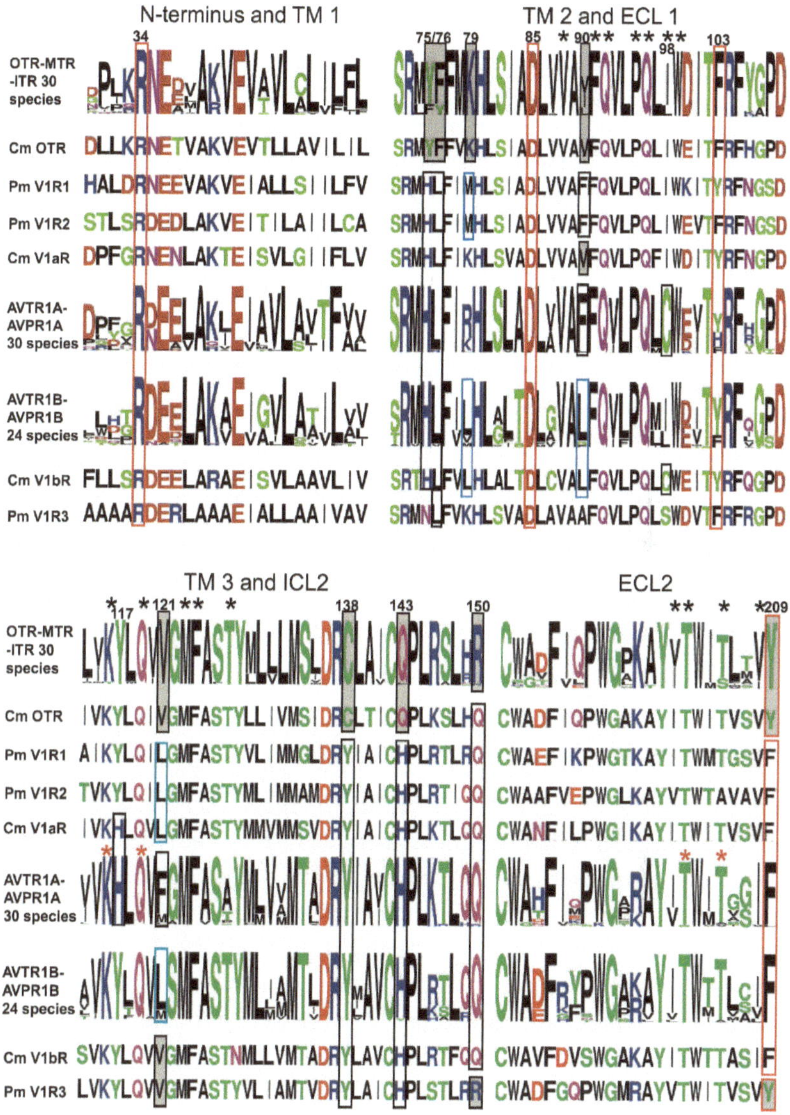

Fig. 3 Sequence logos (https://weblogo.berkeley.edu/logo.cgi) are used to show how differentiation may have occurred from the lamprey (Pm; *Petromyzon marinus*) vasotocin (AVT) receptors V1R1, V1R2, and V1R3 to the oxytocin (OXTR), V1AR, and V1BR types in the elephant shark (Cm; *Callorhinchus milii*) and other jawed vertebrates. Regions of importance are shown as determined by other researchers; putative OXT/AVP ligand binding amino acids in mammalian receptors are marked with black asterisks (Chini & Fanelli, 2000; Koehbach, Stockner, Bergmayr, Muttenthaler, & Gruber, 2013). Red asterisks note important residues for AVT binding (Acharjee, Do-Rego, Oh, Ahn, et al., 2004; Chini & Fanelli, 2000). Red boxes show six residues known to be important for oxytocin ligand selectivity, binding, and signaling (Koehbach et al., 2013). Shaded boxes show matches specific to the gnathostome OXTR type; clear black boxes show matches specific to V1AR, or to both V1AR and V1BR; and clear blue boxes show matches specific to V1BR.

(Continued)

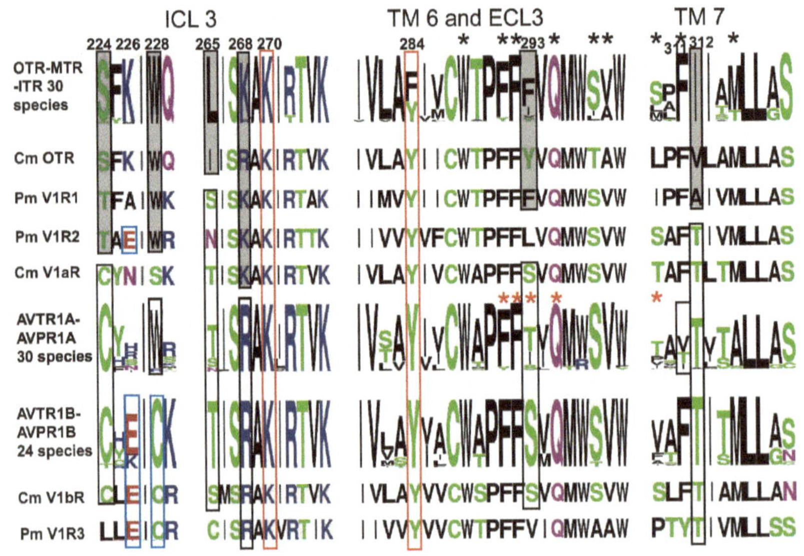

Fig. 3—Cont'd

In addition to these five known or putative ligand binding positions, we manually evaluated other positions within the functional domain regions to identify those that differ distinctly between receptor types. We found 15 positions of interest that have not previously been demonstrated to be involved in ligand binding. We judged exchangeability properties of the amino acids based on a statistical model that predicts probabilities of evolutionary acceptance of missense mutations (i.e., non-synonymous amino acid substitutions) in humans (Yampolsky & Stoltzfus, 2005). Because genomes can differ among species, these data were used as the basis for estimates of high, moderate or low probability of exchange. This method was supported by general knowledge of amino acid physio-chemical similarity and biological consequences of substitution (Betts & Russell, 2003). The 20 positions of interest are summarized and probability of amino acid exchanges and consequences are indicated in Table 2. Conserved amino acids that differed between the OXTR and the V1AR/V1BR had different acid/base, polarity or charge properties, indicating possible specificity in function and potentially revealing an ortholog "signature."

Table 2 Amino acid positions in ligand and G protein binding domains (see text and Fig. 3) manually evaluated and identified as having non-similar amino acids among receptor types, potentially indicating differentiation.

Domain	Position	Amino Acid: Gnath OXTR	Cm OXTR	PmV1R1	PmV1R2	CmV1AR	Gnath V1AR	Gnath V1BR	CmV1BR	PmV1R3	Exchangeability	Properties/consequence of exchange
TM2	75	YF	Y	H	H	H	H	H	H	N	H→Y or F	H has unique charge properties, not particularly exchangeable, often involved in protein interaction.
	76	FY	F	L	L	L	LR	LV	L	L	L→F, Y→L	F and Y aromatic side chains interact with proteins. L may render the position less reactive.
	79	K	K	M	M	K	RK	LVM	L	K	K▶RM, L▶K	Polar, positively charged K and R often involved in protein binding.
	90	VIL	V	F	F	V	FL	LGF	L	A	LI→F-->V, VIL◀▶VIL	I, L, V have non-reactive aliphatic side chains, rarely involved directly in function; I, L, V and M can play role in ligand recognition.
ECL1	98	IL	I	I	I	I	C	IL	C	S	I→C→L	C can form disulfide bonds, cellular locations are important.
	103	F	F	Y	F	Y	YHF	YF	Y	F	F↔Y	F103 is a known ligand binding position in the human OXTR.
TM3	117	Y	Y	Y	Y	H	H	Y	Y	Y	H→Y	Y uncharged; H unique charge properties.
	121	V	V	L	L	L	FM	LM	V	V	M▶L→F→V	V and L may function in ligand recognition; F may function in ligand binding.
ICL2	138	C	C	Y	Y	Y	Y	Y	Y	Y	C→Y	C specificity in OXTR may indicate need for sulfide moeity.
	143	QY	Q	H	H	H	H	HY	H	H	Q↔H	Q uncharged, amine side chain. H has unique charge properties.
	150	RK	Q	Q	Q	Q	QR	Q	Q	R	Q→R	R is a charged, polar, basic amino acid. Q, R and H often involved in protein interactions.
ECL2	209	Y	Y	F	F	F	FL	F	F	Y	F↔Y	Y is polar, F is nonpolar.
ICL3	224	SG	S	T	T	C	C	CT	C	L	T◀▶S→C	S and T are polar, uncharged; C is nonpolar, with sulfide moeity.
	226	K	K	A	E	N	HR SNI	EK	E	E	E→K▶HR	K is a charged, polar, basic amino acid; E is polar, acidic.
	228	W	W	W	W	S	WR	C	C	C	W x C x S	W is aromatic, unique chemistry, large size; exchange may impact protein function.
	265	L	I	S	N	T	TSN	T	S	C	T<-->L	T/S uncharged, polar; L/I nonpolar.
	268	KR	R	K	K	K	R	R	R	R	K▶R	K to R exchange more likely; R highly conserved in gnathostome V1AR/V1BR.
TM6	293	FIVY	Y	F	L	S	TIS	S	S	V	YF<-->T, S→YF	F, I, V nonpolar; S, T, Y uncharged, polar, hydroxyl side chain.
TM7	311	F	F	F	F	F	VTI	F	F	Y	F<-->VT	F conserved except gnathostome V1AR.
	312	I	V	A	T	T	TA	T	T	T	I<-->T↔V	T is polar, uncharged, reactive. I, V nonreactive, ligand recognition. A is small, nonpolar, non-critical for protein interaction.

Font size indicates relative prevalence of an amino acid among species at that position. Shaded positions are known to be involved in ligand binding. Most likely direction of exchange: bold arrows indicate high, narrow arrows moderate, and dashed arrows low exchangeability based on a probability model by Yampolsky and Stoltzfus (2005). Amino acid properties and consequences of exchange are based on information from Betts and Russell (2003).

Our results were similar to those of Indrischek, Prohaska, Gurevich, Gurevich, and Stadler (2017) in a study of arrestins. They found that evolutionary substitutions differentiating function between proteins occurred in positions neighboring the binding site amino acids more often than at the binding residues themselves. The elephant shark results, as in the bit score analysis, showed the receptors' similarities to their putative orthologs. Nineteen of the 20 potential differentiated positions in the shark OXTR, for example, match the amino acids in the OXTR-type in other gnathostomes (Fig. 3; Table 2).

Differentiation resulting in distinction and functional specificity between the OXTR-type sequences and the V1AR and V1BR types is understandable due to the selective pressure of the second ligand. Notably in the human OXTR, Y209 in ECL 2 and F284 in TM6 together interact with residues at positions 2 (Y) and 3 (I) of the OXT hormone (Koehbach et al., 2013). Y and I also occupy positions 2 and 3 in mesotocin and isotocin as well as vasotocin, while vasopressin differs with F in the second position (Table 1). F and Y are exchanged at 209 and 284 in other receptors (Fig. 3, Table 2), which may result in differences in ligand interaction, and potentially conformation of the receptor after ligand activation. Acharjee, Do-Rego, Oh, Ahn, et al. (2004) found that F291Y, I293T and P310T (Fig. 3; Table 2) exchanges in the rat V1AR, and the reverse in the bullfrog, altered selectivity between AVT and AVP ligands. In their further work, Acharjee, Do-Rego, Oh, Moon, et al. (2004) sequenced and characterized an amphibian V1AR and a mesotocin receptor (MTR) and found that Y in the AVTR replaces F in the MTR at position 293, which they propose accounts for a difference in antagonist sensitivity between the two receptors. This may be due to how the antagonist changes the receptor conformation, which in turn results in a difference in the interface presented to the G protein (Acharjee, Do-Rego, Oh, Moon, et al., 2004).

The ICL2 region is known to help provide an interface for G protein coupling (Flock et al., 2017), specifically the Gq/11 protein in V1-type receptors (Gimpl & Fahrenholz, 2001). The Gq/11 protein signaling cascade ultimately results in the release of intracellular calcium. One of the major regions of difference between these receptors is in ICL3, which is of known importance for receptor binding to the stimulatory G protein (Gs; resulting in cAMP release). Differences exist among all receptor types at ICL3 positions 224, 226, 228, 265, 268 and 270 (Fig. 3; Table 2). Position 228 exhibits significant differences in amino acids among receptors, with W in the lamprey V1R1 and V1R2, and gnathostome V1AR and OXTR types; but C in the gnathostome V1BR. W is aromatic with unique chemistry and size, and replacement often has dire consequences for protein function. W is therefore usually highly conserved, and the change may indicate functional differentiation. K270 is a known determinant of Gq/11 protein binding in the human OXTR (Yang et al., 2002), and is fully conserved across all species (Fig. 3). However, although the G protein that normally binds with all three of these receptor types is the Gq/11, only ICL2 containing the DRY motif is required for V1-type receptor binding, while a more complex set of residues including some in the C-terminus are required for OXTR-Gq/11 coupling. It is also known that ICL3 in OXTR is involved in binding to the inhibitory G protein (Gi) in the mammalian myometrium during labor (Gimpl & Fahrenholz, 2001; Strakova & Soloff, 1997). Therefore, specific and less frequent, but not trivial, functions require the ability to bind to more than one G protein. Additionally, the V1BR is also known to activate the Gs protein under certain circumstances (Orcel et al., 2009; Thibonnier et al., 1997). Thibonnier et al. (1997) showed that the human V1b (V3) receptor in the pituitary could utilize either the Gq/11 calcium or the Gs cyclic AMP pathway, depending on the density of receptor expression in the cell. Orcel et al. (2009) found that the specific receptor sequences may contribute to aspects beyond affinity for ligand binding, to the potential for homodimerization. The conservation of some residues in V1BR ICL3 with those in the V2AR is part of a discussion below on the evolution of cAMP signaling and antidiuresis.

In summary, of the 20 positions of interest, the sea lamprey V1R1 receptor has 9 positions with identical or similar residues to the OXTR-types in gnathostomes and 16 overall matches with V1AR-V1BR types (some positions overlap among types). Of four specific matches to the OXTR type (224, 268, 293 and 312) two are in ICL3, with potential for functional difference from V1-type receptors at 224. W228 is also a V1R1-OXTR match with potential functional significance (Fig. 3; Table 2). This may have been enough to tip the balance for the lamprey V1R1 to cluster marginally with gnathostome OXTR-type in our phylogenetic analysis (Fig. 1; Mayasich & Clarke, 2016). The lamprey V1R3 had fewer matches to the gnathostome receptors, but to cluster with the V1BR-type, its two specific matches are to gnathostome V1BR in ICL3 (226 and 228) (Fig. 3, Table 2). These could be "signature" amino acids of early differentiation between the AVTR paralogs that ultimately resulted in neofunctionalization and clues to orthology. With the known potential for the V1BR to bind the Gs protein stimulating cAMP signaling, we further explored changes in the ICL3 domain of the V2-type receptors for their significance in Gs binding.

3. Evolution of the V2A receptor and cAMP signaling

3.1 cAMP and aquaporin expression

A well-studied canonical function of vasopressin in mammals, as discussed in other chapters of this volume, is antidiuresis. This involves both the regulation of aquaporin-2 (AQP2) protein expression and the stimulation of short-term insertion of AQP2-containing vesicles into the apical membrane of renal collecting ducts in the kidney. Based on the work of Kim et al. (2000) regulation of AQP2 is dependent on cAMP. Ocampo Daza, Lewicka, and Larhammar (2012) and Yamaguchi et al. (2012) identified the V2BR group that is structurally V2-like, but found by Yamaguchi et al. (2012) to utilize the calcium second messenger system rather than cAMP. As noted above, cAMP signaling has been found to occur with V1BRs under certain conditions (Orcel et al., 2009; Thibonnier et al., 1997), but was not found to be inducible in V1AR or OXTR types (Terrillon et al., 2003).

Wells, Anderson, and Hazon (2002) demonstrated the antidiuretic function of 10^{-9} M AVT in the cartilaginous dogfish *Scyliorhinus canicula* in vivo and an in situ kidney preparation. Glomerular flow was reduced; however, there was no effect on tubular (collecting duct) diuresis. Birds are the only vertebrates other than mammals that concentrate urine, and aquaporins are involved in water flow in the avian renal collecting ducts. However, Nishimura and Yang (2013) report that the regulation of aquaporins in birds may be through a mechanism other than the AVT hormone-receptor system. Birds apparently lost the V2AR gene, retaining a V2CR as found in the chicken (*Gallus gallus*). Therefore, there is no evidence, to our knowledge, that AVT mediates antidiuresis through the regulation of aquaporins in the absence of a V2AR. Our previous studies (Mayasich & Clarke, 2016) show low or no expression of the sea lamprey V1R3 and V2BR2 and moderate expression of the other three-receptor genes in the adult lamprey kidneys. Yamaguchi et al. (2012) reported similar results in the elephant shark, indicating a minor role for AVT in renal function in these animals.

3.2 Receptor structure for Gs coupling

The evolutionary question of what mutational steps were responsible for driving V2AR genes to evolve cAMP signaling capability may be found in determining what structural components of the V2AR are responsible for Gs protein coupling. Studies in mammals have been focused on conformational changes triggered by ligand binding and ICL/carboxy-terminal tail structures. Rahmeh et al. (2012) showed with selective V2AR agonists that a conformational change and stabilization in the TM6-ICL3 domain is necessary for Gs protein activation. Erlenbach and Wess (1998) replaced the V1AR segment at the TM5-ICL3 junction with the collocated V2AR segment consisting of the amino acid sequence QVLIFREIHASL and shortening the "bottom" of the loop by 10 amino acids. The resulting V1AR-V2AR hybrid had more than 95% of the native V2AR cAMP activity.

We compared amino acid alignments of the ICL3 and carboxy-terminal domain segments among all of the AVPR-OXTR types to find whether these motifs were conserved in V2AR across species, and how they differed from the V2BR and V2CR types as well as the other receptor types. We found that the QVLIFREIHASL motif is generally conserved in the V2AR type, except that the V may be an I or F, the first L may be an R, the R may be a K (FREIH or FKEIH), and the last three letters vary among species (Fig. 4). The Q is present in all V2B/C-type receptors and the first I is conserved among all receptor types (V1, OXTR and V2). Therefore, the critical change in the motif was to arrive at the FREIH or FKEIH sequence. In addition to mammals, activation of the cAMP pathway has been demonstrated for the V2AR type in Japanese tree frog (FREIH; Kohno, Kamishima, & Iguchi, 2003), African lungfish (FKEIH; Konno et al., 2009), and medaka (FREIH) and gray bichir fish (FKEIH; Konno et al., 2010). Neither the chicken V2CR (CKIIK) (Tan et al., 2000) nor, as noted, the shark V2BR1 (CRAIQ) (Yamaguchi et al., 2012), when transfected into mammalian cells, have shown cAMP accumulation upon stimulation with AVT. Thus, we believe the FKEIH or FREIH motif in ICL3 is a critical determinant of Gs protein binding leading to cAMP signaling.

3.3 Evolution of the V2AR sequence

In Fig. 5 we postulate the evolutionary path the V2 receptor line may have followed to arrive at the proposed FREIH/FKEIH cAMP signaling motif. Phylogeny (Fig. 1) indicates that the lamprey V2-like receptor genes are structurally more similar to the V2BR or V2CR group and there is no V2AR gene. Based on analysis of the lamprey V2BR2 (CRIIR) and V2BC-like (CRVIR) ICL3 motifs, (Fig. 4), we would predict no AVT-inducible interaction with the Gs protein, and therefore no cAMP accumulation. However, only four to five nucleotide changes from the lamprey

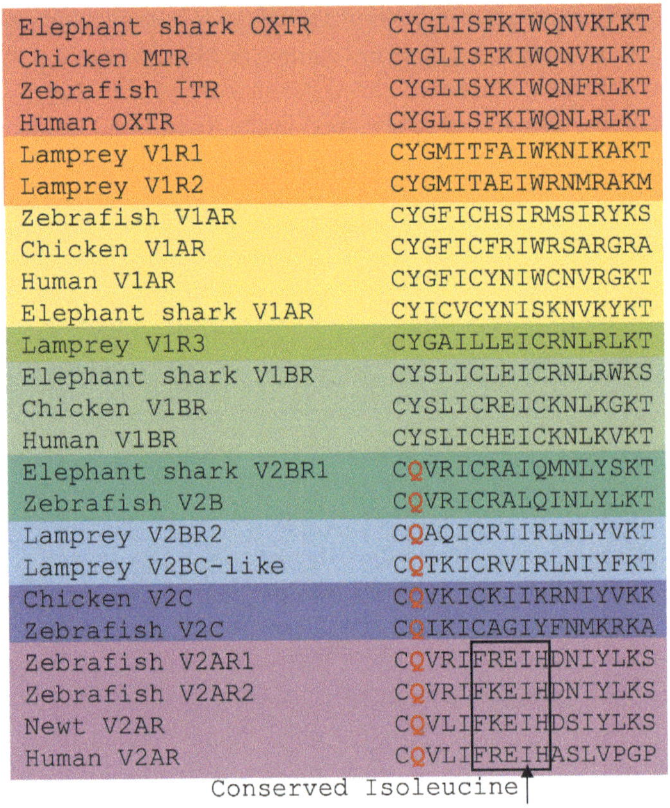

Fig. 4 Comparison of intracellular loop 3 (ICL3) amino acid motifs among receptor types and species. The boxed sequence is proposed to confer the ability to interact with the Gs protein resulting in cAMP accumulation.

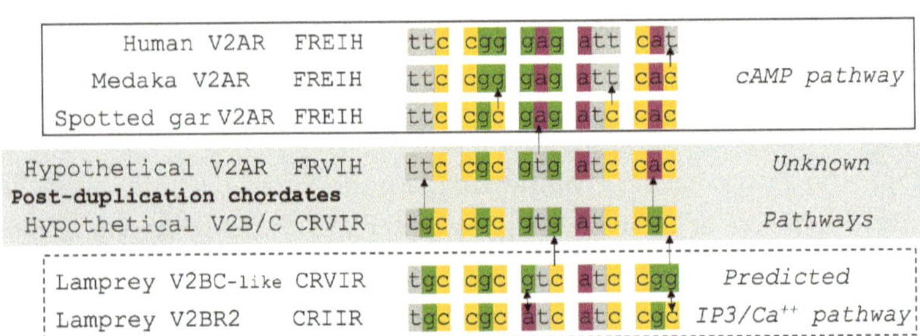

Fig. 5 Evolutionary path of nucleotide mutations (arrows) proposed to result in the putative Gs protein binding motif.

V2BC-like receptor, through hypothetical post-WGD chordate intermediate species, would be required to result in a cAMP-functional V2AR receptor motif such as in the spotted gar (Fig. 5).

This significant differentiation and those of other AVTR/AVPR/OXTR family paralogs after gene duplication and subsequent nucleotide mutations is demonstrably evident in protein function. Less evident are DNA changes in parts of the gene that do not code for protein, although they may be responsible for whether the protein is expressed in a specific tissue, or expressed at all. The AVTR/AVPR/OXTR proteins are pleiotropic, having tissue- and life-stage-specific functions leading to different phenotypic traits. Our relatively recent understanding of non-coding DNA prompts a look into the regulatory regions of these genes as another key to the conservation or differentiation of protein function.

4. Promoter region regulatory elements and evolution of expression

4.1 Conservation and loss of enhancers through WGD and chromosomal duplication

Onimaru and Kuraku (2018) list several unresolvable factors in extrapolating genomic duplications and phenotypic traits. Among these are the specialized traits of basal vertebrates like the lamprey that obfuscate the timing of trait gains or losses. They pointed to the potential loss of enhancer and promoter regions during WGD or chromosomal duplication, which might appear as gene losses. If regulatory regions are lost or rearranged, the gene would be expressed differently or not at all, resulting in an altered phenotype.

4.2 A potential enhancer and promoter region in a lamprey receptor

The concept that specific enhancers remain part and parcel of an operating gene after duplication can also be applied to determining orthology of lamprey and gnathostome genes through the comparison of their regulatory elements. Unfortunately, study of non-coding DNA, though rapidly progressing in mammalian systems, is still very new to non-mammals. Empirical demonstration of regulatory function calls for such methods as transcription factor binding and DNA methylation studies. DNA methylation of cytosine-guanine (CpG) dinucleotides in promoter regions is well known to regulate the binding of transcription factors and therefore gene transcription in mammals. The high GC content in coding regions of lamprey genes, however, raised the question of whether this mechanism operates in lamprey gene transcription. We conducted a study with the hypothesis that differential DNA methylation of CpG dinucleotides in the promoter region of the lamprey V1R1 gene is correlated with gene expression in a manner similar to that in mammalian OXTR genes (Mayasich, Bemis, & Clarke, 2016). Genomic DNA from adult heart and larval liver where the V1R1 gene is transcribed was compared to adult liver where no mRNA expression has been observed. Bisulfite conversion of the genomic DNA followed by high resolution melt (HRM) polymerase chain reaction (PCR) and subsequent sequencing were performed. The results showed for the first time in a specific lamprey gene that DNA methylation was in fact correlated with life-stage- and tissue-specific differences in mRNA transcription. Methylation of CpGs occurred in a specific pattern in a CpG island area (as defined by Li & Dahiya, 2002) from approximately −900 to −1600 nucleotide base pairs (bp) upstream of the transcription start site (TSS). However, because the high levels of methylation in the heart and larval liver appeared to result in high levels of transcription, contrary to the prevalent mode in mammals, it was important to ascertain how this could work.

The putative V1R1 receptor promoter region was searched and found to contain consensus sequences matching those of regulatory elements known to control OXTR gene transcription in mammals. Estrogen response element (ERE) sites and half-sites are present in the V1R1 putative promoter. The estrogen nuclear receptor in lampreys binds the classic estradiol molecule, and the activated receptor binds ERE consensus DNA sequences (Paris et al., 2008). Moreover, recent research actually provides evidence of regulatory element sequences containing CpGs that require methylation in order for certain transcription factors to bind and initiate transcription (Hu et al., 2013; Kaczynski, Cook, & Urrutia, 2003; Liu et al., 2014). These transcription factors include Krüppel-like factor 4 (KLF4) and the related Specificity proteins Sp-1 and Sp-2. Consensus DNA binding site sequences for these transcription factors are found in the V1R1 promoter DNA; the CpGs within these sequence motifs were highly methylated in adult heart and larval liver samples (high transcription), but unmethylated or at low methylation levels in the adult liver (no transcription). The KLF4 and Sp genes exist in both the *P. marinus* and *L. japonicum* genomes, and the DNA binding domain is highly conserved among jawed vertebrates including humans (Mayasich et al., 2016). Similar correlations between CpG methylation levels of Sp-1 and ERE motifs and tissue-specific expression were found in mammalian (mouse) OXTR promoters

(Harony-Nicolas et al., 2014; Mamrut et al., 2013). Although no methylation difference of the ERE sites was found in the V1R1-expressing versus non-expressing lamprey tissues we tested (heart and liver), these motifs may have regulatory significance in estrogen-responsive lamprey tissues (Mayasich et al., 2016).

Other transcription factor binding motifs exhibiting differential methylation correlated with tissue expression (Dido1), and several motifs known to bind gene enhancer elements (Gata3, MYC and RUNX3) were also found in the region at 1500–1375 bp upstream of the V1R1 TSS. The significance of these results is that the lamprey V1R1 possesses a putative enhancer and promoter region that is structurally and functionally similar to that of the OXTR in mammals. Further study is needed to test whether the predicted transcription factor binding in the lamprey V1R1 actually occurs.

Specific promoter region elements are known to regulate the V1AR, including renin-angiotensin system (RAS) responsive elements that bind the Rreb1 transcription factor, and microsatellites (short tandem repeats of a few nucleotides) (Donaldson & Young, 2013). Microsatellites also regulate V1BR gene transcription (Roper, O'Carroll, Young III, & Lolait, 2011). The lamprey V1R3 promoter, for which the putative TSS has also been determined (Mayasich & Clarke, 2016), could therefore be similarly evaluated.

5. Conclusions and future directions

5.1 Duplication scenarios

In this study we conducted in-depth sequence similarity and orthological comparisons of the lamprey AVTRs with the OXTR/V1AR/V1BR types in jawed vertebrates, utilizing elephant shark AVTR and OXTR as an evolutionary intermediate. While our sequence similarity analysis (Fig. 2) showed general paralogy of lamprey and gnathostome receptors, the regional analysis of known and putative ligand and G–protein interaction (Fig. 3) provided an additional increment of support for orthology of the lamprey V1R1 with gnathostome OXTR and V1AR types, and of V1R3 with V1BR.

Synteny analysis previously revealed a mix of neighboring genes on sea and Japanese lamprey scaffolds, with inter-digitated genes showing orthology to either the chromosome containing V1AR or that containing the OXTR gene in gnathostomes (Mayasich & Clarke, 2016). Two alternative scenarios of gene duplication were proposed (Mayasich & Clarke, 2016) that can still be considered as possible courses of evolutionary events. We find the two rounds of WGD scenario (Scenario 1) to be less parsimonious considering that many gene losses would have to occur in all classes of vertebrates. Also, new information from the latest lamprey genome build divides orthology groups into those that had 1R WGD and those with further, smaller-scale duplications (Smith et al., 2018).

Scenario 2 (Mayasich & Clarke, 2016) postulates 1R WGD with subsequent local duplications and translocations, placing the lamprey V1R1-V2BR2 scaffold and gnathostome OXTR-V2BR chromosome in parallel, emanating from the same ancestral 1R WGD chromosome. The lamprey V1R3-V2BC-like scaffold then parallels a jawed-vertebrate chromosome that underwent local duplication resulting in the V1AR-V2CR chromosomal block, with the V2AR and the V1BR on separate chromosomes (Mayasich & Clarke, 2016).

Here we put forth a third scenario that also proposes 1R WGD in the lamprey (Fig. 6). After lamprey divergence, the 1R ancestral chromosome that had resulted in the lamprey V1R1-V2BR2 block is proposed to have undergone chromosome-level duplication to form OXTR-V2BR and V1AR-V2CR-bearing chromosomes. This scenario also avoids multiple

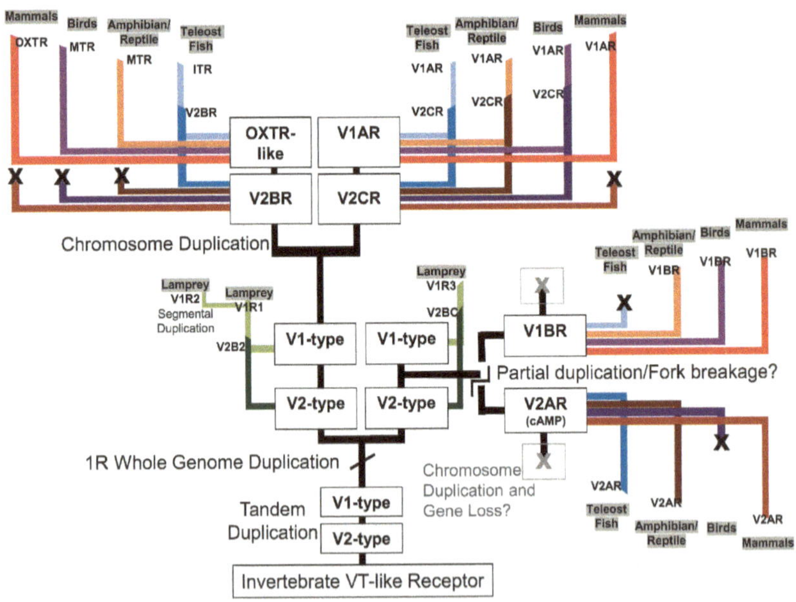

Fig. 6 Possible whole-genome duplication (WGD) and smaller-scale duplications that resulted in the gene compliments of the vasopressin/oxytocin receptor family in vertebrate taxa. An "X" indicates a gene loss. Overlap of a light-colored V1-type and dark-colored V2-type receptor indicate colocation on the same scaffold or chromosome.

gene losses, accounts for the synteny, promoter region and gene differentiation data, and avoids applying specific orthology to the lamprey genes. Although the V1AR and V1BR appear more closely related phylogenetically (Fig. 1), this may be an artifact of having the same ligand and therefore remaining less differentiated from each other than from the OXTR types in jawed vertebrates. The new scenario also accounts for postulated events such as replication fork breakage that could have led the V2BC-like ancestor to become the V2AR through mutation. However, duplication and gene loss could also explain the lack of a V1 partner on the V2AR chromosome and a V2 partner on the V1BR chromosomes in jawed species (Fig. 6). This conundrum cannot be solved until more genomic information is available. For example, in the genome build of the cartilaginous elephant shark (which does not possess a V2AR gene), pairing cannot be resolved because all of the AVTR/OXTR genes are currently on separate scaffolds (NCBI assembly GCA_000165045.2 Callorhinchus_milii-6.1.3; Venkatesh et al., 2014).

5.2 Future directions and the broader importance of molecular evolution studies

Our current findings show that substitutions at ligand-binding residues and neighboring residues, as well as how the binding of an antagonistic chemical may affect function through downstream events such as G protein binding, should all be considered. Human health research into the effects of mutation of this GPCR family can benefit from the findings of evolutionary studies (Pifano et al., 2018). Another critical need is to understand how amino acid substitutions affect the binding not only of ligands across species, but, by extension, to predict cross-species sensitivity to chemicals in the environment (LaLone et al., 2016). These chemicals may act as agonists or antagonists, disrupting endocrine function. Genomics, transcriptomics and proteomics research have become essential tools revolutionizing the field of environmental toxicology (Villeneuve & Garcia-Reyero, 2010). Coupled with high-throughput methods similar to those applied to drug discovery, molecular evolution research can help to minimize the need for in vivo animal testing (Schroeder, Ankley, Houck, & Villeneuve, 2016) while protecting the health not only of humans but a multitude of species.

Acknowledgments

We thank Dr. Carlie LaLone for review and comments on the manuscript.

References

Acharjee, S., Do-Rego, J. L., Oh, D. Y., Ahn, R. S., Choe, H., Vaudry, H., et al. (2004). Identification of amino acid residues that direct differential ligand selectivity of mammalian and nonmammalian V1a type receptors for arginine vasopressin and vasotocin. Insights into molecular coevolution of V1a type receptors and their ligands. *Journal of Biological Chemistry*, *279*(52), 54445–54453.

Acharjee, S., Do-Rego, J. L., Oh, D. Y., Moon, J. S., Ahn, R. S., Lee, K., et al. (2004). Molecular cloning, pharmacological characterization, and histochemical distribution of frog vasotocin and mesotocin receptors. *Journal of Molecular Endocrinology*, *33*(1), 293–313.

Balment, R. J., Lu, W., Weybourne, E., & Warne, J. M. (2006). Arginine vasotocin a key hormone in fish physiology and behaviour: A review with insights from mammalian models. *General and Comparative Endocrinology*, *147*, 9–16.

Betts, M. J., & Russell, R. B. (2003). Amino acid properties and consequences of subsitutions. In M. R. Barnes & I. C. Gray (Eds.), *Bioinformatics for geneticists* (pp. 289–316): Wiley.

Chini, B., & Fanelli, F. (2000). Molecular basis of ligand binding and receptor activation in the oxytocin and vasopressin receptor family. *Experimental Physiology*, *85*, 59S–66S, Spec No.

Crooks, G. E., Hon, G., Chandonia, J. M., & Brenner, S. E. (2004). WebLogo: A sequence logo generator. *Genome Research*, *14*, 1188–1190.

Donaldson, Z. R., & Young, L. J. (2008). Oxytocin, vasopressin, and the neurogenetics of sociality. *Science*, *322*, 900–904.

Donaldson, Z. R., & Young, L. J. (2013). The relative contribution of proximal 5′ flanking sequence and microsatellite variation on brain vasopressin 1a receptor (Avpr1a) gene expression and behavior. *PLoS Genetics*, *9*(8), e1003729.

Ellens, E. R., Kittilson, J. D., Hall, J. A., Sower, S. A., & Sheridan, M. A. (2013). Evolutionary origin and divergence of the growth hormone receptor family: Insight from studies on sea lamprey. *General and Comparative Endocrinology*, *192*, 222–236.

Erlenbach, I., & Wess, J. (1998). Molecular basis of V2 vasopressin receptor/Gs coupling selectivity. *Journal of Biological Chemistry*, *273*, 26549–26558.

Flock, T., Hauser, A. S., Lund, N., Gloriam, D. E., Balaji, S., & Babu, M. M. (2017). Selectivity determinants of GPCR-G-protein binding. *Nature*, *545*(7654), 317–322.

Gimpl, G., & Fahrenholz, F. (2001). The oxytocin receptor system: Structure, function, and regulation. *Physiological Reviews*, *81*, 629–683.

Goodson, J. L. (2005). The vertebrate social behavior network: Evolutionary themes and variations. *Hormones and Behavior*, *48*, 11–22.

Goodson, J. L., & Kabelik, D. (2009). Dynamic limbic networks and social diversity in vertebrates: From neural context to neuromodulatory patterning. *Frontiers in Neuroendocrinology*, *30*, 429–441.

Gwee, P. C., Tay, B. H., Brenner, S., & Venkatesh, B. (2009). Characterization of the neurohypophysial hormone gene loci in elephant shark and the Japanese lamprey: Origin of the vertebrate neurohypophysial hormone genes. *BMC Evolutionary Biology*, *9*, 47.

Haitina, T., Klovins, J., Takahashi, A., Löwgren, M., Ringholm, A., Enberg, J., et al. (2007). Functional characterization of two melanocortin (MC) receptors in lamprey showing orthology to the MC1 and MC4 receptor subtypes. *BMC Evolutionary Biology*, *7*, 101.

Harony-Nicolas, H., Mamrut, S., Brodsky, L., Shahar-Gold, H., Barki-Harrington, L., & Wagner, S. (2014). Brain region-specific methylation in the promoter of the murine oxytocin receptor gene is involved in its expression regulation. *Psychoneuroendocrinology*, *29*, 121–131.

Hu, S., Wan, J., Su, Y., Song, Q., Zeng, Y., et al. (2013). DNA methylation presents distinct binding sites for human transcription factors. *eLife2*, e00726. https://doi.org/10.7554/eLife.00726.

Indrischek, H., Prohaska, S. J., Gurevich, V. V., Gurevich, E. V., & Stadler, P. F. (2017). Uncovering missing pieces: Duplication and deletion history of arrestins in deuterostomes. *BMC Evolutionary Biology*, *17*(1), 163.

Kaczynski, J., Cook, T., & Urrutia, R. (2003). Sp1- and Krüppel-like transcription factors. *Genome Biology*, *4*(206), 1–8.

Kim, S. W., Jeon, Y. S., Lee, J. U., Kang, D. G., Kook, H., Ahn, K. Y., et al. (2000). Diminished adenylate cyclase activity and aquaporin 2 expression in acute renal failure rats. *Kidney International*, *57*, 1643–1650.

Koehbach, J., Stockner, T., Bergmayr, C., Muttenthaler, M., & Gruber, C. W. (2013). Insights into the molecular evolution of oxytocin receptor ligand binding. *Biochemical Society Transactions*, *41*(1), 197–204.

Kohno, S., Kamishima, Y., & Iguchi, T. (2003). Molecular cloning of an anuran V2 type [Arg^8] vasotocin receptor and mesotocin receptor: Functional characterization and tissue expression in the Japanese tree frog (*Hyla japonica*). *General and Comparative Endocrinology*, *132*, 485–498.

Konno, N., Hyodo, S., Yamaguchi, Y., Kaiya, H., Miyazato, M., Matsuda, K., et al. (2009). African lungfish, *Protopterus annectens*, possess an arginine vasotocin receptor homologous to the tetrapod V2-type receptor. *Journal of Experimental Biology*, *212*, 2183–2193.

Konno, N., Kurosawa, M., Kaiya, H., Miyazato, M., Matsuda, K., & Uchiyama, M. (2010). Molecular cloning and characterization of V2-type receptor in two ray-finned fish, gray bichir, *Polypterus senegalus* and medaka, *Oryzias latipes*. *Peptides*, *31*, 1273–1279.

Kuenzel, W. J., Kang, S. W., & Jurkevich, A. (2013). Neuroendocrine regulation of stress in birds with an emphasis on vasotocin receptors (VTRs). *General and Comparative Endocrinology*, *190*, 18–23.

Lagman, D., Ocampo Daza, D., Widmark, J., Abalo, X. M., Sundström, G., & Larhammar, D. (2013). The vertebrate ancestral repertoire of visual opsins, transducin alpha subunits and oxytocin/vasopressin receptors was established by duplication of their shared genomic region in the two rounds of early vertebrate genome duplications. *BMC Evolutionary Biology*, *13*, 238.

LaLone, C. A., Villeneuve, D. L., Lyons, D., Helgen, H. W., Robinson, S. L., Swintek, J. A., et al. (2016). Sequence alignment to predict across species susceptibility (SeqAPASS): A web-based tool for addressing the challenges of cross-species extrapolation of chemical toxicity. *Toxicological Sciences*, *153*(2), 228–245.

Li, L. C., & Dahiya, R. (2002). MethPrimer: Designing primers for methylation PCRs. *Bioinformatics*, *18*, 1427–1431.

Liu, Y., Olanrewaju, Y. O., Zheng, Y., Hashimoto, H., Blumenthal, R. M., et al. (2014). Structural basis for Klf4 recognition of methylated DNA. *Nucleic Acids Research*, *42*, 4859–4867.

Mamrut, S., Harony, H., Sood, R., Shahar-Gold, H., Gainer, H., et al. (2013). DNA methylation of specific CpG sites in the promoter region regulates the transcription of the mouse oxytocin receptor. *PLoS One8*, e56869. https://doi.org/10.1371/journal.pone.0056869.

Mayasich, S. A., Bemis, L. T., & Clarke, B. L. (2016). DNA methylation in a sea lamprey vasotocin receptor gene promoter correlates with tissue- and life-stage-specific mRNA expression. *Comparative Biochemistry and Physiology Part B: Biochemistry and Molecular Biology*, *202*, 56–66.

Mayasich, S. A., & Clarke, B. L. (2016). The emergence of the vasopressin and oxytocin hormone receptor gene family lineage: Clues from the characterization of vasotocin receptors in the sea lamprey (*Petromyzon marinus*). *General and Comparative Endocrinology*, *226*, 88–101.

Mehta, T. K., Ravi, V., Yamasaki, S., Lee, A. P., Lian, M. M., Tay, B.-H., et al. (2013). Evidence for at least six Hox clusters in the Japanese lamprey (*Lethenteron japonicum*). *Proceedings of the National Academy of Sciences of the United States of America*, *110*, 16044–16049.

Nishimura, H., & Yang, Y. (2013). Aquaporins in avian kidneys: Function and perspectives. *American Journal of Physiology: Regulatory, Integrative and Comparative Physiology, 305*, R1201–R1214.

Ocampo Daza, D., Lewicka, M., & Larhammar, D. (2012). The oxytocin/vasopressin receptor family has at least five members in the gnathostome lineage, including two distinct V2 subtypes. *General and Comparative Endocrinology, 175*, 135–143.

Onimaru, K., & Kuraku, S. (2018). Inference of the ancestral vertebrate phenotype through vestiges of the whole-genome duplications. *Briefings in Functional Genomics, 17*, 351–361, ely008.

Orcel, H., Albizu, L., Perkovska, S., Durroux, T., Mendre, C., Ansanay, H., et al. (2009). Differential coupling of the vasopressin V1b receptor through compartmentalization within the plasma membrane. *Molecular Pharmacology, 75*, 637–647.

Paris, M., Pettersson, K., Schubert, M., Bertrand, S., Pongratz, I., et al. (2008). An amphioxus orthologue of the estrogen receptor that does not bind estradiol: Insights into estrogen receptor evolution. *BMC Evolutionary Biology, 8*(219), 1–20.

Pifano, M., Garona, J., Sobol, N. T., Albertó, M., Alonso, D. F., & Ripoll, G. V. (2018). Search of vasopressin analogs with antiproliferative activity on small-cell lung cancer: Drug design based on two different approaches. *Future Medicinal Chemistry, 10*(8), 879–894.

Rahmeh, R., Damian, M., Cottet, M., Orcel, H., Mendre, C., Durroux, T., et al. (2012). Structural insights into biased G protein-coupled receptor signaling revealed by fluorescence spectroscopy. *Proceedings of the National Academy of Sciences of the United States of America, 109*(17), 6733–6738.

Roper, J. A., O'Carroll, A.-M., Young, W. S., III, & Lolait, S. J. (2011). The vasopressin Avpr1b receptor: Molecular and pharmacological studies. *Stress, 14*(1), 98–115.

Schroeder, A. L., Ankley, G. T., Houck, K. A., & Villeneuve, D. L. (2016). Environmental surveillance and monitoring—The next frontiers for high-throughput toxicology. *Environmental Toxicology and Chemistry, 35*(3), 513–525.

Smith, J. J., Antonacci, F., Eichler, E. E., & Amemiya, C. T. (2009). Programmed loss of millions of base pairs from a vertebrate genome. *Proceedings of the National Academy of Sciences of the United States of America, 106*(27), 11212–11217.

Smith, J. J., & Keinath, M. C. (2015). The sea lamprey meiotic map improves resolution of ancient vertebrate genome duplications. *Genome Research, 25*, 1–10.

Smith, J. J., Kuraku, S., Holt, C., Sauka-Spengler, T., Jiang, N., Campbell, M. S., et al. (2013). Sequencing of the sea lamprey (*Petromyzon marinus*) genome provides insights into vertebrate evolution. *Nature Genetics, 45*, 415–421.

Smith, J. J., Timoshevskaya, N., Ye, C., Holt, C., Keinath, M. C., Parker, H. J., et al. (2018). The sea lamprey germline genome provides insights into programmed genome rearrangement and vertebrate evolution. *Nature Genetics, 50*, 270–277.

Sower, S. A., Decatur, W. A., Joseph, N. T., & Freamat, M. (2012). Evolution of vertebrate GnRH receptors from the perspective of a basal vertebrate. *Frontiers in Endocrinology, 3*, 140.

Stafflinger, E., Hansen, K. H., Hauser, F., Schneider, M., Cazzamali, G., Williamson, M., et al. (2008). Cloning and identification of an oxytocin/vasopressin-like receptor and its ligand from insects. *Proceedings of the National Academy of Sciences of the United States of America, 105*, 3262–3267.

Strakova, Z., & Soloff, M. S. (1997). Coupling of oxytocin receptor to G proteins in rat myometrium during labor: Gi receptor interaction. *American Journal of Physiology, 272*(5 Pt. 1), E870–E876.

Tan, F. L., Lolait, S. J., Brownstein, M. J., Saito, N., MacLeod, V., Baeyens, D. A., et al. (2000). Molecular cloning and functional characterization of a vasotocin receptor subtype that is expressed in the shell gland and brain of the domestic chicken. *Biology of Reproduction, 62*, 8–15.

Terrillon, S., Durroux, T., Mouillac, B., Breit, A., Ayoub, M. A., Taulan, M., et al. (2003). Oxytocin and vasopressin V1a and V2 receptors form constitutive homo- and heterodimers during biosynthesis. *Molecular Endocrinology*, *17*, 677–691.

Thibonnier, M., Preston, J. A., Dulin, N., Wilkins, P. L., Berti-Mattera, L. N., & Mattera, R. (1997). The human V3 pituitary vasopressin receptor: Ligand binding profile and density-dependent signaling pathways. *Endocrinology*, *138*(10), 4109–4122.

Venkatesh, B., Lee, A. P., Ravi, V., Maurya, A. K., Lian, M. M., Swann, J. B., et al. (2014). Elephant shark genome provides unique insights into gnathostome evolution. *Nature*, *505*(7482), 174–179.

Villeneuve, D. L., & Garcia-Reyero, N. (2010). Vision & strategy: Predictive ecotoxicology in the 21st century. *Environmental Toxicology and Chemistry*, *30*(1), 1–8.

Wells, A., Anderson, W. G., & Hazon, N. (2002). Development of an in situ perfused kidney preparation for elasmobranch fish: Action of arginine vasotocin. *American Journal of Physiology: Regulatory, Integrative and Comparative Physiology*, *282*, R1636–R1642.

Wilczynski, W., Quispe, M., Muñoz, M. I., & Penna, M. (2017). Arginine vasotocin, the social neuropeptide of amphibians and reptiles. *Frontiers in Endocrinology*, *8*, 186, 1–17.

Yamaguchi, Y., Kaiya, H., Konno, N., Iwata, E., Miyazato, M., Uchiyama, M., et al. (2012). The fifth neurohypophysial hormone receptor is structurally related to the V2-type receptor but functionally similar to V1-type receptors. *General and Comparative Endocrinology*, *178*, 519–528.

Yampolsky, L. Y., & Stoltzfus, A. (2005). The exchangeability of amino acids in proteins. *Genetics*, *170*(4), 1459–1472.

Yang, M., Wang, W., Zhong, M., Philippi, A., Lichtarge, O., & Sanborn, B. M. (2002). Lysine 270 in the third intracellular domain of the oxytocin receptor is an important determinant for G α(q) coupling specificity. *Molecular Endocrinology*, *16*(4), 814–823.

CHAPTER TWO

Oxytocin/vasopressin-like neuropeptide signaling in insects

Edin Muratspahić[a,†], Emilie Monjon[b,†], Leopold Duerrauer[a], Stephen M. Rogers[c], Darron A. Cullen[b], Jozef Vanden Broeck[b], Christian W. Gruber[a,*]

[a]Center for Physiology and Pharmacology, Medical University of Vienna, Vienna, Austria
[b]Molecular Developmental Physiology and Signal Transduction Group, Zoological Institute, KU Leuven, Leuven, Belgium
[c]Department of Zoology, University of Cambridge, Cambridge, United Kingdom
*Corresponding author: e-mail address: christian.w.gruber@meduniwien.ac.at

Contents

Abstract

The origin of the oxytocin (OT)/vasopressin (VP) signaling system is thought to date back more than 600 million years. OT/VP-like peptides have been identified in numerous invertebrate phyla including molluscs, annelids, nematodes and insects. However, to date we only have a limited understanding of the biological role(s) of this GPCR-mediated signaling system in insects. This chapter presents the current knowledge of OT/VP-like neuropeptide signaling in insects by providing a brief overview of insect OT/VP-like neuropeptides, their genetic and structural commonalities, and their experimentally tested and proposed functions. Despite their widespread occurrence across insect orders these peptides (and their endogenous receptors) appear to be absent in common insect model species, such as flies and bees. We therefore explain the known functionalities of this signaling system in three different insect model systems: beetles, locusts, and ants. Additionally, we review the phylogenetic distribution of

† These authors contributed equally.

Vitamins and Hormones, Volume 113
ISSN 0083-6729
https://doi.org/10.1016/bs.vh.2019.08.011

the OT/VP signaling system in arthropods as obtained from extensive genome/transcriptome mining. Finally, we discuss the unique challenges in the development of selective OT/VP ligands for human receptors and share our perspective on the possible application of insect- and other non-mammalian-derived OT/VP-like peptide ligands in pharmacology.

1. Introduction

Neuropeptides acting as neurotransmitters or neurohormones are key players in the regulation of physiological, behavioral, and developmental processes in animals (Gäde, 2004). In mammals, oxytocin (OT) and vasopressin (VP) exert their effects *via* four G protein-coupled receptors (GPCRs): the oxytocin receptor, OTR, and three vasopressin receptors, $V_{1a}R$, $V_{1b}R$ and V_2R (Barberis, Mouillac, & Durroux, 1998; Gimpl & Fahrenholz, 2001). GPCRs are proteins with seven membrane-spanning α-helical segments. They transduce extracellular stimuli into intracellular signals and hence regulate various cellular processes (Gruber, Muttenthaler, & Freissmuth, 2010; Kroeze, Sheffler, & Roth, 2003; Vanden Broeck, 1996). Agonist binding results in a stabilized active receptor conformation that in turn recruits and regulates intracellular transducers, such as heterotrimeric G-proteins and arrestins (Wacker, Stevens, & Roth, 2017). GPCR agonists are diverse and can include neurotransmitters, hormones, growth factors, odorants, photons, small molecules, peptides and proteins (Gruber et al., 2010). The human genome contains more than 800 GPCR genes, making them the largest family of membrane proteins (Fredriksson, Lagerstrom, Lundin, & Schioth, 2003). OT and VP are synthesized by neurons in the paraventricular and supraoptic nuclei of the hypothalamus that project to and have release sites in the pituitary gland, from which both hormones are released into peripheral circulation and the brain. In addition, parvocellular neurons originating in the paraventricular nucleus also express OT and VP and project to discrete brain regions (Meyer-Lindenberg, Domes, Kirsch, & Heinrichs, 2011). Central and peripheral functions of these two peptides include reproduction, osmoregulation, memory and learning (Donaldson & Young, 2008; McCall & Singer, 2012). In mammals and other vertebrates the two neuropeptides OT and VP exhibit distinct physiological roles. By contrast, invertebrates generally have only a single homolog of the OT/VP precursor and a single OT/VP-like receptor (Gruber, 2014; Liutkevičiūtė & Gruber, 2016).

The OT/VP-like signaling system has an ancient origin and is found in groups whose last common ancestor lived over 600 million years ago, including insects, molluscs, annelids and nematodes (Gruber, 2014; Liutkevičiūtė, Koehbach, Eder, Gil-Mansilla, & Gruber, 2016). It has been suggested that the vertebrate OT/VP-like peptides originated from a gene duplication of the ancestral peptide vasotocin (Liutkevičiūtė & Gruber, 2016); refer to Chapter 4 of this volume).

All OT/VP-like molecules are nonapeptides possessing high sequence homology. They comprise a six residue N-terminal ring formed by a disulfide bond between the conserved cysteines in positions 1 and 6, and an amidated three residue C-terminal tail (Gruber, Koehbach, & Muttenthaler, 2012). They derive from larger precursor proteins, which contain a short N-terminal endoplasmic reticulum signal peptide followed by the mature nonapeptide domain and a canonical amidation processing sequence. The C-terminal domain is characterized by a protein called neurophysin. Neurophysins are primarily known to act as carrier proteins for OT and VP, but there is evidence that they have other OT/VP-independent biological functions as well (Elphick, 2010).

Despite a high degree of sequence similarity, OT/VP-like peptides are known by different names in different taxa such as annetocin (annelids, OT-like), conopressin (molluscs, VP-like), nematocin (nematodes, OT-like), inotocin (insects, OT/VP-like) or vasotocin (vertebrate ancestral form, OT/VP-like) (Gruber, 2014). These names mask both fundamental differences and similarities between these molecules across species. The current names do not uniquely describe the peptide sequence (there are several different peptides in insects, for example, which are all referred to as "inotocin") and the use of the suffixes "-pressin" or "-tocin" are arbitrary and not related to underlying functional similarities. In this chapter we use the terms "OT/VP-like" or "inotocin signaling" as equivalent or synonymous when referring to insects.

OT/VP-like signaling appears to be involved in neurotransmission, metabolism, and osmoregulation in molluscs and annelids (Gruber, 2014); in the nematode *Caenorhabditis elegans*, it regulates gustatory associative learning and coordinates reproductive behavior (Beets et al., 2012; Garrison et al., 2012). Until very recently there has been relatively little information available on the role of this signaling in insects.

In this chapter we will provide an overview of the current knowledge of OT/VP-like signaling in beetles, locusts and ants as widely-studied insect model systems (summarized in Fig. 1). Interestingly, OT/VP-like signaling is absent in the European honey bee, *Apis mellifera,* and the fruit fly,

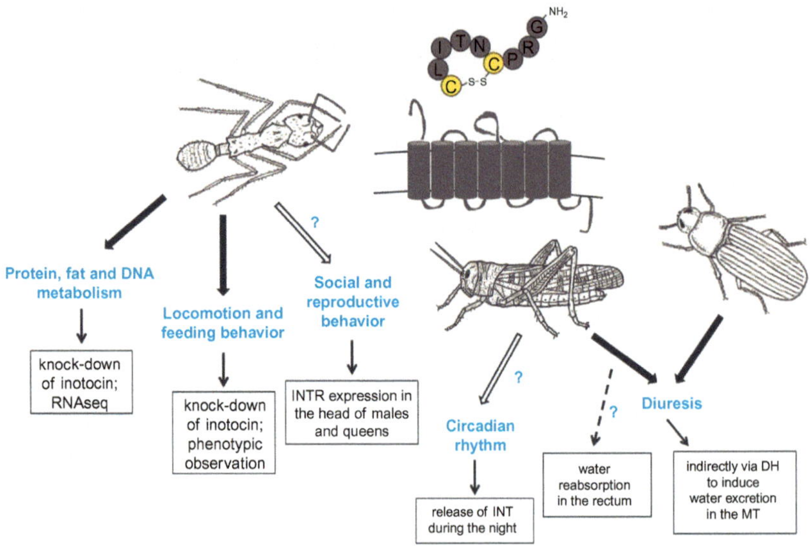

Fig. 1 Summary of functional role of the inotocin signaling in ants, locusts and beetles. INT (inotocin), INTR (inotocin receptor), DH (diuretic hormone), MT (Malpighian tubules). Black arrows indicate experimentally tested functions and white arrows indicate suggested functions (based on weaker experimental evidence). *Insect drawings were prepared by E. Gil-Mansilla.*

Drosophila melanogaster. We provide an overview of the phylogenetic distribution of OT/VP signaling and highlight recent advances in the discovery of ligand and receptor genes in insects. Lastly, we will summarize available information about the possible application of insect OT/VP-like ligands in human pharmacology.

2. OT/VP-like signaling in beetles

The first evidence for the existence of an OT/VP-like signaling system in beetles was obtained in the early 1980s after immunostaining of the central nervous system (CNS) of the Colorado potato beetle, *Leptinotarsa decemlineata* with anti-vasopressin and anti-oxytocin antibodies (Veenstra, 1984; Veenstra, Romherg-Privee, & Schooneveld, 1984). More recently, the presence of an OT/VP-like peptide, designated as inotocin, was predicted from the genome sequence data of the red flour beetle, *Tribolium castaneum*, and cloned (Aikins et al., 2008; Stafflinger et al., 2008). Inotocin-encoding genes are likely present in many other beetle species based on genome or transcriptome predictions, while *L. decemlineata* even appears to have two different copies of this gene (Liutkevičiūtė et al., 2016). On available evidence, CLITNCP**RG***[a] is the most common peptide sequence in beetles, while *L. decemlineata* has both CLITNCP**KG*** and CLITNCP**IG*** as putative inotocin-like peptides (Liutkevičiūtė et al., 2016).

An inotocin receptor has been predicted in several beetle species based on available genome or transcriptome sequences (Liutkevičiūtė et al., 2016), but has only been cloned and characterized in *T. castaneum* (Aikins et al., 2008; Stafflinger et al., 2008). As expected, this receptor is a GPCR and is activated by the monomeric form of inotocin (CLITNCPRG*). The *T. castaneum* receptor could also be activated by two other analogs, which were tested because they had been previously identified by high performance liquid chromatography (HPLC) in the migratory locust, *Locusta migratoria* (Proux et al., 1987). These are an antiparallel and a parallel dimer of the nonapeptide, which activated the receptor as partial agonists and had ~10- and ~80-fold lower potency than the monomer, respectively (Aikins et al., 2008; Keov, Liutkevičiūtė, Hellinger, Clark, & Gruber, 2018). Furthermore, oxytocin, vasotocin, isotocin (OT-like peptide in fishes; Godwin & Thompson, 2012) and conopressin could all induce a response (but only at higher concentrations than any of the inotocin forms), whereas vasopressin and other insect neuropeptides [such as crustacean cardioactive peptide (CCAP), adipokinetic hormones I and II (AKH I and AKH II), and corazonin (CRZ)] could not. From these results, it seems that the inotocin receptor from *T. castaneum* is more OT-like than VP-like (*cf.* the preference for isoleucine at position 3 of the peptide). This is supported by the amino acid sequence homology of the beetle receptor to its human counterparts (~40% similarity with the human OT receptor compared to 37–38% with the human VP receptors; Aikins et al., 2008; Keov et al., 2018; Stafflinger et al., 2008).

In situ hybridization and immunohistochemistry studies on CNS tissues from *T. castaneum*, *L. decemlineata* and the mealworm *Tenebrio molitor* showed that inotocin is synthesized by a single pair of neurons with cell bodies situated in the subesophageal ganglion (Aikins et al., 2008; Veenstra, 1984; Veenstra et al., 1984). These two neurons project symmetrically throughout the entire CNS, from the brain to the terminal abdominal ganglion. Interestingly, there was no immunostaining observed in peripheral nerves in *T. castaneum*, suggesting a major central role and possibly a minor one

[a] *C-terminal amidation.

peripherally (Aikins et al., 2008). Moreover, expression analysis of the inotocin receptor in this species revealed that it is highly expressed in the CNS (mainly in the head), while there is little expression in the Malpighian tubules and the hindgut, further supporting this idea (Aikins et al., 2008; Stafflinger et al., 2008). Although little is known about the *in vivo* roles of inotocin in beetles, there is an indication that it could play a role in diuresis in both *T. castaneum* and *T. molitor*. Injecting monomeric inotocin in adult *T. castaneum* induced an immediate diuretic response, as measured in an evaporation chamber, whereas neither dimer was active. In *T. molitor*, an *ex vivo* experiment was performed by incubating Malpighian tubules with different tissues and the inotocin peptides. The only condition under which inotocin (monomer) induced an increase in excretion was when incubated with the CNS together with the corpora cardiaca/corpora allata (CC/CA) complex. Without the CC/CA complex (only the CNS) or when incubated only with inotocin, Malpighian tubules did not excrete any water. Again, the antiparallel dimer of inotocin was not able to induce any response. Diuretic hormone, which has been shown to induce diuresis in the Malpighian tubules (Aikins et al., 2008; Furuya, Schegg, Wang, Kingt, & Schooley, 1995), is stored and released by the CC/CA complex of *T. molitor* (Wiehart, Torfs, Van Lommel, Nicolson, & Schoofs, 2002). Hence, this experimental result suggested an indirect role of inotocin on diuresis (for instance *via* central signaling instead of peripheral action). Indeed, it was hypothesized that inotocin could be acting on the CC/CA of *T. molitor* to induce the release of diuretic hormone that in turn triggers water excretion from the Malpighian tubules (Aikins et al., 2008) and the same could occur in *T. castaneum* (Fig. 1).

Knockdown studies of OT/VP-like signaling using RNA interference (RNAi) have also been performed in *T. castaneum*, but downregulating the precursor and receptor transcripts in this animal did not influence any of the processes monitored: general mortality, developmental time in the pupal stage, egg-laying behavior, egg hatching, and mortality in high salt loads (Aikins et al., 2008). Although these knockdown experiments had no observable effect, there is still the possibility that inotocin has additional roles in beetles other than diuresis. Other approaches that could be used to study OT/VP-like signaling include modulation of the receptor by inotocin analogs, which can be injected into the animal. Potentially useful analogs that could be used for this purpose are atosiban, a well-known synthetic oxytocin analog, and kalata B7, a plant-derived cyclotide (Keov et al., 2018). In addition, their structures may also serve as leads for development of additional receptor ligands.

3. OT/VP-like signaling in locusts

Locusts are swarm-forming insects, which can be distinguished from other grasshoppers (Orthoptera:Acrididae) by their ability to switch between solitarious and gregarious phases upon changes in population density, a process known as phase polyphenism (Simpson & Sword, 2008). The first evidence for an OT/VP-like peptide in insects came from the migratory locust, *L. migratoria*, nearly 40 years ago (Rémy, Girardie, & Dubois, 1979), when immunostaining of the subesophageal ganglion with an anti-mammalian VP antibody revealed a single pair of immunoreactive neurons. Inotocin was subsequently isolated using HPLC and identified as being a nonapeptide with the sequence CLITNCPRG (Proux et al., 1987), which is the most common inotocin sequence across insect taxa (Liutkevičiūtė et al., 2016). Proux et al. (1987) isolated two forms of inotocin: a monomer and its antiparallel dimer; later a parallel dimer was also found (Baines & Bacon, 1994). The anti-mammalian VP antibody was effective in insects because VP and inotocin share a common C-terminus (residues in positions 6–9). The N-terminals, however, differ (residues at positions 2, 3 and 4; Donaldson & Young, 2008).

The widespread presence of inotocin in other grasshoppers is supported by immunostaining studies of the whole CNS in 16 grasshopper species across several subfamilies with the same AVP antibody used previously by Rémy and co-workers (Evans & Cournil, 1990; Rémy et al., 1979; Tyrer, Davis, Arbas, Thompson, & Bacon, 1993). Unfortunately, the peptide has not been isolated and sequenced in any species other than *L. migratoria,* hence these immunostaining results are the only evidence for widespread OT/VP-like signaling in grasshoppers, and it is unknown whether there are multiple variants of the inotocin sequence across the Orthoptera.

In common with other animal species, inotocin in grasshoppers is formed by the post-translational modification of a large precursor hormone containing a signal peptide, followed by the nonapeptide itself, and finally a neurophysin domain (Aikins et al., 2008; Caldwell & Young, 2006; Di Giglio et al., 2017; Liutkevičiūtė et al., 2016). The presence of a neurophysin co-expressed with inotocin was supported by Rémy et al. (1979), who discovered that a mammalian anti-neurophysin II antibody stained the same pair of neurons in the subesophageal ganglion that expressed inotocin. It was only with the recent publication of the genome of *L. migratoria*, that this finding was confirmed and the full precursor sequence determined (Wang et al., 2014). A single OT/VP-like peptide receptor has also been predicted in the genome of *L. migratoria*, based on sequence similarities with other known OT/VP-like peptide receptors. As with all other OT/VP-like peptide receptors so far identified, the predicted *L. migratoria* OT/VP-like peptide receptor belongs to the GPCR family (Liutkevičiūtė et al., 2016). To our knowledge, this inotocin receptor has yet to be cloned or characterized.

The single pair of neurons that biosynthesize inotocin in grasshoppers closely resemble those found in beetles, having cell bodies of about 50–60 μm diameter lying near the ventral midline of the subesophageal ganglion (Evans & Cournil, 1990; Rémy & Girardie, 1980; Rémy et al., 1979; Thompson, Tyrer, May, & Bacon, 1991; Tyrer et al., 1993). These two neurons extend parallel axons that project both anteriorly into the brain and posteriorly to the terminal abdominal ganglion. The exact pattern of arborization appears to vary between different subfamilies of grasshopper. In the Cyrtacanthacridinae, which includes the desert locust, *Schistocerca gregaria*, there is extensive central branching within the brain and in each ganglion of the ventral nerve cord but no fibers enter the optic lobes or the peripheral nerves (except in males, see below). In the Gomphocerinae, such as the field grasshopper, *Chorthippus brunneus*, there is likewise extensive arborization within the CNS but there is also some branching within the lobulae of the optic lobes. Conversely, grasshoppers in the Oedipodinae, which includes *L. migratoria*, have relatively sparse branching in the central neuropils of the brain and ventral nerve cord, predominately in dorsal regions, but have extensive arborizations throughout the entire optic lobe. A distinctive feature of the neuron in this subfamily is a proliferation of fine fibers that exit the CNS to form plexuses in most peripheral nerves (Evans & Cournil, 1990; Thompson et al., 1991; Tyrer et al., 1993).

Sexually dimorphic branching of the inotocin expressing neurons has so far only been found in *S. gregaria*, which occurs in both the seventh and terminal (fused segments 8–11) abdominal ganglia. In females no neuronal processes exit any of these ganglia, while in males some fibers enter nerves innervating the genitals and accessory reproductive structures.

The contrasting arborization patterns in different subfamilies of grasshoppers raise important questions as to the extent to which inotocin acts as a neurotransmitter or neuromodulator within the CNS, is released onto peripheral effectors, or enters into general circulation in the hemolymph in different acridid lineages (Evans & Cournil, 1990; Tyrer et al., 1993).

Hemolymph concentrations of inotocin are far below what would be expected for a circulating hormone (10^{-12}M vs 10^{-9}M), supporting a central role or targeted peripheral delivery (Baines, Thompson, Rayne, & Bacon, 1995).

Several studies in *L. migratoria* have suggested that inotocin expression varies with the light-dark cycle over the day (Picquot & Proux, 1987; Thompson & Bacon, 1991) (Fig. 1). In mammals, VP expression varies according to an intrinsic circadian rhythm, which is entrained to the external light intensity (Li et al., 2008; Li, Burton, Zhang, Hu, & Zhou, 2009; Reppert, Schwartz, & Uhl, 1987). In locusts the inotocin-containing neurons are constantly active under dark conditions firing at 2–8 spikes s^{-1}, but silent in daylight, an effect seen even when the optic lobes and ocellar nerves were detached from the rest of the brain (Thompson & Bacon, 1991). This led Thompson and Bacon (1991) to propose that locusts, like some other insect species, may possess extraocular photoreceptors on the surface of the brain that detect light independently from the ocelli and compound eyes. The authors suggest that these extraocular photoreceptors synapse with a descending interneuron in the pars intercerebralis of the brain that descends *via* the circumesophageal connectives to synapse with the inotocin-expressing neurons in the subesophageal ganglion.

Neither the extraocular receptors nor descending neuron have been positively identified, but their properties can be inferred from their effects on the inotocin-expressing neurons. Physiological saline with high Ca^{2+} concentrations promotes continuous synaptic release and under these conditions the inotocin-expressing neurons are active even under strong light conditions, suggesting that their activity is driven by a continuous excitatory drive from the descending neuron (Baines & Bacon, 1994). In dark conditions, the descending interneuron is spontaneously active and continually excites the inotocin-expressing neurons (Baines & Bacon, 1994; Lundquist, Baines, & Bacon, 1996; Thompson & Bacon, 1991). Pharmacological experiments suggest that the descending neuron is cholinergic, and that the inotocin-expressing neurons have both fast nicotinic and slow muscarinic receptors leading to transient and long-lasting (several seconds) depolarizations, respectively (Baines & Bacon, 1994). Pressure injections of histamine, an inhibitory neurotransmitter of insect primary photoreceptors (Schlemermeyer, Schütte, & Ammermüller, 1989), into the brain near the putative site of the extraocular receptors led to the suppression of spontaneous firing in the inotocin expressing neurons, mimicking the effect of light conditions, with the inference that the extraocular photoreceptors inhibit the firing of the descending neuron which in turn removes the excitatory drive to the inotocin-expressing neurons (Lundquist et al., 1996).

The inotocin-expressing neurons in non-oedipodinine grasshoppers do not show the light-dark response characteristic of *L. migratoria* (Lundquist, Baines, Thompson, & Bacon, 1998). In these grasshoppers the inotocin expressing neurons are tonically active under both light and dark conditions. Mechanosensory stimulation from anywhere on the body is a potent stimulus of the inotocin-expressing neurons in all grasshopper species, and appears to act *via* similar cholinergic descending interneurons as mediate the light response in *L. migratoria* (Lundquist et al., 1998).

As in beetles, the *in vivo* function of inotocin in locusts is unclear. There is some evidence supporting a role in diuresis in *L. migratoria*, even though this is controversial (Fig. 1). Diuresis occurs essentially at two sites: at the Malpighian tubules where water is excreted and forms the primary urine, and in the rectum where there is some water, ion and metabolite reabsorption (Spring, 1990). It is likely that inotocin acts in an indirect way to promote water excretion at the level of the Malpighian tubules. Incubation of the Malpighian tubules attached to the posterior part of the midgut with inotocin or an extract of the subesophageal ganglion led to an increase in excretion, in a concentration-dependent manner (Proux et al., 1987; Proux & Herault, 1988; Proux, Picquot, Herault, & Fournier, 1988; Proux, Rougon, & Cupo, 1982), whereas if only the Malpighian tubules were put in the incubation medium, nothing was observed (Coast et al., 1993). It is now known that the posterior midgut and the ampullae possess endocrine cells containing diuretic hormone and that this neuropeptide induces strong diuresis by elevating cAMP levels in the Malpighian tubules (Montuenga et al., 1996). Hence, it is possible that inotocin could affect diuresis indirectly by acting on the endocrine cells of the midgut and the ampullae to induce the release of diuretic hormone, which in turn, induces a rise in cAMP that triggers water excretion. There is some additional evidence supporting a possible role of inotocin in diuresis. Relative atmospheric humidity has some influence on inotocin concentration in the subesophageal ganglion and ventral nerve cord. Decreasing relative humidity to 30–40% induced a significant decrease in inotocin expression, but no significant differences were observed following an increase to 80–100% relative humidity (Picquot & Proux, 1989; Proux & Rougon-Rapuzzi, 1980), supporting a role in the reduction of water loss in dry environments. Changing the concentration of physiological saline, mimicking the effect of altered water balance in the hemolymph had no effect on the firing rate of the inotocin-expressing neurons, however (Thompson & Bacon, 1991). Furthermore, there was only a loose correlation between variation in the concentration of inotocin in the hemolymph and diuresis in locusts over a 24h period, with the peak of both occurring 4h after feeding in the morning, but being otherwise unrelated (Picquot & Proux, 1987).

A role of inotocin in diuresis of locusts is consistent with a conserved function of the VP-like system across the animal kingdom and is similar to the role attributed to inotocin in *T. castaneum* and *T. molitor*, (see above), although in mammals VP is an anti-diuretic rather than diuretic hormone (Caldwell & Young, 2006). In mammals, annelids and chordates OT/VP-like peptides play a more extensive role in regulating osmoregulatory function (Caldwell & Young, 2006; Gruber, 2014). However, it is still possible that inotocin also plays a role in water reabsorption in the rectum of insects similar to the role of VP in the kidneys of mammals. This was suggested by Proux and co-workers (Proux et al., 1988; Proux & Herault, 1988) in early studies, where they observed that low inotocin concentrations were capable of inducing fluid reabsorption in the rectum.

From the three identified forms of inotocin, only one was shown to be effective in inducing diuretic effects (or water reabsorption) in locusts, the antiparallel dimer (Proux et al., 1987, 1988; Proux & Herault, 1988). The biological function of the monomer form is not known, but it was hypothesized that it is just a precursor of the antiparallel dimer. Serine proteases were able to transform the monomer into the antiparallel dimer and induce diuresis, and the dimer could be degraded by enzymes extracted from the rectum. When those degradation products were analyzed by HPLC, a product similar to the monomer could be detected (Picquot & Proux, 1990). It is possible that the monomer has a role as a neurotransmitter or neuromodulator in the CNS: in a coarse biochemical assay, the monomer induced a net decrease in cAMP levels throughout the thoracic CNS, but neither dimer had any measurable effect (Baines et al., 1995). It is clear that much remains to be done to understand the roles of inotocin in locusts, both as a potential hormone and neurochemical, and how the tactile and visual stimuli that drive the activity of the inotocin expressing neurons affect physiology or behavior.

4. OT/VP-like signaling in ants

Gruber and Muttenthaler (2012) reported the discovery of the precursor protein encoding genes and mature OT/VP-like peptide sequences using a simple genome-mining approach in three ant species from the subfamilies Myrmicinae (*Atta cephalotes*), Formicinae (*Camponotus floridanus*) and Ponerinae (*Harpegnathos saltator*). The identified OT/VP-related peptides with the amino acid sequences CLITNCPRG* (in *A. cephalotes* and *H. saltator*) and CLIVNPRG* (in *C. floridanus*), exhibited high similarity to vasotocin and contained elements of both mammalian OT (CYIQNCPLG*) as well as VP (CYFQNCPRG*) peptides. Further analysis showed that the precursor proteins shared considerable sequence homology to the precursors of the other insect OT/VP-like proteins as well as mammalian OT and VP precursors. Apart from the mature nonapeptide sequences, each precursor protein contained a canonical amidation sequence and a neurophysin domain (Gruber & Muttenthaler, 2012). In addition to the discovered inotocin nonapeptides, Gruber and Muttenthaler (2012) also identified putative receptor sequences in ant genomes that are similar to inotocin receptors of other insects such as the red flour beetle *T. castaneum*. Another important finding was the absence of inotocin signaling system in another hymenopteran, the honey bee (*A. mellifera*) (Stafflinger et al., 2008). A subsequent transcriptomic study confirmed these initial findings and identified inotocin peptides and receptor sequences in 17 ant species (Liutkevičiūtė et al., 2016).

These observations not only provided an exciting opportunity to advance our understanding of the evolution of inotocin signaling but led to the elucidation of the molecular mechanisms of pharmacological and physiological actions of inotocin peptides and their receptors in ants, as an important group of social insects. Di Giglio et al. (2017) cloned the inotocin peptide precursor and its putative receptor from the black garden ant *Lasius niger*, before determining the pharmacological properties of a synthetic inotocin ligand on the receptor. The results provided evidence that the endogenous inotocin peptide was indeed the cognate ligand of the identified receptor. Shortly afterwards, Chérasse and Aron (2017) studied the gene expression levels of the inotocin receptor in *L. niger* ant queens during reproductive, social and aggressive behavior. Using qPCR, they observed that the inotocin receptor is expressed in the heads, with the highest expression levels occurring early in the queens' lives under crowded and mated conditions (Chérasse & Aron, 2017). The authors suggested that the inotocin receptor could play an important role in the control of social and reproductive behaviors in ants, which has been further studied by Gruber and colleagues. Liutkevičiūtė et al. (2018) studied the functionality of inotocin signaling in two *Lasius* species, *L. neglectus* and *L. niger* using qPCR, gene-knockdown (RNAi), transcriptomics, as well as phenotypic screening of physiology and behavior. They initially characterized inotocin peptide-receptor signaling in *L. neglectus* and observed that inotocin is a potent agonist of its cognate receptor for the G_q- and G_s-signaling pathways (Liutkevičiūtė et al., 2018). Radioligand binding studies showed that a tritiated inotocin analog was able to bind to membranes transiently expressing the inotocin receptor. The authors measured the expression levels in both ant species in different developmental stages, castes, parts of the body and organs (Liutkevičiūtė et al., 2018). The results demonstrated that the expression of the inotocin peptide and receptor are low in larvae and were higher in pupae, with the highest expression levels detected in adult ants. Compared to workers, the expression of the inotocin precursor in mated queens was about 10-fold lower. When comparing all castes of ants used in this study, the expression of the inotocin precursor was most pronounced in the heads, while the receptor was expressed throughout the body with the highest expression found in the heads of males (Liutkevičiūtė et al., 2018). These gene expression data were confirmed using immunostaining of the mature peptide. The authors analyzed cellular localization of the inotocin peptide in both ant species and revealed that inotocin localizes in the brain, and notably in the subesophageal ganglion. Further experiments aimed to determine changes in expression driven by external stimuli. The expression profile of the inotocin peptide and receptor were affected by seasonal changes. The data suggest that inotocin peptide and receptor expression in *L. neglectus* is lower in winter compared to summer conditions, and lower in nursing ants compared to foragers (Liutkevičiūtė et al., 2018). Furthermore, starvation led to the downregulation of the inotocin peptide mRNA levels. RNAi was used to downregulate the inotocin peptide mRNA in an attempt to alter phenotypes and gene expression. An mRNA-sequencing analysis uncovered approximately 100 differentially expressed genes related to lipid, protein and DNA metabolism. Given that OT and VP are involved in the regulation of social behavior in humans, Liutkevičiūtė et al. (2018) then conducted experiments with the RNAi-knockdown ants to examine whether inotocin signaling affected behavior. The knockdown of inotocin peptide in ants resulted in higher walking activity, enhanced self-grooming in the brood chamber, and a higher motivation for feeding (Fig. 1) (Liutkevičiūtė et al., 2018).

In summary, these data support a role for inotocin signaling in ants, particularly with respect to metabolism, activity and feeding behavior.[b]

[b] During production of this chapter, the following manuscript was published (Koto et al., 2019), which adds further information to the function of inotocin signalling in ants.

Additionally, inotocin in ants could play an important role in the control of energy status linking appetite and locomotion. Ants have provided valuable insights into the roles of inotocin signaling, but there are limitations: for example, due to their small size, ant tissues cannot be readily dissected and employed in physiological and biochemical experiments. Moreover, ants do not possess a very robust RNAi response compared to several other insects. Nevertheless, in our view these studies constitute an excellent step toward better understanding of inotocin signaling in insects.

5. Mining of ligand and receptor genes in other insects

Driven by the intriguing distribution of the OT/VP encoding genes in insects—present in ants, beetles and grasshoppers, but absent in flies and bees—Liutkevičiūtė et al. (2016) further investigated the phylogenetic distribution of the inotocin signaling system in arthropods using the extensive library of insect sequence data compiled by Misof et al. (2014). Liutkevičiūtė et al. (2016) searched for genes encoding components of the inotocin GPCR signaling system in genomes and/or transcriptomes of more than 260 species of insects and other arthropods. They found that inotocin signaling-related genes are present in all subphyla of the Arthropoda, and in 25 out of the 32 orders of the subphylum Hexapoda (Fig. 2). Closer analysis, however, revealed a more complex evolutionary history of OT/VP-like signaling in these groups. For instance, all 20 beetle (Coleoptera) species analyzed had the inotocin signaling system (receptor and/or precursor),[c] whereas in the Hymenoptera (and as outlined above), this pathway appeared to be present in sawflies (Symphyta), parasitoid wasps (Parasitica), social wasps (Vespidae) and ants (Formicoidea) but has seemingly been lost in bees (Apoidea). A similar pattern is observed in the Hemiptera: inotocin being apparently present in just some Heteroptera, widely present in plant, leaf and froghoppers (Auchenorrhyncha, but apparently absent in Cicadas), and totally absent in aphids (Sternorrhynca: Aphidiformes). Likewise, in the Chelicerata, OT/VP-like signaling appears to be present in mites (Acari) and scorpions (Scorpiones) but absent in spiders (Aranea; Liutkevičiūtė et al., 2016). The seven insect orders that appeared to have completely lost inotocin signaling are the Mantophasmatodea, Strepsiptera, Trichoptera, Lepidoptera, Siphonaptera, Mecoptera and

[c] A functional signaling system requires the expression of precursor and receptor; lack of either may be due to low quality of transcriptome data.

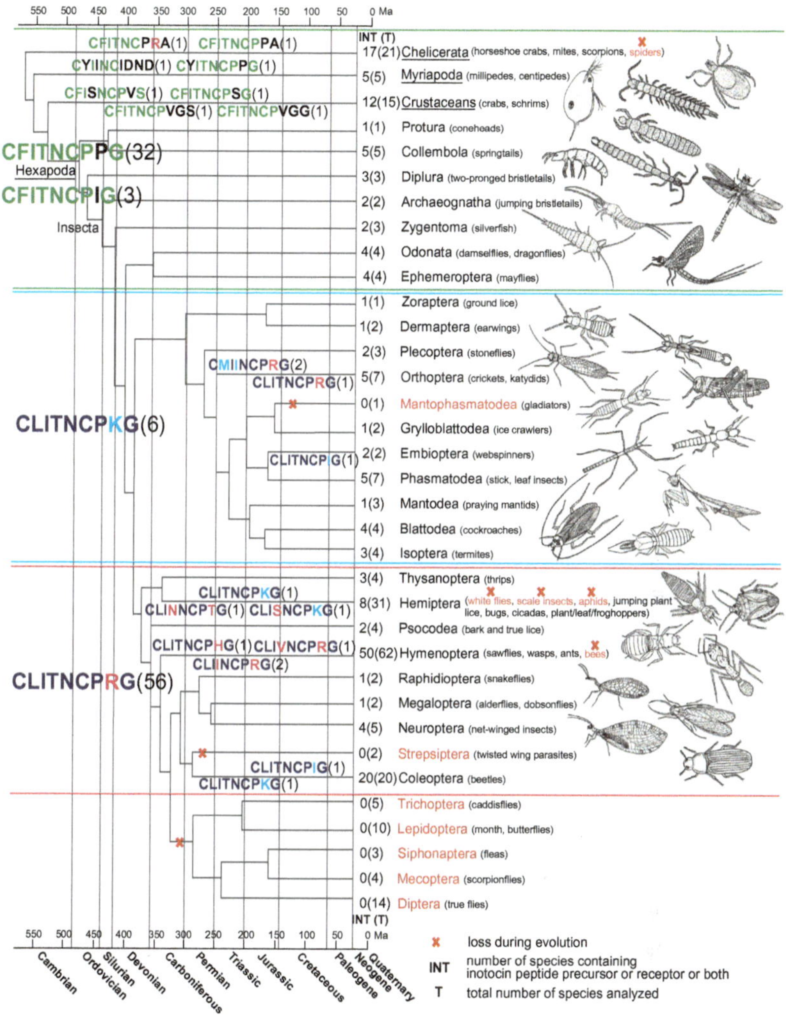

Fig. 2 Map of inotocin peptide GPCR signaling in arthropods. A phylogenetic map based on the recently established insect phylogeny (Misof et al., 2014) is shown. The absence of the inotocin signaling system is highlighted in red. Different putative peptide sequences are shown in different colors to indicate the diversity and distribution throughout the phylogeny. The numbers in brackets next to the peptide sequences indicate the frequencies of occurrence (from a total of 121 analyzed precursor sequences that contained a mature peptide domain). Number of species where the inotocin signaling system (receptor and/or precursor) is present (INT) as well as the total number of sampled species (T) has been indicated next to the tree branches. The inotocin signaling system is confined to specific groups of arthropods. For clarity of this phylogenetic illustration, the upper three groups of Arthropoda (Chelicerata, Myriapoda and Crustaceans; underlined) represent subphyla. All other groups denote orders of the subphylum Hexapoda; the class of Insecta comprises the orders Archaeognatha to Diptera. *Reproduced from Liutkevičiūtė, Z., Koehbach, J., Eder, T., Gil-Mansilla, E., & Gruber, C. W. (2016). Global map of oxytocin/vasopressin-like neuropeptide signalling in insects.* Scientific Reports, 6, 39177. *10.1038/srep39177 with permission under a Creative Commons Attribution 4.0 International License. To view a copy of this license, visit http:/creativecommons.org/licenses/by/4.0/.*

Diptera (although both propeptide and receptor sequences somewhat resembling those of OT/VP have been found in the New Zealand glow worm, *Arachnocampa luminosa*, which belongs to a basal family of Diptera; Sharpe, Dearden, Gimenez, & Krause, 2015). These results support the hypothesis that inotocin signaling was present in ancestral arthropod lineages but has since been lost multiple times (Liutkevičiūtė et al., 2016). In addition to their phylogenetic study, Liutkevičiūtė et al. (2016) further analyzed the sequence variability of inotocin peptides and receptors. Most analyzed species have one gene encoding the inotocin receptor and one gene encoding the cognate inotocin peptide. Liutkevičiūtė and colleagues (Liutkevičiūtė et al., 2016) observed that all identified inotocin precursor proteins exhibit a high degree of homology. There are several exceptions to this general observation: for instance, in the crustacean, *Xibalbanus tulumensis* (formerly known as *Speleonectes tulumensis*), belonging to the Remipedes, a sister group to the Hexapoda, no amidation processing signal could be identified. Additionally, while all other beetle data comprise only one precursor copy encoding the mature inotocin peptide CLITNCP**R**G*, the Colorado potato beetle (*L. decemlineata*) contains two copies encoding two distinct inotocin-like peptides (CLITNCP**K**G* and CLITNCP**I**G*). Liutkevičiūtė et al. (2016) also identified inotocin-like decapeptides in the centipede (Myriapoda), *Scolopendra subspinipes*, and the Copepod sea-lice (Crustacea), *Lepeophtheirus salmonis* and *Caligus rogercresseyi*. Interestingly, six of the identified multi-copy receptor species were discovered in the Chelicerata and one in Myriapoda, but not a single species within the Hexapoda. Liutkevičiūtė et al. (2016) also analyzed four different OT/VP-like receptor transcripts from the crustacean water flea (Cladocera) *Daphnia magna* and found that they were almost identical except for minor deletions and insertions present in the intracellular loop 3 and transmembrane domains 3 and 5.

Overall, these data have considerably extended our knowledge regarding the phylogenetic distribution and molecular structure of inotocin GPCR signaling in insects and other arthropods. Nevertheless, sequence similarity between different insect species does not always correspond to functional similarity and *vice versa*, so it is of crucial importance that we continue to functionally characterize and compare inotocin GPCR signaling in different insect species. Such comparative analysis may enable the identification of common, as well as distinct, features of OT/VP physiology across the animal kingdom.

6. Possible applications of insect inotocin ligands in pharmacology

Following this review of the discovery, distribution and physiology of the insect OT/VP signaling system, we will now provide a few examples of how these discoveries could be exploited. Amongst the diversity of insect inotocin ligands, there could be novel chemical probes or pharmacological ligands for human OT/VP receptors. The OT/VP signaling system plays a role in a number of physiological functions and behaviors, such as anti-diuretic action, vasoconstriction and lactation (summarized in Gruber et al., 2010; Jurek & Neumann, 2018), aggression (Bosch, Meddle, Beiderbeck, Douglas, & Neumann, 2005; Wersinger, Ginns, O'Carroll, Lolait, & Young, 2002), attachment (Insel & Young, 2001), anxiety-related behavior (Landgraf, 2006), fear conditioning (Eckstein et al., 2015), labor (Gruber & O'Brien, 2011), social recognition (Kirsch et al., 2005), and memory and learning (Alescio-Lautier & Soumireu-Mourat, 1998). Consequently, malfunctioning of this signaling system may contribute to various disorders and diseases, including autism, social anxiety disorder, schizophrenia, borderline personality disorder, depression, preterm-labor (Gruber & O'Brien, 2011), cardiovascular diseases and cancer (Goldman, Marlow-O'Connor, Torres, & Carter, 2008; Gruber & O'Brien, 2011; Keri, Kiss, & Kelemen, 2009; McQuaid, McInnis, Abizaid, & Anisman, 2014; Meyer-Lindenberg et al., 2011; Scantamburlo et al., 2007). Despite progress in the development of probes for the modulation of OT/VP receptors, assigning specific functions to particular receptors is still challenging. This problem could be addressed by the development of novel and more receptor-specific ligands, which could then lead to highly targeted therapeutics for diseases associated with different OT/VP receptors.

There has been sustained industrial and academic research over the past half-century to design and synthesize novel OT and VP analogs, resulting in several commercial OT/VP agonists and antagonists. The main challenge in the development of selective ligands is the high sequence similarity of the OT/VP receptors, particularly in the extracellular region, where the approximately 80% homology is one of the main obstacles in drug development (Koehbach, Stockner, Bergmayr, Muttenthaler, & Gruber, 2013). Receptor selectivity should ideally be greater than 2 orders of magnitude (defined as the difference in affinity of a compound to bind to a single receptor isoform over the remaining ones). In reality this has only been achieved by very few VP analogs (e.g., desmopressin, d[Cha4]AVP, d[Leu4]AVP; Manning et al., 2012) and selectivity is generally lost at higher doses. Additionally, receptor selectivity is species dependent: a compound selective for mouse receptors is not necessarily selective in humans, which poses an additional obstacle in the use of animal models for drug development. Furthermore, *in vivo* selectivity is enhanced through various other factors such as receptor up- and downregulation, local ligand production and degradation (Gruber et al., 2012). Finally, there have been reports of functional OT and VP receptor homo- and heterodimers, which make it difficult to predict the signaling profile and selectivity of any given ligand (Terrillon et al., 2003).

All of the above have hampered the development of specific small molecule agonists and antagonists, to the point where companies such as Merck, Pfizer and Sanofi have abandoned their development programs of non-peptide based compounds (Gruber et al., 2012; Manning et al., 2012). Peptide ligands, however, have a unique potential to overcome these constraints. Despite their handicaps of a short half-life, and difficulty crossing the blood-brain barrier, an advantage of peptides over small molecules is their potential to interact with a larger surface area of the receptor, as well as their greater stereochemical and structural complexity.

Rational design of peptide ligands has yielded several semi-selective therapeutics. For example, atosiban is a competitive OT/VP receptor antagonist first described in 1985. It is a nonapeptide, desamino-oxytocin analog most commonly used intravenously to halt preterm labor (Akerlund, Carlsson, Melin, & Trojnar, 1985). Another example, desmopressin (1-desamino-8-D-arginine vasopressin), is a synthetic human VP derivative. In contrast to VP, the first cysteine residue has been deaminated *via* replacement with β-mercaptopropionic acid and the arginine residue in position eight has been switched from levorotation to dextrorotation. It is relatively selective for the human V_2 receptor, where it is used as an agonist in the treatment of diabetes insipidus (Vavra et al., 1968).

An alternative to the rational design of ligands is the screening of naturally occurring compounds, utilizing the evolutionary conservation of peptide-ligands as a starting point. This approach has recently been successfully used to develop a selective human V_{1a} receptor antagonist. Inotocin was isolated from the black garden ant, *L. niger*, and investigated for its pharmacological properties on human OT/VP receptors. It was discovered that inotocin activates the human V_{1b} receptor while inhibiting the human V_{1a} receptor. This functional dichotomy was then exploited by replacing the arginine residue in position eight of inotocin with its stereoisomer D-arginine, thereby creating a stable, competitive V_{1a} receptor-antagonist with a 3000-fold increased binding selectivity toward the V_{1a} receptor compared to the other human receptor subtypes (Di Giglio et al., 2017). Another rich source of bioactive peptides are so-called conus-toxins, derived from cone snails (*Conus* spp.). Conopressin-T, a peptide isolated from *C. tulipa* venom, is a selective antagonist of the human V_{1a} receptor. Substitution of the valine in position nine with glycine switches the peptide from antagonism to agonism (Dutertre et al., 2008).

The distribution of OT/VP-like peptides of pharmacological interest is therefore not limited to vertebrates but also extends to invertebrates, which provide a considerable pool of candidates and vantage points for the development of superior, selective molecular probes for both OT and VP receptors. The survey by Liutkevičiūtė et al. (2016) revealed 11 different OT/VP-like peptide sequences just within insects, and more probably remain to be discovered. Despite limited knowledge about the function of this system in insects, where there is evidence for regulation of metabolism and locomotor activity, inotocin-like peptide precursors and cognate receptors have been identified in hundreds of arthropod species (Liutkevičiūtė et al., 2016, 2018), such as arachnotocin in the mite *Varroa destructor*. Moreover, distinct OT-, VP- and inotocin-like peptides have been identified in a growing number of other invertebrates, including molluscs (octopressin and cephalotocin in *Octopus vulgaris*), annelids (annetocin in oligochaetes), nematodes (nematocin) and starfish (asterotocin) where they generally modulate social and reproductive behavior (Donaldson & Young, 2008; Gruber, 2014), thus indicating the presence of an even wider pool of neuropeptide candidates.

Plants are another potential source of therapeutic lead compounds. Ethnopharmacological investigations of uterotonic plant preparations have identified the cyclotide kalata B7, which has been successfully employed as a template for the design of synthetic OT-like peptides with nanomolar affinity and selectivity for the oxytocin receptor (Koehbach, O'Brien, et al., 2013).

In conclusion, we believe that invertebrate OT/VP-like signaling systems are worthwhile targets for future research. Due to the functional conservation of the system, any new knowledge gleaned from insects and other invertebrates might be applicable for translational therapeutic research in humans. In addition, studying insect OT/VP signaling systems could also provide insights and valuable lead compounds for the development of peptide-derived OT/VP ligands. In contrast to small molecule ligands, these compounds could overcome the problem of receptor selectivity, an important limitation in the pursuit of highly effective OT/VP therapeutics and probes. Breakthroughs in this area will result in a deeper understanding of this fascinating neuroendocrine system both in humans and other animals, and could also provide novel treatments for a diverse range of disorders and diseases.

Acknowledgments

The study of insect OT/VP-like signaling in the laboratories of Christian Gruber and Jozef Vanden Broeck is supported by the Austrian Science Fund FWF (I3243) and the Fund for Scientific Research—Flanders FWO (G0F2417N).

References

Aikins, M. J., Schooley, D. A., Begum, K., Detheux, M., Beeman, R. W., & Park, Y. (2008). Vasopressin-like peptide and its receptor function in an indirect diuretic signaling pathway in the red flour beetle. *Insect Biochemistry and Molecular Biology*, *38*, 740–748. https://doi.org/10.1016/j.ibmb.2008.04.006.

Akerlund, M., Carlsson, A. M., Melin, P., & Trojnar, J. (1985). The effect on the human uterus of two newly developed competitive inhibitors of oxytocin and vasopressin. *Acta Obstetricia et Gynecologica Scandinavica*, *64*(6), 499–504.

Alescio-Lautier, B., & Soumireu-Mourat, B. (1998). Role of vasopressin in learning and memory in the hippocampus. *Progress in Brain Research*, *119*, 501–521.

Baines, R. A., & Bacon, J. P. (1994). Pharmacological analysis of the cholinergic input to the locust VPLI neuron from an extraocular photoreceptor system. *Journal of Neurophysiology*, *72*, 2864–2874. https://doi.org/10.1152/jn.1994.72.6.2864.

Baines, R. A., Thompson, K. S. J., Rayne, R. C., & Bacon, J. P. (1995). Analysis of the peptide content of the locust vasopressin-like immunoreactive (VPLI) neurons. *Peptides*, *16*, 799–807. https://doi.org/10.1016/0196-9781(95)00038-L.

Barberis, C., Mouillac, B., & Durroux, T. (1998). Structural bases of vasopressin/oxytocin receptor function. *The Journal of Endocrinology*, *156*(2), 223–229.

Beets, I., Janssen, T., Meelkop, E., Temmerman, L., Suetens, N., Rademakers, S., et al. (2012). Vasopressin/oxytocin-related signaling regulates gustatory associative learning in *C. elegans*. *Science*, *338*(6106), 543–545. https://doi.org/10.1126/science.1226860.

Bosch, O. J., Meddle, S. L., Beiderbeck, D. I., Douglas, A. J., & Neumann, I. D. (2005). Brain oxytocin correlates with maternal aggression: Link to anxiety. *The Journal of Neuroscience*, *25*(29), 6807–6815. https://doi.org/10.1523/JNEUROSCI.1342-05.2005.

Caldwell, H. K., & Young, W. S. (2006). *Oxytocin and vasopressin: Genetics and behavioral implications* (*Vol. 495*) (3rd ed.). Springer-Verlag New York Inc.

Chérasse, S., & Aron, S. (2017). Measuring inotocin receptor gene expression in chronological order in ant queens. *Hormones and Behavior*, *96*, 116–121. https://doi.org/10.1016/j.yhbeh.2017.09.009.

Coast, G. M., Rayne, R. C., Hayes, T. K., Mallet, A. I., Thompson, K. S. J., & Bacon, J. P. (1993). A comparison of the effects of two putative diuretic hormones from Locusta migratoria on isolated locust malpighian tubules. *The Journal of Experimental Biology*, *175*, 1–14.

Di Giglio, M. G., Muttenthaler, M., Harpsøe, K., Liutkeviciute, Z., Keov, P., Eder, T., et al. (2017). Development of a human vasopressin V_{1a}-receptor antagonist from an evolutionary-related insect neuropeptide. *Scientific Reports*, 7, 41002. https://doi.org/10.1038/srep41002.

Donaldson, Z. R., & Young, L. J. (2008). Oxytocin, vasopressin, and the neurogenetics of sociality. *Science*, *322*, 900–904.

Dutertre, S., Croker, D., Daly, N. L., Andersson, A., Muttenthaler, M., Lumsden, N. G., et al. (2008). Conopressin-T from Conus tulipa reveals an antagonist switch in vasopressin-like peptides. *Journal of Biological Chemistry*, *283*(11), 7100–7108. https://doi.org/10.1074/jbc.M706477200.

Eckstein, M., Becker, B., Scheele, D., Scholz, C., Preckel, K., Schlaepfer, T. E., et al. (2015). Oxytocin facilitates the extinction of conditioned fear in humans. *Biological Psychiatry*, *78*(3), 194–202. https://doi.org/10.1016/j.biopsych.2014.10.015.

Elphick, M. R. (2010). NG peptides: A novel family of neurophysin-associated neuropeptides. *Gene*, *458*(1–2), 20–26. https://doi.org/10.1016/j.gene.2010.03.004.

Evans, P. D., & Cournil, I. (1990). Co-localization of FLRF- and vasopressin-like immunoreactivity in a single pair of sexually dimorphic neurones in the nervous system of the locust. *The Journal of Comparative Neurology*, *292*, 331–348. https://doi.org/10.1002/cne.902920302.

Fredriksson, R., Lagerstrom, M. C., Lundin, L. G., & Schioth, H. B. (2003). The G-protein-coupled receptors in the human genome form five main families. Phylogenetic analysis, paralogon groups, and fingerprints. *Molecular Pharmacology*, *63*(6), 1256–1272. https://doi.org/10.1124/mol.63.6.1256.

Furuya, K., Schegg, K. M., Wang, H., Kingt, D. S., & Schooley, D. A. (1995). Isolation and identification of a diuretic hormone from the mealworm Tenebrio molitor (water balance/corticotropin-releasing factor/cyclic AMP/Malpighian tubules). *Proceedings of the National Academy of Sciences*, *92*, 12323–12327.

Gäde, G. (2004). Regulation of intermediary metabolism and water balance of insects by neuropeptides. *Annual Review of Entomology*, *49*, 93–113. https://doi.org/10.1146/annurev.ento.49.061802.123354.

Garrison, J. L., Macosko, E. Z., Bernstein, S., Pokala, N., Albrecht, D. R., & Bargmann, C. I. (2012). Oxytocin/vasopressin-related peptides have an ancient role in reproductive behavior. *Science*, *338*(6106), 540–543. https://doi.org/10.1126/science.1226201.

Gimpl, G., & Fahrenholz, F. (2001). The oxytocin receptor system: Structure, function, and regulation. *Physiological Reviews*, *81*(2), 629–683. https://doi.org/10.1152/physrev.2001.81.2.629.

Godwin, J., & Thompson, R. (2012). Nonapeptides and social behavior in fishes. *Hormones and Behavior*, *61*(3), 230–238. https://doi.org/10.1016/j.yhbeh.2011.12.016.

Goldman, M., Marlow-O'Connor, M., Torres, I., & Carter, C. S. (2008). Diminished plasma oxytocin in schizophrenic patients with neuroendocrine dysfunction and emotional deficits. *Schizophrenia Research*, *98*(1–3), 247–255. https://doi.org/10.1016/j.schres.2007.09.019.

Gruber, C. W. (2014). Physiology of invertebrate oxytocin and vasopressin neuropeptides. *Experimental Physiology*, *99*(1), 55–61. https://doi.org/10.1113/expphysiol.2013.072561.

Gruber, C. W., Koehbach, J., & Muttenthaler, M. (2012). Exploring bioactive peptides from natural sources for oxytocin and vasopressin drug discovery. *Future Medicinal Chemistry*, *4*(14), 1791–1798. https://doi.org/10.4155/fmc.12.108.

Gruber, C. W., & Muttenthaler, M. (2012). Discovery of defense- and neuropeptides in social ants by genome-mining. *PLoS One*, 7(3), e32559. https://doi.org/10.1371/journal.pone.0032559.

Gruber, C. W., Muttenthaler, M., & Freissmuth, M. (2010). Ligand-based peptide design and combinatorial peptide libraries to target G protein-coupled receptors. *Current Pharmaceutical Design*, *16*(28), 3071–3088.

Gruber, C. W., & O'Brien, M. (2011). Uterotonic plants and their bioactive constituents. *Planta Medica*, 77(3), 207–220. https://doi.org/10.1055/s-0030-1250317.

Insel, T. R., & Young, L. J. (2001). The neurobiology of attachment. *Nature Reviews. Neuroscience*, *2*(2), 129–136. https://doi.org/10.1038/35053579.

Jurek, B., & Neumann, I. D. (2018). The oxytocin receptor: From intracellular signaling to behavior. *Physiological Reviews*, *98*(3), 1805–1908. https://doi.org/10.1152/physrev.00031.2017.

Keov, P., Liutkevičiūtė, Z., Hellinger, R., Clark, R. J., & Gruber, C. W. (2018). Discovery of peptide probes to modulate oxytocin-type receptors of insects. *Scientific Reports*, *8*, 10020. https://doi.org/10.1038/s41598-018-28380-3.

Keri, S., Kiss, I., & Kelemen, O. (2009). Sharing secrets: Oxytocin and trust in schizophrenia. *Social Neuroscience*, *4*(4), 287–293. https://doi.org/10.1080/17470910802319710.

Kirsch, P., Esslinger, C., Chen, Q., Mier, D., Lis, S., Siddhanti, S., et al. (2005). Oxytocin modulates neural circuitry for social cognition and fear in humans. *The Journal of Neuroscience*, *25*(49), 11489–11493. https://doi.org/10.1523/JNEUROSCI.3984-05.2005.

Koehbach, J., O'Brien, M., Muttenthaler, M., Miazzo, M., Akcan, M., Elliott, A. G., et al. (2013). Oxytocic plant cyclotides as templates for peptide G protein-coupled receptor ligand design. *Proceedings of the National Academy of Sciences of the United States of America*, *110*(52), 21183–21188. https://doi.org/10.1073/pnas.1311183110.

Koehbach, J., Stockner, T., Bergmayr, C., Muttenthaler, M., & Gruber, C. W. (2013). Insights into the molecular evolution of oxytocin receptor ligand binding. *Biochemical Society Transactions*, *41*(1), 197–204. https://doi.org/10.1042/BST20120256.

Koto, et al. (2019). Oxytocin/vasopressin-like peptide inotocin regulates cuticular hydrocarbon synthesis and water balancing in ants. *Proceedings of the National Academy of Sciences of the United States of America*, *116*(12), 5597–5606.

Kroeze, W. K., Sheffler, D. J., & Roth, B. L. (2003). G-protein-coupled receptors at a glance. *Journal of Cell Science*, *116*(Pt. 24), 4867–4869. https://doi.org/10.1242/jcs.00902.

Landgraf, R. (2006). The involvement of the vasopressin system in stress-related disorders. *CNS & Neurological Disorders Drug Targets*, *5*(2), 167–179.

Li, J. D., Burton, K. J., Zhang, C., Hu, S. B., & Zhou, Q. Y. (2009). Vasopressin receptor V1a regulates circadian rhythms of locomotor activity and expression of clock-controlled genes in the suprachiasmatic nuclei. *American Journal of Physiology Regulatory, Integrative and Comparative Physiology*, *296*(3), R824–R830. https://doi.org/10.1152/ajpregu.90463.2008.

Li, B., Predel, R., Neupert, S., Hauser, F., Tanaka, Y., Cazzamali, G., et al. (2008). Genomics, transcriptomics, and peptidomics of neuropeptides and protein hormones in the red flour beetle *Tribolium castaneum*. *Genome Research*, *18*, 113–122. https://doi.org/10.1101/gr.6714008.

Liutkevičiūtė, Z., Gil-Mansilla, E., Eder, T., Casillas-Pérez, B., Di Giglio, M. G., Muratspahić, E., et al. (2018). Oxytocin-like signaling in ants influences metabolic gene expression and locomotor activity. *The FASEB Journal*, *32*, fj201800443. https://doi.org/10.1096/fj.201800443.

Liutkevičiūtė, Z., & Gruber, C. W. (2016). Evolutionary aspects of physiological function and molecular diversity of the oxytocin/vasopressin signaling system. In D. Murphy & H. Gainer (Eds.), *Molecular neuroendocrinology: From genome to physiology* (pp. 5–23). Wiley-Blackwell.

Liutkevičiūtė, Z., Koehbach, J., Eder, T., Gil-Mansilla, E., & Gruber, C. W. (2016). Global map of oxytocin/vasopressin-like neuropeptide signalling in insects. *Scientific Reports*, *6*, 39177. https://doi.org/10.1038/srep39177.

Lundquist, C., Baines, R. A., & Bacon, J. P. (1996). Evidence that histamine is a neurotransmitter in an insect extraocular photoreceptor pathway. *Journal of Experimental Biology, 199*, 1973–1982.

Lundquist, C. T., Baines, R. A., Thompson, K. S. J., & Bacon, J. P. (1998). Differential regulation of a homologous peptidergic neuron in grasshoppers: Evolution of a neural circuit. *Journal of Comparative Physiology A, 182*, 755–765.

Manning, M., Misicka, A., Olma, A., Bankowski, K., Stoev, S., Chini, B., et al. (2012). Oxytocin and vasopressin agonists and antagonists as research tools and potential therapeutics. *Journal of Neuroendocrinology, 24*(4), 609–628. https://doi.org/10.1111/j.1365-2826.2012.02303.x.

McCall, C., & Singer, T. (2012). The animal and human neuroendocrinology of social cognition, motivation and behavior. *Nature Neuroscience, 15*(5), 681–688. https://doi.org/10.1038/nn.3084.

McQuaid, R. J., McInnis, O. A., Abizaid, A., & Anisman, H. (2014). Making room for oxytocin in understanding depression. *Neuroscience and Biobehavioral Reviews, 45*, 305–322. https://doi.org/10.1016/j.neubiorev.2014.07.005.

Meyer-Lindenberg, A., Domes, G., Kirsch, P., & Heinrichs, M. (2011). Oxytocin and vasopressin in the human brain: Social neuropeptides for translational medicine. *Nature Reviews. Neuroscience, 12*(9), 524–538. http://www.nature.com/nrn/journal/v12/n9/suppinfo/nrn3044_S1.html.

Misof, B., Liu, S., Meusemann, K., Peters, R. S., Donath, A., Mayer, C., et al. (2014). Phylogenomics resolves the timing and pattern of insect evolution. *Science, 346*(6210), 763–767. https://doi.org/10.1126/science.1257570.

Montuenga, L. M., Zudaire, E., Prado, M. A., Audsley, N., Burrell, M. A., & Coast, G. M. (1996). Presence of Locusta diuretic hormone in endocrine cells of the ampullae of locust Malpighian tubules. *Cell and Tissue Research, 285*, 331–339. https://doi.org/10.1007/s004410050650.

Picquot, M., & Proux, J. P. (1987). Relationship between excretion of primary urine and haemolymph level of diuretic hormone in the migratory locust. *Physiological Entomology, 12*, 455–460. https://doi.org/10.1111/j.1365-3032.1987.tb00772.x.

Picquot, M., & Proux, J. P. (1989). Influence of hydration on the concentration of two arginine-vasopressin-like neuropeptides coexisting in the ventral nerve cord of the migratory locust, Locusta migratoria. *Journal of Insect Physiology, 35*, 571–577.

Picquot, M., & Proux, J. P. (1990). Biosynthesis and degradation of the arginine-vasopressin-like insect diuretic hormone, a neurohormone in the migratory locust. *Regulatory Peptides, 31*, 139–156. https://doi.org/10.1016/0167-0115(90)90001-D.

Proux, J. P., & Herault, J.-P. (1988). Cyclic AMP: A second messenger of the newly characterized AVP-like insect diuretic hormone, the migratory locust diuretic hormone. *Neuropeptides, 12*, 7–12. https://doi.org/10.1016/S0143-4179(98)90003-8.

Proux, J. P., Miller, C. A., Li, J. P., Carney, R. L., Girardie, A., Delaage, M., et al. (1987). Identification of an arginine vasopressin-like diuretic hormone from Locusta migratoria. *Biochemical and Biophysical Research Communications, 149*, 180–186. https://doi.org/10.1016/0006-291X(87)91621-4.

Proux, J. P., Picquot, M., Herault, J.-P., & Fournier, B. (1988). Diuretic activity of a newly identified neuropeptide—The arginine-vasopressin-like insect diuretic hormone: Use of an improved bioassay. *Journal of Insect Physiology, 34*, 919–927. https://doi.org/10.1016/0022-1910(88)90127-8.

Proux, J. P., Rougon, G., & Cupo, A. (1982). Enhancement of excretion across locust Malpighian tubules by a diuretic vasopressin-like hormone. *General and Comparative Endocrinology, 47*, 449–457. https://doi.org/10.1016/0016-6480(82)90123-X.

Proux, J. P., & Rougon-Rapuzzi, G. (1980). Evidence for vasopressin-like molecule in migratory locust. Radioimmunological measurements in different tissues: Correlation with various states of hydration. *General and Comparative Endocrinology*, *42*, 378–383. https://doi.org/10.1016/0016-6480(80)90168-9.

Rémy, C., & Girardie, J. (1980). Anatomical organization of two vasopressin-neurophysin-like neurosecretory cells throughout the central nervous system of the migratory locust. *General and Comparative Endocrinology*, *40*, 27–35. https://doi.org/10.1016/0016-6480(80)90092-1.

Rémy, C., Girardie, J., & Dubois, M. P. (1979). Vertebrate neuropeptide-like substances in the subesophageal ganglion of two insects: Locusta migratoria R. and F. (Orthoptera) and Bombyx mori L. (Lepidoptera) immunocytological investigation. *General and Comparative Endocrinology*, *37*, 93–100. https://doi.org/10.1016/0016-6480(79)90050-9.

Reppert, S. M., Schwartz, W. J., & Uhl, G. R. (1987). Arginine vasopressin: A novel peptide rhythm in cerebrospinal fluid. *Trends in Neurosciences*, *10*, 76–80. https://doi.org/10.1016/0166-2236(87)90029-4.

Scantamburlo, G., Hansenne, M., Fuchs, S., Pitchot, W., Marechal, P., Pequeux, C., et al. (2007). Plasma oxytocin levels and anxiety in patients with major depression. *Psychoneuroendocrinology*, *32*(4), 407–410. https://doi.org/10.1016/j.psyneuen.2007.01.009.

Schlemermeyer, E., Schütte, M., & Ammermüller, J. (1989). Immunohistochemical and electrophysiological evidence that locust ocellar photoreceptors contain and release histamine. *Neuroscience Letters*, *99*, 73–78. https://doi.org/10.1016/0304-3940(89)90267-X.

Sharpe, M. L., Dearden, P. K., Gimenez, G., & Krause, K. L. (2015). Comparative RNA seq analysis of the New Zealand glowworm Arachnocampa luminosa reveals bioluminescence-related genes. *BMC Genomics*, *16*, 825. https://doi.org/10.1186/s12864-015-2006-2.

Simpson, S. J., & Sword, G. A. (2008). Locusts. *Current Biology*, *18*, R364–R366. https://doi.org/10.1016/j.cub.2008.02.029.

Spring, J. H. (1990). Endocrine regulation of diuresis in insects. *Journal of Insect Physiology*, *36*, 13–22. https://doi.org/10.1016/0022-1910(90)90146-7.

Stafflinger, E., Hansen, K. K., Hauser, F., Schneider, M., Cazzamali, G., Williamson, M., et al. (2008). Cloning and identification of an oxytocin/vasopressin-like receptor and its ligand from insects. *Proceedings of the National Academy of Sciences of the United States of America*, *105*, 3262–3267. https://doi.org/10.1073/pnas.0710897105.

Terrillon, S., Durroux, T., Mouillac, B., Breit, A., Ayoub, M. A., Taulan, M., et al. (2003). Oxytocin and vasopressin V1a and V2 receptors form constitutive homo- and heterodimers during biosynthesis. *Molecular Endocrinology*, *17*(4), 677–691. https://doi.org/10.1210/me.2002-0222.

Thompson, K. S. J., & Bacon, J. P. (1991). The vasopressin-like immunoreactive (VPLI) neurons of the locust, Locusta migratoria. II. Physiology. *Journal of Comparative Physiology. A, Sensory, Neural, and Behavioral Physiology*, *168*, 619–630.

Thompson, K. S., Tyrer, N. M., May, S. T., & Bacon, J. P. (1991). The vasopressin-like immunoreactive (VPLI) neurons of the locust, Locusta migratoria. I. Anatomy. *Journal of Comparative Physiology. A, Sensory, Neural, and Behavioral Physiology*, *168*, 605–617.

Tyrer, N. M., Davis, N. T., Arbas, E. A., Thompson, K. S. J., & Bacon, J. P. (1993). Morphology of the vasopressin-like immunoreactive (VPLI) neurons in many species of grasshopper. *The Journal of Comparative Neurology*, *329*, 385–401. https://doi.org/10.1002/cne.903290309.

Vanden Broeck, J. J. (1996). G-protein-coupled receptors in insect cells. *International Review of Cytology*, *164*, 189–268.

Vavra, I., Machova, A., Holecek, V., Cort, J. H., Zaoral, M., & Sorm, F. (1968). Effect of a synthetic analogue of vasopressin in animals and in patients with diabetes insipidus. *Lancet*, *1*(7549), 948–952.

Veenstra, J. A. (1984). Immunocytochemical demonstration of a homology in peptidergic neurosecretory cells in the suboesophageal ganglion of a beetle and a locust with antisera to bovine pancreatic polypeptide, FMRFamide, vasopressin and α-MSH. *Neuroscience Letters*, *48*, 185–190. https://doi.org/10.1016/0304-3940(84)90017-X.

Veenstra, J. A., Romherg-Privee, H. M., & Schooneveld, H. (1984). Immunocytochemical localization of peptidergic cells in the neuro-endocrine system of the Colorado potato beetle Leptinotarsa decemlineata, with antisera against vasopressin, vasotocin and oxytocin. *Histochemistry*, *81*, 29–34.

Wacker, D., Stevens, R. C., & Roth, B. L. (2017). How ligands illuminate GPCR molecular pharmacology. *Cell*, *170*(3), 414–427. https://doi.org/10.1016/j.cell.2017.07.009.

Wang, X., Fang, X., Yang, P., Jiang, X., Jiang, F., Zhao, D., et al. (2014). The locust genome provides insight into swarm formation and long-distance flight. *Nature Communications*, *5*, 2957. https://doi.org/10.1038/ncomms3957.

Wersinger, S. R., Ginns, E. I., O'Carroll, A. M., Lolait, S. J., & Young, W. S., 3rd. (2002). Vasopressin V1b receptor knockout reduces aggressive behavior in male mice. *Molecular Psychiatry*, 7(9), 975–984. https://doi.org/10.1038/sj.mp.4001195.

Wiehart, U. I. M., Torfs, P., Van Lommel, A., Nicolson, S. W., & Schoofs, L. (2002). Immunocytochemical localization of a diuretic hormone of the beetle Tenebrio molitor, Tenmo-DH(37), in nervous system and midgut. *Cell and Tissue Research*, *308*, 421–429. https://doi.org/10.1007/s00441-002-0552-9.

CHAPTER THREE

Amyloid-like aggregation of provasopressin

Martin Spiess*, Nicole Beuret, Cristina Prescianotto Baschong, Jonas Rutishauser
Biozentrum, University of Basel, Basel, Switzerland
*Corresponding author: e-mail address: martin.spiess@unibas.ch

Contents

Abstract

The antidiuretic hormone vasopressin is synthesized as a longer precursor protein. After folding in the endoplasmic reticulum (ER), provasopressin is transported through the secretory pathway, forms secretory granules in the trans-Golgi network (TGN), is processed, and finally secreted into the circulation. Mutations in provasopressin cause autosomal dominant diabetes insipidus. They prevent native protein folding and cause fibrillar, amyloid-like aggregation in the ER, which eventually results in cell death. Secretory granules of peptide hormones were proposed to constitute functional amyloids and thus might be the cause of amyloid formation of misfolded mutant protein in the ER. Indeed, the same two segments in the precursor—vasopressin and a C-terminal glycopeptide—were found to be responsible for pathological aggregation in the ER and physiological aggregation in granule formation in the TGN. Furthermore, even wild-type provasopressin

Vitamins and Hormones, Volume 113
ISSN 0083-6729
https://doi.org/10.1016/bs.vh.2019.08.014

tends to aggregate in the ER, but is controlled by ER-associated degradation. When essential components thereof, Sel1L or Hrd1, were inactivated, wild-type provasopressin accumulated as fibrillar aggregates in vasopressinergic neurons in mice, causing diabetes insipidus. Evolution of amyloidogenic sequences for granule formation thus made provasopressin dependent on ER quality control mechanisms. These principles may similarly apply to other peptide hormones.

1. Biosynthesis and secretion of vasopressin

1.1 The nonapeptide hormone vasopressin regulates water homeostasis

While more than 90% of the originally ~150–180 L/day of primary glomerular filtrate of the kidney is reabsorbed constitutively, the hormone vasopressin (also called arginine vasopressin [AVP] or antidiuretic hormone [ADH]) regulates reabsorption of the last ~12 L/day, typically limiting daily urine output to 0.8–2 L/day. Vasopressin is expressed mainly in vasopressinergic neurons of the supraoptic and paraventricular nuclei of the hypothalamus with axonal extensions to the neurohypophysis (posterior pituitary), from where the hormone is released into the blood stream in a regulated manner. Vasopressin secretion is triggered by increased osmolality of plasma and interstitium, reduced circulating blood volume, and reduced blood pressure (Robertson, 2001). By binding to type 2 vasopressin receptors in collecting duct principal cells of the kidney, vasopressin induces the insertion of aquaporin-2 water channels into the apical membrane, resulting in increased water reabsorption and thus water conservation.

The final processed hormone is a nonapeptide with an amidated C-terminus and an intramolecular disulfide bond connecting the two cysteines C1 and C6 (Fig. 1).

1.2 Precursor modification and folding in the endoplasmic reticulum (ER)

Vasopressin is initially synthesized as preprovasopressin, a larger precursor protein of 164 amino acids, that allows cotranslational targeting to the secretory pathway and translocation into the endoplasmic reticulum (ER). Preprovasopressin consists of (i) an N-terminal signal sequence for targeting to the ER, (ii) the vasopressin nonapeptide hormone, (iii) neurophysin II (NPII) of 93 residues, and (iv) a 39-amino acid C-terminal glycopeptide (also called copeptin; Fig. 2). As provasopressin is translocated into the

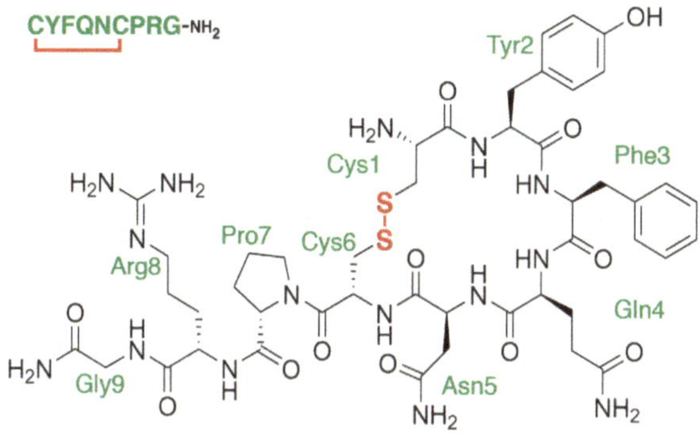

Fig. 1 Chemical structure of vasopressin.

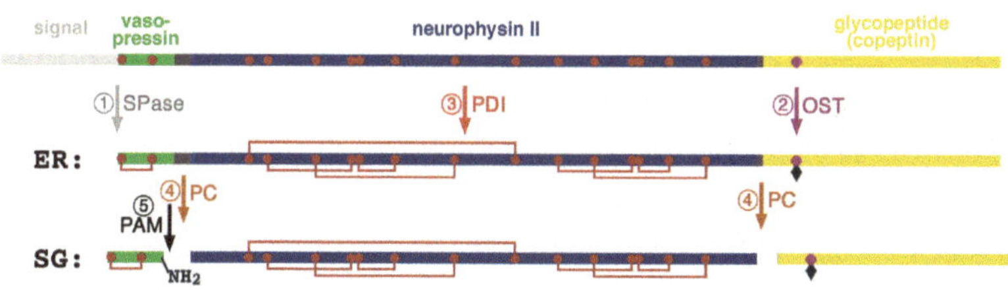

Fig. 2 Preprovasopressin domains and processing. The different segments of the precursor are shown in different colors with cysteines indicated as red dots, the site for N-glycosylation as a pink dot, and the glycan as a black diamond. In the ER, (1) the signal peptide is removed by signal peptidase (SPase), (2) the glycopeptide is glycosylated by oligosaccharyl transferase(OST), and (3) the 16 cysteines are oxidized to disulfide bonds (lines connecting cysteines) by protein disulfide isomerase (PDI). In secretory granules (SG), (4) vasopressin and glycopeptide are separated from NPII by prohormone convertase (PC) cleavage and (5) vasopressin is amidated by peptidyl-glycine α-amidating monooxygenase (PAM).

ER lumen, the signal sequence is cleaved off by signal peptidase, and an N-linked glycan is transferred onto the glycopeptide. Vasopressin and NPII fold together into a single globular domain with the glycopeptide as a separate extension.

Protein folding in the ER lumen is assisted by molecular chaperones such as the Hsp70 family member BiP (immunoglobulin Binding Protein or GRP-78). In the oxidative environment of the ER, disulfide bonds are formed between the 16 cysteines of vasopressin–NPII to produce 8 disulfide bonds, 1 in vasopressin and 7 in NPII. Disulfide bond formation and shuffling of incorrectly formed bonds is primarily catalyzed by protein disulfide isomerase (PDI). Chaperones associate with unfolded or incompletely folded polypeptide segments that expose hydrophobic sequences of the interior of the final (native) folded state, to keep them in solution. Concomitant with folding, correct disulfide bonds become inaccessible to PDI inside the fold.

BiP and PDI are also part of ER quality control mechanisms. While these chaperones are bound to a substrate protein, it is retained in the ER and prevented from exiting to the Golgi apparatus. A high load of chaperone-associated substrate proteins triggers two distinct processes to maintain protein homeostasis in the ER: the unfolded protein response (UPR) and ER-associated degradation (ERAD). For UPR, increased BiP binding to substrates reduces the concentration of free BiP and thereby activates three separate stress sensors, IRE1α, PERK, and ATF6, to trigger transcriptional upregulation of the ER folding capacity, particularly by increasing the amount of chaperones, PDI, and ERAD components, and translational downregulation to reduce the ER substrate load (Schröder & Kaufman, 2005; Wang & Kaufman, 2012). Upon prolonged failure to fold, misfolded proteins are targeted to cytosolic degradation by proteasomes. ERAD is initiated by different chaperones, including BiP, and involves the Sel1L–Hrd1 protein complex, consisting of the E3 ubiquitin ligase Hrd1 and its cofactor Sel1L (Ruggiano, Foresti, & Carvalho, 2014). Hrd1 is likely to mediate retrotranslocation through the ER membrane, energized by the p97/CDC48 ATPase.

1.3 Sorting into secretory granules, precursor processing, and regulated secretion

Folded provasopressin exits ER, is transported to the Golgi apparatus, and migrates from the *cis*- to the *trans*-side, where the relatively compact stack of cisternae end in a tubulo-vesicular cluster of membranes called the *trans*-Golgi network (TGN). From the TGN, membrane carriers emanate to transport cargo proteins to the endosomal-lysosomal system as well as to the plasma membrane (Guo, Sirkis, & Schekman, 2014). Peptide hormones in endocrine cells are concentrated in this compartment to produce secretory granules with characteristic dense-core appearance in electron microscopy. The mechanism of granule formation is not entirely clear. While the involvement of sorting receptors have been implicated in the process (Kim, Gondré-Lewis, Arnaoutova, & Loh, 2006), they have remained controversial. However, there is general agreement that self-aggregation of regulated cargo proteins in the specific conditions in the TGN is essential for granule formation. This notion is supported by the observation that expression of provasopressin and of other regulated cargo proteins (prohormones and granins) was found to be sufficient to produce dense granule-like structures in fibroblasts and other cell types lacking endocrine-specific machinery (Beuret, Stettler, Renold, Rutishauser, & Spiess, 2004). The environment within the TGN is characterized by a reduced pH (Paroutis, Touret, & Grinstein, 2004), high concentrations of Ca^{2+} and possibly other divalent cations such as Zn^{2+} and Mg^{2+} (Hutton, Penn, & Peshavaria, 1983; Thorlacius-Ussing, 1987), and the presence of glycosaminoglycans (Kolset, Prydz, & Pejler, 2004) that may contribute to self-aggregation, directly and/or via conformational changes in the protein.

In addition to prohormones, secretory granules also accumulate granins, acidic secretory proteins that coaggregate with peptide hormones (Taupenot, Harper, & O'Connor, 2003). Furthermore, enzymes for prohormone processing are included. In the case of provasopressin, the prohormone convertases PC1/3 and PC2 are the most likely candidates to separate vasopressin (residues 1–10 of provasopressin, to be exact) from NPII (residues 13–105) and from the glycopeptide (residues 107–145) (Birch, Hakes, Dixon, & Mezey, 1994). The C-terminal glycine of vasopressin 1–10 is converted by peptidyl-glycine α-amidating monooxygenase to a C-terminal amide and free glyoxylate (Kim & Seong, 2001).

The maturing secretory granules are transported along actin cables through the axon to the nerve endings in the neurohypophysis (the posterior pituitary), where they dock to the cell membrane. Vesicle fusion is stimulated by action potentials and results in the regulated release of granule contents into the blood stream. Secreted NPII and glycopeptide are eventually degraded. The glycopeptide/copeptin is useful as a diagnostic surrogate of vasopressin release (e.g., Fenske et al., 2018).

1.4 Structure of vasopressin–NPII

The structure of mature, processed NPII with and without bound vasopressin (Fig. 3) was solved by crystallography (Chen et al., 1991; Wu et al., 2001). The N-terminus of vasopressin in bound into a pocket in NPII. The C-terminus of the hormone as well as the N-terminus of NPII are not recovered in the structure most likely because they are not fixed in the crystal. The original connecting loop is thus flexible and exposed for cleavage by prohormone convertases. There is no structure available for the released glycopeptide.

Two segments in the NPII sequence (residues 12–31 and 60–76) show significant homology, indicating partial gene duplication. The disulfide

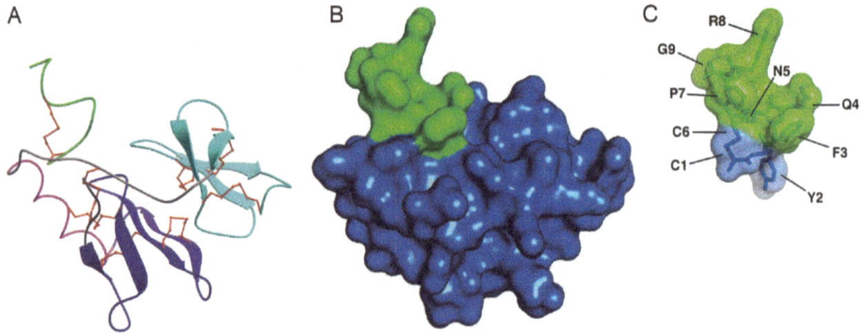

Fig. 3 Structure of vasopressin–NPII (Protein Data Bank 1JK4; Wu:2001gp). (A): The backbone structure of vasopressin is shown in green and of NPII with the N- and C-terminal β-sheets in dark and light blue, respectively, and the connecting α-helix in purple. Disulfide bonds are indicated in red. (B): Surface representation of NPII (blue) with bound vasopressin (green). Residues P7, R8, and G9, which were not resolved in the structure, are presented in an arbitrary conformation. (C): Vasopressin is shown with the portion interacting with NPII in blue. *Panels B and C from Beuret, N., Hasler, F., Prescianotto-Baschong, C., Birk, J., Rutishauser, J., & Spiess, M. (2017). Amyloid-like aggregation of provasopressin in diabetes insipidus and secretory granule sorting. BMC Biology, 15(1), 5. http:/doi.org/10.1186/s12915-017-0347-9.*

pairing and the secondary structure also indicate a bipartite organization, dividing the molecule into an N-terminal and a C-terminal subdomain. Each of them forms a four-stranded β-sheet with three disulfide bridges. They are linked by a single α-helix. Vasopressin inserts itself into a binding pocket in NPII. The α-amino group of vasopressin forms an ion bridge to the side chain carboxylate of E47 of NPII at its bottom. The disulfide bond between C1 and C6 and Y2 interact with a largely hydrophobic surface in the cavity. F3 contacts the surface NPII. Importantly, vasopressin is necessary for the folding of NPII (De Bree et al., 2003). In addition, vasopressin binding to mature NPII increases its propensity to dimerize (reducing the dissociation constant of 0.2 mM 10- to 100-fold; De Bree & Burbach, 1998). Since vasopressin remains tethered to NPII in provasopressin before processing in granules (producing a very high local concentration), provasopressin is likely dimeric in the secretory pathway.

2. Autosomal dominant neurohypophyseal diabetes insipidus (ADNDI) caused by amyloid-like provasopressin aggregation in the ER

Inactivating mutations anywhere in the signaling chain regulating water reabsorption—from preprovasopressin, the vasopressin type 2 receptor and the components of its intracellular signaling cascade, to aquaporin-2—lead to diabetes insipidus, i.e., polyuria and polydipsia. More than 70 different mutations in preprovasopressin have so far been described to cause the neurohypophyseal form of diabetes insipidus resulting from the absence of vasopressin in the circulation (Human gene mutation database HGMD http://www.hgmd.cf.ac.uk/). Most strikingly, almost all of them are dominant (Fig. 4) and eventually result in the absence of any circulating vasopressin. Heterozygous patients develop diabetes insipidus gradually within months to years after birth, but with essentially complete penetrance. In post mortem biopsies, marked loss of magnocellular AVP-producing neurons, and in some cases gliosis of vasopressinergic nuclei, were noted (Bergeron, Kovacs, Ezrin, & Mizzen, 1991; Green, Buchan, Alvord, & Swanson, 1967; Nagai, Li, Hsieh, Kizaki, & Urano, 1984). This phenotype was reproduced in knock-in mice with heterozygous expression of the human mutant C67X (codon for C67 of NPII altered to a stop codon): the mice gradually lost circulating vasopressin, suffered from diabetes insipidus, and lost vasopressinergic neurons detectable by immunostaining (Hayashi et al., 2009; Russell et al., 2003). Cells still producing vasopressin mRNA, however, could still be detected in one study (Hayashi et al., 2009). Only when vasopressin production was stimulated by intermittent water deprivation, cell death was also observed in these mice (Hagiwara et al., 2014).

ADNDI mutations are found across NPII, in the sequence of vasopressin and the signal peptide, but not in the glycopeptide. As a common feature of all these mutations, they are likely to interfere with folding of the vasopressin–NPII domain. This is particularly obvious for the many

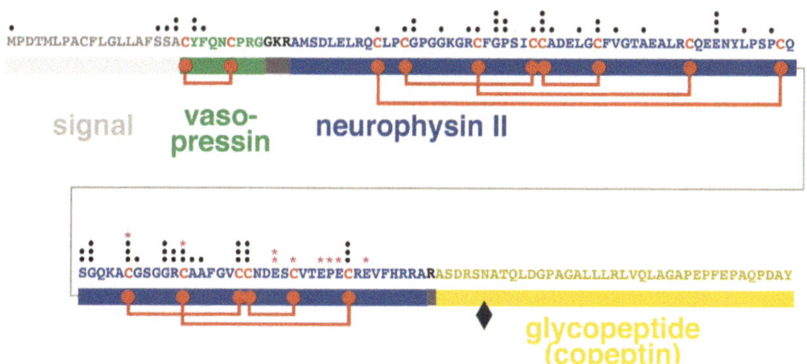

Fig. 4 Dominant mutations in the sequence of preprovasopressin causing ADNDI. Each black dot represents a different missense or deletion mutation, each pink asterisk a nonsense mutation (stop codon or frame shift).

mutations of cysteine residues involved in disulfide bonds inside NPII. The mutations in vasopressin include Y2, which inserts deeply into the pocket in NPII in the native structure, and the deletion of F3, which likely alters the ring structure of vasopressin and prevents concerted folding together with NPII. The mutations in the signal peptide prevent its removal by signal peptidase (Beuret, Rutishauser, Bider, & Spiess, 1999; Ito et al., 1993); as a result, the N-terminus of vasopressin is blocked and insertion into the pocket in NPII and thus proper folding of the vasopressin–NPII domain is impossible. In support of this notion, the preprovaspressin mutants expressed in cultured cell lines were not secreted into the medium, but were retained in the ER (Beuret et al., 1999; Friberg, Spiess, & Rutishauser, 2004). Inhibition of proteasomal degradation resulted in accumulation of full-size mutant proteins and deglycosylated intermediates, demonstrating retrotranslocation for cytosolic degradation (Friberg et al., 2004). Interestingly, even wild-type provasopressin, most of which was secreted, was intracellularly stabilized by proteasome inhibition, suggesting that some wild-type precursor did not successfully fold to its native structure and was delivered to ERAD (Friberg et al., 2004) (see Section 5).

By immunofluorescence, ADNDI mutants were found to produce heterogeneous aggregations in a majority of the expressing cells, colocalizing with ER markers (Birk, Friberg, Prescianotto-Baschong, Spiess, & Rutishauser, 2009). By immunogold electron microscopy, these accumulations revealed a fibrillar appearance strongly suggesting an amyloid-like aggregation of the mutant protein accumulating in the ER (Fig. 5). These aggregates contained ER chaperones, such as calreticulin, as expected for an aggregating misfolded protein. Large aggregates were also detected in the C67X knock-in mice in the brain cells identified by in situ hybridization to be vasopressinergic neurons (Hagiwara et al., 2014; Hayashi et al., 2009; Hiroi et al., 2010). However, these large inclusions were too densely packed to recognize any fibrillar substructure and could not be stained with antibodies.

Formation of fibrillar aggregates could also be observed *in vitro* with bacterially expressed and purified provasopressin (Birk et al., 2009). In this purified system, fibers can be directly attributed to provasopressin, where as *in vivo* they likely consist of provasopressin with associated chaperones.

Newly synthesized mutant proteins—and to a small extent also wild-type provasopressin—formed disulfide-linked homo-oligomers (Birk et al., 2009), consistent with intermolecular disulfide bonds being formed among the accumulating misfolded polypeptides. Mutagenesis experiments showed that the presence of cysteines, but no specific single cysteine, is essential for

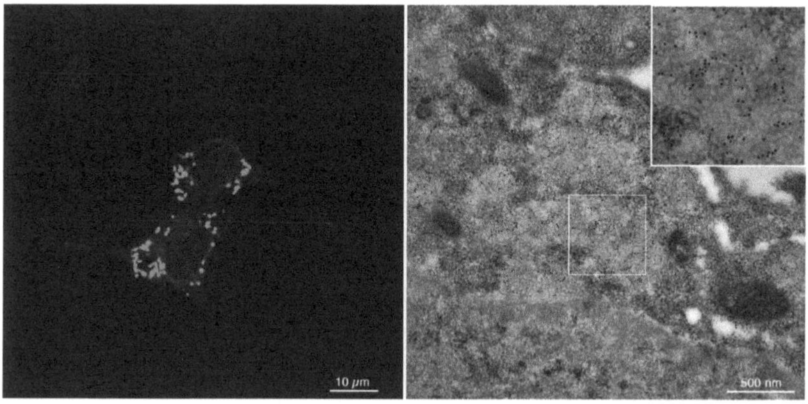

Fig. 5 ER aggregates formed by the ADNDI mutant provasopressin ΔE47 (deletion of codon 47 of NPII) in transiently transfected neuronal cell lines. Left: Immunofuorescence staining of ΔE47 expressed in HN10 cells in green. Nuclei stained with DAPI (4′,6-diamidino-2-phenylindole) in blue. Right: Immunogold electron microscopy of ΔE47 expressed in Neuro2a cells. (Inset: enlarged twofold.)

disulfide oligomerization and aggregation in transfected cultured cells (Birk et al., 2009). These findings indicate that the basis for the dominant loss of vasopressin hormone production was amyloid-like, fibrillar aggregation of the folding-deficient precursor protein in the ER, stabilized by disulfide crosslinking. When mutant and wild-type provasopressin were coexpressed in transfected cultured cells, wild-type protein was also significantly retained (Ito, Yu, & Jameson, 1999). Similarly, aggregation of the mutant protein in the C67X knock-in mice resulted in loss of circulating hormone(Hayashi et al., 2009; Russell et al., 2003), suggesting co-aggregation of both wild-type and mutant precursors.

ADNDI can thus be attributed to the group of neurodegenerative diseases associated with the formation of amyloid-like protein aggregates. In contrast to the extensively studied aggregations associated with Alzheimer's, Parkinson's and Huntington's disease that are extracellular, cytoplasmic or intranuclear, the aggregations formed by dominant provasopressin mutants are produced within the lumen of the ER. Like in most other aggregate diseases, it is not clear how aggregates cause cell toxicity. Only a single study reported cytotoxicity of ADNDI mutant precursors in expressing cell lines (Ito, Jameson, & Ito, 1997). Indeed, formation of aggregates or inclusion bodies may represent a mechanism of protection by concentrating the protein with low surface area. In the mouse model, autophagy was shown to be induced to eliminate the C67X aggregates and was proposed to be involved in cell death (Hagiwara et al., 2014).

3. Granule sorting by functional amyloid aggregation in the TGN

Peptides and proteins producing amyloids in living cells or organisms have long been associated exclusively with disease. In recent years, however, so-called functional amyloids have been discovered that perform physiological roles (Blanco, Evans, Smith, Badtke, & Chapman, 2012; Chiti & Dobson, 2006; Fowler, Koulov, Balch, & Kelly, 2007). Human Pmel17, for instance, organizes melanin in melanosomes, and the bacterial curli protein is involved in biofilm formation. In all these cases, amyloid aggregation has a rather permanent function.

The structural hallmark of amyloids is a repetitive assembly of peptides in a cross-β organization. They consist of β-sheets with their β-strands arranged perpendicular to the axis of the fiber and the sheets extending along its length. This geometry produces the dominant inter-strand distances of 4.7 Å defined by the backbone hydrogen bonds and the inter-sheet distances of ~10 Å due to the side chain packaging of parallel β-sheets (Chuang, Hori, Hesketh, & Shorter, 2018). Characteristically, amyloids are extremely stable and therefore insoluble. This is in contrast to aggregation of regulated secretory cargo in the TGN, which is only transient followed by rapid solubilization upon stimulated release into the blood stream. It therefore came as a surprise, when Riek and colleagues proposed that secretory granules of pituitary peptide hormones constitute functional amyloids (Maji et al., 2009).

This proposal was based first on the observation that a large number among 42 peptide hormones tested produced fibrillar aggregates *in vitro* under slightly acidic conditions (pH 5.5) in the presence of glycosaminoglycans (Maji et al., 2009), approximating the environment in the TGN. These *in vitro* aggregates stained positive with Thioflavin T and Congo Red, two fluorescent dyes empirically used to detect amyloids. The peptides' circular dichroism spectra frequently indicated the presence of β-sheet secondary structure. X-ray diffraction of these *in vitro* aggregates in at least a few cases revealed the typical reflections at 4.7 and ~10 Å. These analyses indicated that peptide hormones were frequently able to form amyloids *in vitro*.

As to naturally formed secretory granules, Maji et al. (2009) showed that the material contained in purified secretory granules from AtT20 cells or from rat pituitary stained positive for the amyloid-specific antibody OC, produced fluorescence with Thioflavin T, and diffracted at least weakly in the region of 4.7 Å and near 10 Å, consistent with and indicating the presence of amyloids. Most significantly, immunohistochemical staining of sections of mouse pituitary co-stained the regions with secretory granules containing prolactin, adrenocorticotropic hormone, growth hormone, oxytocin, or vasopressin with Thioflavin S, further strengthening the concept that granules are functional amyloids.

The resulting model suggested that peptide hormones undergo amyloid aggregation under TGN conditions to form dense granule structures. Upon stimulated release into the blood stream, these amyloids must be able to dissolve sufficiently fast to exert their function at the target cells. *In vitro* aggregates were shown to be able to redissolve within several hours (Maji et al., 2009), demonstrating that they are much less insoluble than the typical pathogenic amyloids. Further *in vitro* studies showed that somatostatin-14, a 14-amino acid peptide hormone with an intramolecular disulfide bond between C3 and C14, can form fibrillar aggregates that stain positive with Thioflavin S and Congo Red. Removal of the disulfide bond by mutation of one of the cysteines resulted in more rapid aggregation and reduced rate of dissolution (Anoop et al., 2014). An intramolecular disulfide loop cannot form a canonical β-sheet and may therefore hinder the formation of a "perfect," i.e., extremely stable, amyloid. The functional amyloids in granules may thus be a "poor" amyloid by design, rendering the aggregates readily reversible.

4. Sequence elements mediating ER aggregation and granule sorting

The notion that hormone aggregation in secretory granules corresponds to amyloid formation might explain why polypeptide hormones are over-represented among the proteins known to form amyloid fibers in human disease (Westermark, 2005a). Examples are amylin (islet amyloid polypeptide, IAPP) associated with type 2 diabetes mellitus (Westermark, 2005b), calcitonin associated with medullary carcinoma of the thyroid (Chiti & Dobson, 2006), atrial natriuretic factor (ANF) associated with atrial amyloidosis (Chiti & Dobson, 2006), and indeed—provasopressin associated with familial neurohypophyseal diabetes insipidus.

In this light, it appears improbable that fibrillar amyloid-like aggregation of mutant provasopressins in the ER is accidental, but rather likely that it is the consequence of the evolved ability of the protein to form amyloids under the specific conditions of the TGN. As misfolded protein accumulates in the ER, its properties might result in premature amyloid aggregation in the wrong compartment. This hypothesis predicts that the same peptide segments required for physiological aggregation of provasopressin in granule formation at the TGN are also responsible for pathological aggregation of mutant precursors in the ER.

4.1 Two amyloidogenic sequences for ER aggregation: Vasopressin and glycopeptide

To identify sequences responsible for aggregation in the ER, the vasopressin precursor was mutated either by deleting parts of the protein or by scanning mutagenesis (Beuret et al., 2017). For this, segments of 10 amino acids throughout the vasopressin–NPII domain were replaced by a proline-glycine-rich sequence that is incompatible with β-sheets (or any other secondary structure) and thus expected to disrupt amyloidogenic elements. The mutant proteins were expressed in differentiated neuronal HN10 cells and analyzed for ER aggregation by immunofluorescence microscopy. Surprisingly, this analysis revealed that provasopressin contains two independent peptide segments able to cause ER aggregation: the vasopressin hormone sequence and the glycopeptide (Beuret et al., 2017).

Several ADNDI mutants are C-terminally truncated, the shortest one at position 61 of NPII (position 73 of provasopressin, consisting only of the N-terminal half of the protein). Since these mutants lack the glycoprotein, aggregation via the vasopressin sequence is thus sufficient to cause disease. Whether this is also the case for aggregation via the glycopeptide cannot be definitely concluded, since there are no naturally occurring pathogenic mutants lacking vasopressin. For the ADNDI mutations Y2H, Y2S, and ΔF3 in vasopressin, it is not known whether they abolish vasopressin's ability to aggregate and cause disease.

Electron microscopy showed fibrillar substructure in the aggregates formed by truncation constructs lacking the glycopeptide, i.e., aggregating via vasopressin. The fibers appeared to be packed more densely than observed for the full-size mutant ΔE47 in Neuro2a cells, but similar to ΔE47 expressed in COS cells (Birk et al., 2009). The density of the ER aggregates may therefore depend on the cell type and/or expression levels. The aggregates seen in the C67X expressing transgenic mouse were indeed so compacted that they appeared homogeneous with no substructure (Hagiwara et al., 2014; Hayashi et al., 2009). Importantly, constructs lacking the vasopressin sequence and thus aggregating via the glycopeptide also showed a fibrillar appearance when expressed in HN10 cells (Beuret et al., 2017), indicating an amyloid nature. These aggregates displayed various degrees of compactness both between and within accumulations.

ER aggregates were shown to be highly crosslinked by intermolecular disulfide bonds, since under nonreducing conditions the protein accumulating in expressing cells did not enter the gel for SDS-polyacrylamide gel electrophoresis and immunoblotting. Even after reduction by boiling cells in SDS-sample buffer with dithiothreitol (DTT), a large fraction of provasopressin mutant protein was recovered as SDS- and DTT-resistant homo-oligomers (Beuret et al., 2017), again a manifestation of amyloid formation. Strikingly, only constructs lacking the vasopressin sequence could be completely reduced to monomers, indicating that it is responsible for this phenomenon. Disulfide crosslinking is a likely mechanism stabilizing aggregate formation. A mutant lacking any cysteines was found not to produce ER aggregates (Birk et al., 2009). This suggests that the glycopeptide is not sufficient to produce aggregates under ER conditions without stabilization by disulfide crosslinks (assuming vasopressin was inactivated as an aggregation sequence by mutation of its two cysteines).

4.2 Vasopressin and glycopeptide also mediate granule aggregation

To identify sequences necessary or sufficient to mediate granule formation (or sorting) by mutagenesis is difficult, because it requires that the mutant proteins are able to fold and pass ER quality control to reach the TGN. This could explain why few "granule sorting signals" have been convincingly identified so far. Examples include the small N-terminal fold of chromogranin A (Taupenot et al., 2002), which is otherwise an unstructured protein that could be truncated without misfolding and ER retention, and two short segments in secretogranin II (Courel, Vasquez, Hook, Mahata, & Taupenot, 2008), an equally extended granule cargo.

Another difficulty is that soluble cargo is not efficiently excluded from budding secretory granules. Depending on the overexpression level, sufficient amounts of constitutive secretory proteins may be included nonspecifically into granules to be detectable by immunofluorescence. It is therefore important to keep expression levels low by producing stable cell lines expressing similar amounts for proteins to be compared, rather than transient transfections. These difficulties are probably why in one previous study using green fluorescent protein (GFP) fusions, residues 1–27 of provasopressin, but not 1–10 or 13–27 alone, were identified in one study

(De Bree, Knight, Howell, & Murphy, 2000) and NPII in another (Cool, Jackson, & Waddell, 2008) to mediate granule localization.

The glycopeptide is structurally separate from the globular vasopressin–NPII domain and can therefore be deleted without affecting folding and ER exit of the rest of the protein. In contrast, NPII cannot fold in the absence of vasopressin and is retained in the ER (De Bree et al., 2003). To retain protein folding while inactivating vasopressin as an amyloidogenic sequence, the five amino acids (besides the terminal glycine) in vasopressin that are not in contact with NPII (Fig. 3C) were mutated to alanines (CYFQNCPRG$\geq$CYAAACAAG) (Beuret et al., 2017). Interestingly, the ability of the resulting sequence V5xA to mediate ER aggregation was reduced, but not eliminated, suggesting that it does not depend strongly on individual amino acid side chains. Most importantly, V5xA was able to fold with NPII and to exit the ER. Stable AtT20 cell lines expressing wild-type provasopressin, provasopressin without the glycopeptide or with the V5xA sequence, or with both mutations were analyzed for protein accumulation in granules relative to an endogenous granule marker and for stimulated secretion (by measuring the ratio of stimulated vs constitutive release into the medium). The results demonstrated that both sequences significantly contribute to granule sorting (Fig. 6; Beuret et al., 2017).

The finding that the same amyloidogenic sequences that cause pathogenic ER aggregation also confer granule sorting and regulated secretion provides biological support for the concept that secretory granules are functional amyloids of peptide hormones.

4.3 TGN-specificity of amyloid-like aggregation

A critical issue with the amyloid hypothesis for granule formation is that proteins normally are folded when they reach the TGN, but require an exposed extended polypeptide segment to form a β-strand for assembly in a multimeric β-sheet amyloid. In the case of provasopressin, the vasopressin part is initially protected from aggregation by folding into its pocket in NPII. It is an attractive hypothesis that under TGN conditions vasopressin binds more weakly and is released from its binding pocket. There is no direct evidence for this. Indeed, hormone binding to mature NPII shows a broad maximum between pH 5.5 and 6.3; the drop in pH in the TGN alone thus cannot account for vasopressin release. Alternatively, TGN conditions favor aggregation and thereby trap vasopressin from its binding equilibrium. The glycopeptide seems to be always exposed. Aggregation is probably favored by the environment in

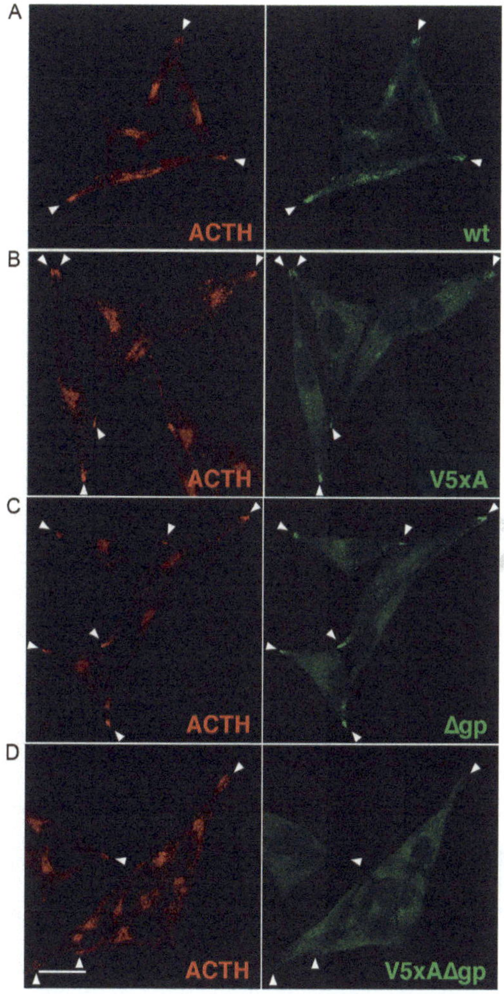

Fig. 6 Both vasopressin and the glycopeptide are required for sorting into secretory granules. AtT20 cell lines stably expressing provasopressin wild-type (wt; panel A), with the V5xA mutant vasopressin (V5xA; panel B), without the glycopeptide (Δgp; panel C), or with both mutations (V5xAΔgp; panel D) were stained by immunofluorescence for the endogenous granule cargo adrenocorticotropic hormone (ACTH; red) and provasopressin (green). Arrowheads point at secretory granules accumulated in the tips of cellular extensions.

the TGN. As mentioned above, the glycopeptide cannot aggregate in the ER without stabilization by disulfide crosslinks (Birk et al., 2009).

Other peptide hormones, such as growth hormone and prolactin, are compact globular structures without apparent exposed peptide segments available to engage in amyloid β-sheets. This has been put forward as an argument against the amyloid hypothesis, at least for these two hormones (Dannies, 2012). In contrast, Jacob et al. (2016) proposed that growth hormone may significantly unfold under TGN conditions to expose an internal segment for amyloid formation. After release into the blood stream, the amyloid would dissolve and the hormone refold. Obviously, more direct *in vivo* evidence will be needed to establish the amyloid proposal for different hormones.

5. Provasopressin is a physiological substrate of ERAD

Cellular quality control mechanisms like UPR and ERAD are typically considered most important to deal with exogenous stresses like temperature shock or chemical damage, or with endogenous errors of transcription and translation. However, even under optimal conditions, the core components of these pathways are essential, since their deletion results in embryonic lethality in mice. For example, knockouts of the ERAD components Hrd1 (E3 ligase and putative ER retrotranslocation pore) and Sel1L (adaptor of Hrd1) were embryonic lethal (Francisco et al., 2010; Yagishita et al., 2005). Induced full-body knockout of Sel1L resulted in progressive weight loss starting after 1 week resulting in premature death after ~3 weeks (Sun et al., 2014). Sel1L was found to be indispensable for Hrd1 stability and ER homeostasis. The exocrine pancreas was most prominently affected, where the ER was dilated, UPR was induced, cell death was increased, and the size of exocrine zymogen granules was reduced. Other cell types have also been reported to be severely affected by Sel1L or Hrd1 deficiency (Qi, Tsai, & Arvan, 2017).

Interestingly, mice with globally induced Sel1L deficiency rapidly developed diabetes insipidus (Shi et al., 2017): within very few days after induction, water intake and urine output increased, and urine osmolality decreased. Vasopressin levels in the serum were found to be reduced and administration of a vasopressin analog restored normal urine osmolality, demonstrating a neurohypophyseal origin of the condition. Specific Sel1L deficiency induced exclusively in vasopressinergic neurons presented these same phenotypes as in the full-body knockout.

Vasopressin-expressing cells in the paraventricular and supraoptic nucleus of the hypothalamus express particularly high levels of Hrd1, which is strongly reduced upon Sel1L knockout (Shi et al., 2017). In wild-type mice, immunofluorescence staining for provasopressin decorated structures (beads-on-a-string) along the axons emanating from these nuclei toward the pituitary, corresponding the secretory granules in transit. Upon Sel1L knockout, however, axonal staining was abolished, while staining of the cell bodies was enhanced, indicating that provasopressin was retained, most likely in the ER. This could also be reproduced in cell culture: Neuro2a cells deleted for Sel1L by CRISPR/Cas9 not only accumulated ADNDI mutant provasopressin in fibrillar ER aggregates, but also the wild-type protein. Most importantly, while provasopressin in brain sections of the paraventricular nucleus of a wild-type mice was detected by immunogold electron microscopy in typical small dense-core secretory granules, it was found in Sel1L$^{-/-}$ brains in large aggregations with a dense meshwork appearance, consistent with amyloid-like aggregation in the ER (Fig. 7). These results indicate that even wild-type provasopressin is normally delivered by ERAD involving Sel1L/Hrd1 to proteasomes. If this pathway is not functional, misfolded and aggregated material accumulates in the ER, to finally capture essentially all provasopressin and to eliminate granule formation and hormone production.

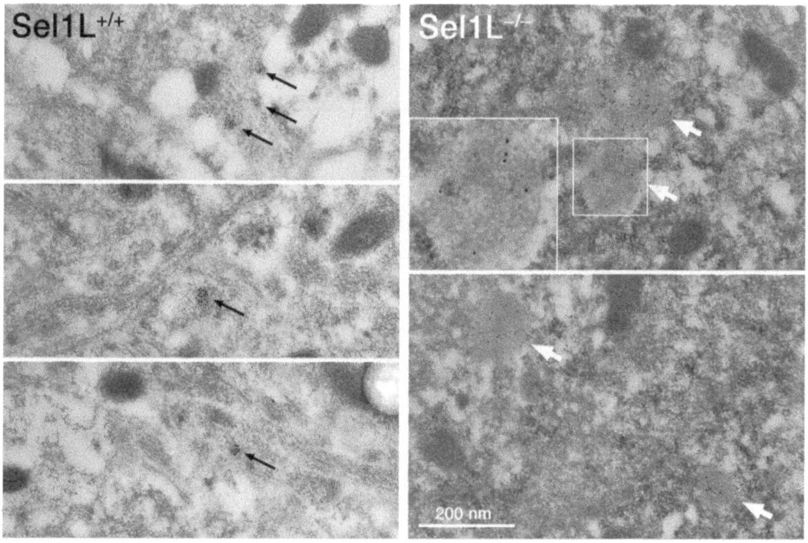

Fig. 7 Immunogold electron microscopy identified small provasopressin-positive structures in brain sections of the paraventricular nucleus of a wild-type (Sel1L+/+) mouse (black arrows) and large fibrillar ER aggregates from a mouse deficient in ERAD in vasopressinergic neurons (Sel1L−/−; white arrows). Inset: enlarged twofold. (In collaboration with Ling Shi and colleagues, Ann Arbor, MI).

Cell culture experiments with functional ERAD, confirmed that wild-type provasopressin is partially ubiquitinated and can be co-immunoprecipitated with Hrd1, demonstrating that provasopressin is an endogenous substrate of Sel1L/Hrd1 (Shi et al., 2017). Furthermore, the majority of wild-type provasopressin in Sel1L- or Hrd1-deficient cells existed as disulfide-linked multimers, just like ADNDI mutants in wild-type cells. These observations demonstrate that wild-type and ADNDI mutants of provasopressin share the tendency to aggregate in the ER. Cells can cope with the amount of misfolded and aggregating wild-type protein by functional ERAD, whose components are already induced to higher levels than in non-secretory cells. Since mutant protein cannot fold, remains chaperone-bound, and cannot leave the ER, the capacity of ERAD is exceeded. Aggregation eventually predominates and coexpressed wild-type is also trapped in the aggregates, ceasing production of functional hormone. This mechanism explains the dominant phenotype of ADNDI mutants, which manifests itself even before the expressing cells have died (Fig. 8).

This scenario may similarly apply to other peptide hormones that also have evolved the ability to form amyloid-like aggregates for granule sorting and whose tendency to aggregate must be kept in check in the ER. Indeed, Sel1L/Hrd1-dependent ERAD was recently shown to be important also for the maturation and function of pro-opiomelanocortin (POMC) (Kim et al., 2018). Mice with specific Sel1L deficiency in POMC neurons developed age-associated obesity due to ER retention of POMC. A fraction of nascent POMC protein is normally ubiquitinated and targeted to proteasomal degradation, preventing accumulation of misfolded and aggregated prohormone. Sel1L deficiency reproduced the result of an autosomal dominant POMC mutation, C28F, causing early onset obesity (Creemers et al., 2008).

One may thus speculate more generally that regulated secretory proteins are "dangerous goods": they employ amyloid-like aggregation to sort and concentrate themselves in the TGN into secretory granules at the risk of premature aggregation in the ER. Normally, low levels of aggregates in the ER are constantly removed by ERAD. When the capacity of ERAD to eliminate them is exceeded, they accumulate to the point where essentially all prohormone products are captured into many large fibrillar aggregates, eventually causing cell death.

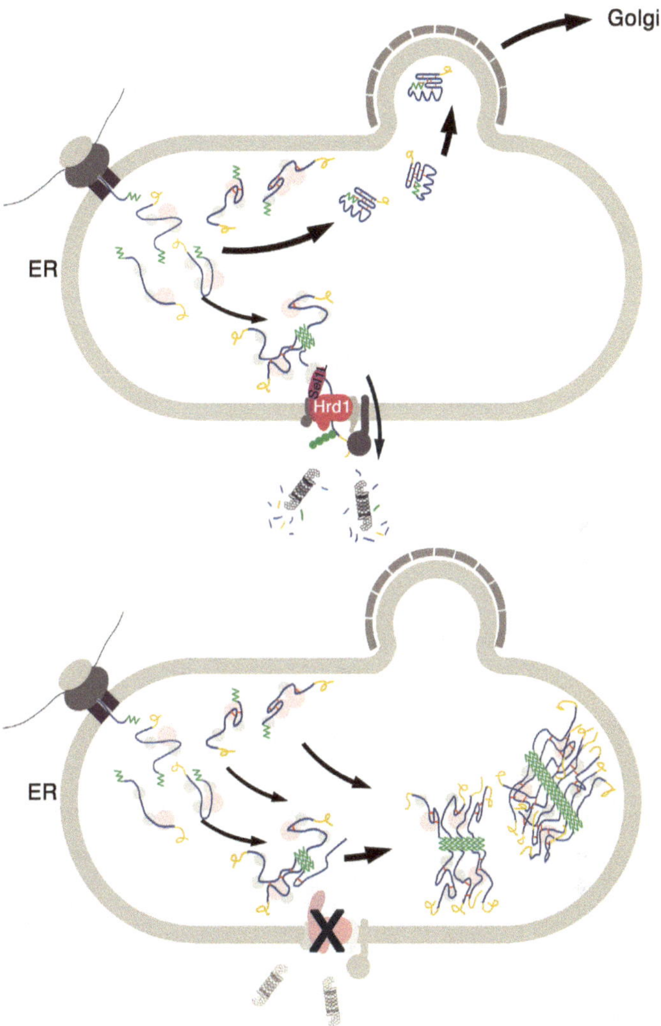

Fig. 8 Provasopressin is a physiological substrate of ERAD. Top: Newly synthesized wild-type provasopressin, assisted by chaperones and PDI, folds and exits the ER to the Golgi. A fraction of the proteins aggregates, but is eliminated by Sel1L- and Hrd1-dependent ERAD by the proteasome. Bottom: In the absence of ERAD (or when at least one allele encodes a folding-deficient mutant protein), aggregation predominates and captures most proteins into fibrillar aggregates.

Acknowledgments

Our work was supported by grant 31003A-162643 (to M.S.) and 32–061978 (to J.R.) from the Swiss National Science Foundation, and by the University of Basel. We are grateful to all our collaborators in the vasopressin project over many years, in particular more recently Dr. Ling Qi and his coworkers (University of Michigan, Ann Arbor, MI).

References

Anoop, A., Ranganathan, S., Dhaked Das, B., Jha, N. N., Pratihar, S., Ghosh, S., et al. (2014). Elucidating the role of disulfide bond on amyloid formation and fibril reversibility of somatostatin-14: Relevance to its storage and secretion. *The Journal of Biological Chemistry*, *289*(24), 16884–16903. https://doi.org/10.1074/jbc.M114.548354.

Bergeron, C., Kovacs, K., Ezrin, C., & Mizzen, C. (1991). Hereditary diabetes insipidus: An immunohistochemical study of the hypothalamus and pituitary gland. *Acta Neuropathologica*, *81*(3), 345–348.

Beuret, N., Hasler, F., Prescianotto-Baschong, C., Birk, J., Rutishauser, J., & Spiess, M. (2017). Amyloid-like aggregation of provasopressin in diabetes insipidus and secretory granule sorting. *BMC Biology*, *15*(1), 5. https://doi.org/10.1186/s12915-017-0347-9.

Beuret, N., Rutishauser, J., Bider, M. D., & Spiess, M. (1999). Mechanism of endoplasmic reticulum retention of mutant vasopressin precursor caused by a signal peptide truncation associated with diabetes insipidus. *The Journal of Biological Chemistry*, *274*(27), 18965–18972.

Beuret, N., Stettler, H., Renold, A., Rutishauser, J., & Spiess, M. (2004). Expression of regulated secretory proteins is sufficient to generate granule-like structures in constitutively secreting cells. *The Journal of Biological Chemistry*, *279*(19), 20242–20249. https://doi.org/10.1074/jbc.M310613200.

Birch, N. P., Hakes, D. J., Dixon, J. E., & Mezey, E. (1994). Distribution and regulation of the candidate prohormone processing enzymes SPC2 and SPC3 in adult rat brain. *Neuropeptides*, *27*(6), 307–322. https://doi.org/10.1016/0143-4179(94)90057-4.

Birk, J., Friberg, M. A., Prescianotto-Baschong, C., Spiess, M., & Rutishauser, J. (2009). Dominant pro-vasopressin mutants that cause diabetes insipidus form disulfide-linked fibrillar aggregates in the endoplasmic reticulum. *Journal of Cell Science*, *122*(Pt. 21), 3994–4002. https://doi.org/10.1242/jcs.051136.

Blanco, L. P., Evans, M. L., Smith, D. R., Badtke, M. P., & Chapman, M. R. (2012). Diversity, biogenesis and function of microbial amyloids. *Trends in Microbiology*, *20*(2), 66–73. https://doi.org/10.1016/j.tim.2011.11.005.

Chen, L. Q., Rose, J. P., Breslow, E., Yang, D., Chang, W. R., Furey, W. F., et al. (1991). Crystal structure of a bovine neurophysin II dipeptide complex at 2.8 Å determined from the single-wavelength anomalous scattering signal of an incorporated iodine atom. *Proceedings of the National Academy of Sciences of the United States of America*, *88*(10), 4240–4244.

Chiti, F., & Dobson, C. M. (2006). Protein misfolding, functional amyloid, and human disease. *Annual Review of Biochemistry*, *75*, 333–366. https://doi.org/10.1146/annurev.biochem.75.101304.123901.

Chuang, E., Hori, A. M., Hesketh, C. D., & Shorter, J. (2018). Amyloid assembly and disassembly. *Journal of Cell Science*, *131*(8), jcs189928. https://doi.org/10.1242/jcs.189928.

Cool, D., Jackson, S., & Waddell, K. (2008). Structural requirements for sorting pro-vasopressin to the regulated secretory pathway in a neuronal cell line. *Open Neuroendocrinology Journal*, *1*, 1–8. https://doi.org/10.2174/1876528900801010001.

Courel, M., Vasquez, M. S., Hook, V. Y., Mahata, S. K., & Taupenot, L. (2008). Sorting of the neuroendocrine secretory protein secretogranin II into the regulated secretory

pathway: Role of N- and C-terminal {alpha}-helical domains. *The Journal of Biological Chemistry*, *283*(17), 11807–11822. https://doi.org/10.1074/jbc.M709832200.
Creemers, J. W. M., Lee, Y. S., Oliver, R. L., Bahceci, M., Tuzcu, A., Gokalp, D., et al. (2008). Mutations in the amino-terminal region of proopiomelanocortin (POMC) in patients with early-onset obesity impair POMC sorting to the regulated secretory pathway. *The Journal of Clinical Endocrinology & Metabolism*, *93*(11), 4494–4499. https://doi.org/10.1210/jc.2008-0954.
Dannies, P. S. (2012). Prolactin and growth hormone aggregates in secretory granules: The need to understand the structure of the aggregate. *Endocrine Reviews*, *33*(2), 254–270. https://doi.org/10.1210/er.2011-1002.
De Bree, F. M., & Burbach, J. P. (1998). Structure-function relationships of the vasopressin prohormone domains. *Cellular and Molecular Neurobiology*, *18*(2), 173–191.
De Bree, F. M., Knight, D., Howell, L., & Murphy, D. (2000). Sorting of the vasopressin prohormone into the regulated secretory pathway. *FEBS Letters*, *475*(3), 175–180.
De Bree, F. M., Van Der Kleij, A. A. M., Nijenhuis, M., Zalm, R., Murphy, D., & Burbach, J. P. H. (2003). The hormone domain of the vasopressin prohormone is required for the correct prohormone trafficking through the secretory pathway. *Journal of Neuroendocrinology*, *15*(12), 1156–1163.
Fenske, W., Refardt, J., Chifu, I., Schnyder, I., Winzeler, B., Drummond, J., et al. (2018). A copeptin-based approach in the diagnosis of diabetes insipidus. *New England Journal of Medicine*, *379*(5), 428–439.
Fowler, D. M., Koulov, A. V., Balch, W. E., & Kelly, J. W. (2007). Functional amyloid–from bacteria to humans. *Trends in Biochemical Sciences*, *32*(5), 217–224. https://doi.org/10.1016/j.tibs.2007.03.003.
Francisco, A. B., Singh, R., Li, S., Vani, A. K., Yang, L., Munroe, R. J., et al. (2010). Deficiency of suppressor enhancer Lin12 1 like (SEL1L) in mice leads to systemic endoplasmic reticulum stress and embryonic lethality. *The Journal of Biological Chemistry*, *285*(18), 13694–13703. https://doi.org/10.1074/jbc.M109.085340.
Friberg, M. A., Spiess, M., & Rutishauser, J. (2004). Degradation of wild-type vasopressin precursor and pathogenic mutants by the proteasome. *The Journal of Biological Chemistry*, *279*(19), 19441–19447. https://doi.org/10.1074/jbc.M310249200.
Green, J. R., Buchan, G. C., Alvord, G. C., & Swanson, A. G. (1967). Hereditary and idiopathic types of diabetes insipidus. *Brain*, *90*(3), 707–714.
Guo, Y., Sirkis, D. W., & Schekman, R. (2014). Protein sorting at the trans-Golgi network. *Annual Review of Cell and Developmental Biology*, *30*(1), 169–206. https://doi.org/10.1146/annurev-cellbio-100913-013012.
Hagiwara, D., Arima, H., Morishita, Y., Wenjun, L., Azuma, Y., Ito, Y., et al. (2014). Arginine vasopressin neuronal loss results from autophagy-associated cell death in a mouse model for familial neurohypophysial diabetes insipidus. *Cell Death & Disease*, *5*, e1148. https://doi.org/10.1038/cddis.2014.124.
Hayashi, M., Arima, H., Ozaki, N., Morishita, Y., Hiroi, M., Ozaki, N., et al. (2009). Progressive polyuria without vasopressin neuron loss in a mouse model for familial neurohypophysial diabetes insipidus. *American Journal of Physiology Regulatory, Integrative and Comparative Physiology*, *296*(5), R1641–R1649. https://doi.org/10.1152/ajpregu.00034.2009.
Hiroi, M., Morishita, Y., Hayashi, M., Ozaki, N., Sugimura, Y., Nagasaki, H., et al. (2010). Activation of vasopressin neurons leads to phenotype progression in a mouse model for familial neurohypophysial diabetes insipidus. *American Journal of Physiology Regulatory, Integrative and Comparative Physiology*, *298*(2), R486–R493. https://doi.org/10.1152/ajpregu.00529.2009.
Hutton, J. C., Penn, E. J., & Peshavaria, M. (1983). Low-molecular-weight constituents of isolated insulin-secretory granules. Bivalent cations, adenine nucleotides and inorganic phosphate. *The Biochemical Journal*, *210*(2), 297–305.

Ito, M., Jameson, J. L., & Ito, M. (1997). Molecular basis of autosomal dominant neurohypophyseal diabetes insipidus. Cellular toxicity caused by the accumulation of mutant vasopressin precursors within the endoplasmic reticulum. *The Journal of Clinical Investigation, 99*(8), 1897–1905. https://doi.org/10.1172/JCI119357.

Ito, M., Oiso, Y., Murase, T., Kondo, K., Saito, H., Chinzei, T., et al. (1993). Possible involvement of inefficient cleavage of preprovasopressin by signal peptidase as a cause for familial central diabetes insipidus. *The Journal of Clinical Investigation, 91*(6), 2565–2571. https://doi.org/10.1172/JCI116494.

Ito, M., Yu, R. N., & Jameson, J. L. (1999). Mutant vasopressin precursors that cause autosomal dominant neurohypophyseal diabetes insipidus retain dimerization and impair the secretion of wild-type proteins. *The Journal of Biological Chemistry, 274*(13), 9029–9037. https://doi.org/10.1074/jbc.274.13.9029.

Jacob, R. S., Das, S., Ghosh, S., Anoop, A., Jha, N. N., Khan, T., et al. (2016). Amyloid formation of growth hormone in presence of zinc: Relevance to its storage in secretory granules. *Scientific Reports, 6*, 23370. https://doi.org/10.1038/srep23370.

Kim, T., Gondré-Lewis, M. C., Arnaoutova, I., & Loh, Y. P. (2006). Dense-core secretory granule biogenesis. *Physiology (Bethesda, Md.), 21*, 124–133. https://doi.org/10.1152/physiol.00043.2005.

Kim, K.-H., & Seong, B. L. (2001). Peptide amidation: Production of peptide hormones in vivo and in vitro. *Biotechnology and Bioprocess Engineering, 6*, 244–251.

Kim, G. H., Shi, G., Somlo, D. R., Haataja, L., Song, S., Long, Q., et al. (2018). Hypothalamic ER-associated degradation regulates POMC maturation, feeding, and age-associated obesity. *The Journal of Clinical Investigation, 128*(3), 1125–1140. https://doi.org/10.1172/JCI96420.

Kolset, S. O., Prydz, K., & Pejler, G. (2004). Intracellular proteoglycans. *The Biochemical Journal, 379*(Pt. 2), 217–227. https://doi.org/10.1042/BJ20031230.

Maji, S. K., Perrin, M. H., Sawaya, M. R., Jessberger, S., Vadodaria, K., Rissman, R. A., et al. (2009). Functional amyloids as natural storage of peptide hormones in pituitary secretory granules. *Science (New York, N.Y.), 325*(5938), 328–332. https://doi.org/10.1126/science.1173155.

Nagai, I., Li, C. H., Hsieh, S. M., Kizaki, T., & Urano, Y. (1984). Two cases of hereditary diabetes insipidus, with an autopsy finding in one. *Acta Endocrinologica, 105*(3), 318–323.

Paroutis, P., Touret, N., & Grinstein, S. (2004). The pH of the secretory pathway: Measurement, determinants, and regulation. *Physiology (Bethesda, Md.), 19*(4), 207–215. https://doi.org/10.1152/physiol.00005.2004.

Qi, L., Tsai, B., & Arvan, P. (2017). New insights into the physiological role of endoplasmic reticulum-associated degradation. *Trends in Cell Biology, 27*(6), 430–440. https://doi.org/10.1016/j.tcb.2016.12.002.

Robertson, G. L. (2001). Antidiuretic hormone. Normal and disordered function. *Endocrinology and Metabolism Clinics of North America, 30*(3), 671–694, vii.

Ruggiano, A., Foresti, O., & Carvalho, P. (2014). ER-associated degradation: Protein quality control and beyond. *The Journal of Cell Biology, 204*(6), 869–879. https://doi.org/10.1083/jcb.201312042.

Russell, T. A., Ito, M., Ito, M., Yu, R. N., Martinson, F. A., Weiss, J., et al. (2003). A murine model of autosomal dominant neurohypophyseal diabetes insipidus reveals progressive loss of vasopressin-producing neurons. *The Journal of Clinical Investigation, 112*(11), 1697–1706. https://doi.org/10.1172/JCI200318616.

Schröder, M., & Kaufman, R. J. (2005). The mammalian unfolded protein response. *Annual Review of Biochemistry, 74*, 739–789. https://doi.org/10.1146/annurev.biochem.73.011303.074134.

Shi, G., Somlo, D. R., Kim, G. H., Prescianotto-Baschong, C., Sun, S., Beuret, N., et al. (2017). ER-associated degradation is required for vasopressin prohormone processing and systemic water homeostasis. *The Journal of Clinical Investigation*, *127*(10), 3897–3912. https://doi.org/10.1172/JCI94771.

Sun, S., Shi, G., Han, X., Francisco, A. B., Ji, Y., Mendonça, N., et al. (2014). Sel1L is indispensable for mammalian endoplasmic reticulum-associated degradation, endoplasmic reticulum homeostasis, and survival. *Proceedings of the National Academy of Sciences of the United States of America*, *111*(5), E582–E591. https://doi.org/10.1073/pnas.1318114111.

Taupenot, L., Harper, K. L., Mahapatra, N. R., Parmer, R. J., Mahata, S. K., & O'Connor, D. T. (2002). Identification of a novel sorting determinant for the regulated pathway in the secretory protein chromogranin A. *Journal of Cell Science*, *115*(Pt. 24), 4827–4841.

Taupenot, L., Harper, K. L., & O'Connor, D. T. (2003). The chromogranin-secretogranin family. *The New England Journal of Medicine*, *348*(12), 1134–1149. https://doi.org/10.1056/NEJMra021405.

Thorlacius-Ussing, O. (1987). Zinc in the anterior pituitary of rat: A histochemical and analytical work. *Neuroendocrinology*, *45*(3), 233–242.

Wang, S., & Kaufman, R. J. (2012). The impact of the unfolded protein response on human disease. *The Journal of Cell Biology*, *197*(7), 857–867. https://doi.org/10.1083/jcb.201110131.

Westermark, G. T. (2005a). Endocrine amyloid. In J. D. Sipe (Ed.), *Amyloid proteins: The beta sheet conformation and disease* (pp. 723–754). Weinheim: Wiley-VCH Verlag GmbH.

Westermark, P. (2005b). Aspects on human amyloid forms and their fibril polypeptides. *The FEBS Journal*, *272*(23), 5942–5949. https://doi.org/10.1111/j.1742-4658.2005.05024.x.

Wu, C. K., Hu, B., Rose, J. P., Liu, Z. J., Nguyen, T. L., Zheng, C., et al. (2001). Structures of an unliganded neurophysin and its vasopressin complex: Implications for binding and allosteric mechanisms. *Protein science: A Publication of the Protein Society*, *10*(9), 1869–1880. https://doi.org/10.1110/ps.10601.

Yagishita, N., Ohneda, K., Amano, T., Yamasaki, S., Sugiura, A., Tsuchimochi, K., et al. (2005). Essential role of synoviolin in embryogenesis. *The Journal of Biological Chemistry*, *280*(9), 7909–7916. https://doi.org/10.1074/jbc.M410863200.

CHAPTER FOUR

V2 vasopressin receptor mutations

Noriko Makita[a,*], Katsunori Manaka[a], Junichiro Sato[a], Taroh Iiri[a,b,*]
[a]Department of Endocrinology and Nephrology, The University of Tokyo, Tokyo, Japan
[b]Department of Pharmacology, St. Marianna University School of Medicine, Kawasaki, Japan
*Corresponding authors: e-mail address: norimaki-tky@umin.ac.jp; tiiri-tky@umin.ac.jp

Contents

Abstract

V2 vasopressin receptor (V2R) is a member of the G protein-coupled receptor (GPCR) family in which many disease-causing mutations have been identified and thus generated much interest. Loss-of-function V2R mutations cause nephrogenic diabetes insipidus (NDI) whereas gain-of-function mutations cause nephrogenic syndrome of inappropriate antidiuresis (NSIAD). The mechanisms underlying a V2R loss-of-function can be theoretically classified as either protein expression, localization (ER retention) or functional disorders. Functional analyses have revealed however that these mechanisms are likely to be complex. Strikingly, V2R mutations at the same site can result in opposite phenotypes, e.g., R137H and R137L/C cause NDI and NSIAD, respectively. These findings support the notion that the constitutive activation of GPCRs might be often associated with their instability and denaturation. Thus, functional analysis of disease-causing V2R mutations may not only reveal potential new treatment strategies using pharmacochaperones for NDI and inverse agonists for NSIAD, but also provide a greater understanding of the physiological functions of GPCRs and highlight the new paradigms, i.e., biased agonism and protean agonism.

Vitamins and Hormones, Volume 113
ISSN 0083-6729
https://doi.org/10.1016/bs.vh.2019.08.012

1. Short introduction: V2 receptor and its physiological role in the kidney

V2 vasopressin receptor (V2R), a member of the class A G protein-coupled receptors (GPCRs), is mainly expressed in the principal cells of the renal collecting tubules where it accelerates water reabsorption upon stimulation by its agonist, arginine vasopressin (AVP). AVP is secreted from the posterior pituitary in response to hypovolemia or high osmolality and binds to V2R in the basolateral membrane of these principal cells. The binding of AVP is assumed to occur at transmembrane helix II–IV (88–96, 119–127, 284–291, 311–317), and thereby stimulate adenylyl cyclase via the stimulatory G protein (Gs). This promotes the cAMP/PKA-mediated trafficking of AQP-2 as a water channel into the luminal membrane, allowing water to freely enter via the medullary osmotic gradient (Moeller, Rittig, & Fenton, 2013). In general, water reabsorption at the collecting ducts comprises 10% of the total water reabsorption by the kidney. AVP augments water absorption by increasing the expression of AQP-2 in a long-term manner. In addition, it has been reported that the exchange protein directly activated by cAMP (Epac)/Ca-calmodulin plays a key role in this regulation and not PKA/CREB (Kortenoeven et al., 2012; Yip, 2006). Recently, it was reported that ERK1/2 plays a role in the translocation of AQP-2 (Gao et al., 2015) (Fig. 1).

Inactivating or activating mutations of V2 receptor cause X-linked congenital nephrogenic diabetes insipidus (NDI) or nephrogenic syndromes of inappropriate antidiuresis (NSIAD) as described further below in this chapter. Efforts to analyze the functional consequences of V2R mutations and identify possible new therapeutics have helped to not only to uncover the mechanisms of the relevant diseases and but also to better understand key features of the physiological function and highlight the conceptual paradigm shift of GPCRs.

2. X-linked congenital nephrogenic diabetes insipidus (NDI)

The human AVPR2 gene located on chromosome region Xq28 and encoding the V2 receptor was cloned in 1992 and mutations in this gene were subsequently identified in patients with X-linked nephrogenic diabetes insipidus (NDI) (Lolait et al., 1992; Pan, Metzenberg, Das, Jing, & Gitschier, 1992; Rosenthal et al., 1992; van den Ouweland et al., 1992). In general,

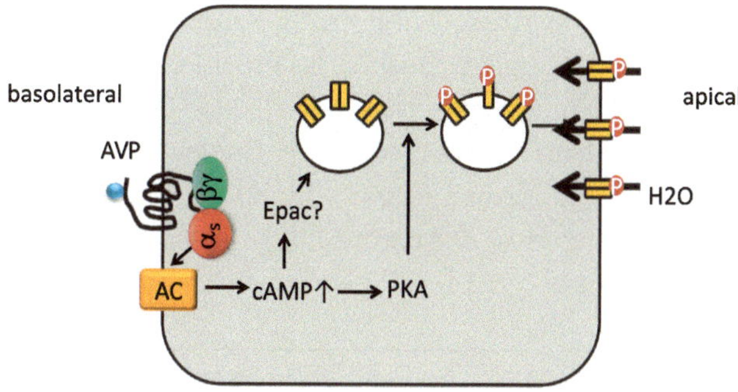

Fig. 1 The V2 receptor signaling cascade in principal cells of the renal collecting duct. AVP is secreted from the pituitary gland following stimulation by hyperosmolality or hypovolemia and then binds to V2R expressed in the basolateral membrane of principal cells of the collecting duct. This then stimulates adenylyl cyclase via Gs and promotes cAMP-mediated trafficking of AQP-2 as a water channel into the luminal membrane. Water is thereby permitted to enter freely through the medullary osmotic gradient. A potential role of an alternative pathway via Epac has also been reported.

loss-of-function (inactivating) mutations of the AVPR2 gene affect men only. In the province of Quebec, the incidence of X-linked NDI was estimated in a previous report to be approximately 8.8 per 1,000,000 male live births (Arthus et al., 2000). However, heterozygous females can also exhibit variable degrees of polyuria and polydipsia because of a skewed X chromosome inactivation (Nomura et al., 1997; Puck & Willard, 1998; Satoh, Ogikubo, & Yoshizawa-Ogasawara, 2008).

It is now known that almost 90% of congenital NDI cases are caused by V2R mutations (Bichet & Bockenhauer, 2016; Bockenhauer & Bichet, 2015). Most male NDI cases caused by V2R mutations show complete impairment of water reabsorption at the renal collecting ducts, which is referred to as complete NDI. For example, an NDI patient's GFR of 144L per day (100mL per minute) will result in about 14L of urine per day (10% of the GFR). Apart from vigorous water intake, only non-specific therapeutic options are currently available for complete NDI. These include sodium restriction, thiazides, or cyclooxygenase inhibitors. Sodium restriction and thiazides accelerate the reabsorption of sodium and water at the proximal tubules as a result of volume depletion. Cyclooxygenase inhibitors decrease the GFR due via the vasoconstriction of afferent arterioles. Both interventions decrease the urinary flow to the distal tubules, resulting in decreased urine volume. In spite of these treatments however, previous instances of mental retardation and renal failure have developed in children due to repeated episodes of hypernatremia, dehydration and a dilated bladder. Although specific effective treatments for NDI are highly desirable, they have proved elusive. New therapeutic strategies using pharmacochaperones (Milano, Carmosino, Gerbino, Svelto, & Procino, 2017; Robben, Sze, Knoers, & Deen, 2007) or that bypass the V2R-Gs-cAMP pathway (Ando et al., 2018; Bockenhauer & Bichet, 2015; Bouley et al., 2011; Bouley, Hasler, Lu, Nunes, & Brown, 2008; Li et al., 2009, 2011; Olesen, Rutzler, Moeller, Praetorius, & Fenton, 2011; Procino et al., 2014) might be possible in the future.

3. Loss-of-function mutations of V2R that cause NDI

3.1 Mechanism of loss-of-function

More than 250 different V2R mutations have now been reported in more than 300 NDI families (Bichet & Bockenhauer, 2016; Spanakis, Milord, & Gragnoli, 2008). Through functional analysis, these variants have been roughly categorized into three classes in accordance with their apparent effect on V2R biosynthesis (Robben, Knoers, & Deen, 2005), i.e., protein expression disorder (type I), localization disorder (ER retention; type II) or functional disorder (type III) (Fig. 2, Table 1). The most common is type II which results in impaired intracellular trafficking of the variant V2R protein, and occurs in up to 70% of NDI cases. For this reason, NDI caused by a V2R mutation is often referred to as a folding disease. The remaining types result in either null, decreased or truncated protein expression due to defects in transcription, mRNA processing, and translation (type I) or functional disorder, which is theoretically classified into ligand-binding disorder, G-protein coupling disorder or constitutive endocytosis (type III) (Deen, Marr, Kamsteeg, & van Balkom, 2000). However, this sub-classification is most likely over simplistic most V2R mutations are likely associated with combined disorders.

Most GPCRs, including V2R, contain a conserved pair of extracellular cysteine residues linking the first and second extracellular loops via a disulfide bond. Substitutions of cysteine for various residues in these loops (R104C, R106C, R181C, G185C, R202C, R204C, Y205C) cause loss-of-function mutations (Fujiwara, Morgan, & Bichet, 1995). This suggests that the presence of an additional cysteine in these loops may cause incorrect disulfide bond formation resulting in receptor dysfunction by interfering with ligand binding (Schulz, Grosse, Schultz, Gudermann, & Schoneberg, 2000). The question that arises however is whether the loss-of-function caused by these V2R mutations is caused only by reduced ligand-binding capacity. In this regard, the R181C mutation has been

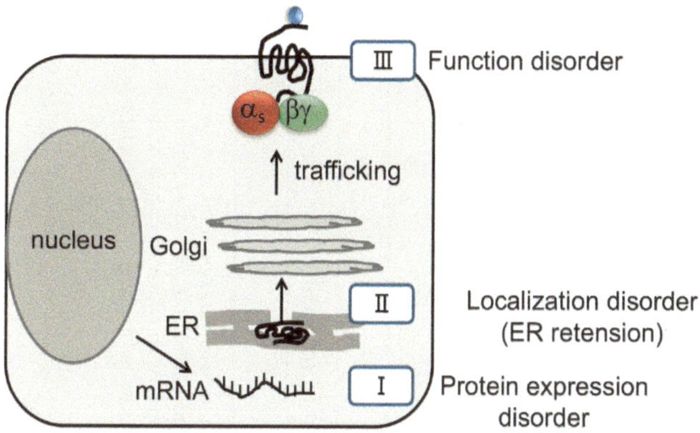

Fig. 2 Classification of V2 receptor loss-of-function mutations by functional analysis. Loss-of-function mutations in V2R have been roughly categorized into three classes based on their effect on protein biosynthesis, i.e., protein expression disorder (type I), localization disorder (ER retention; type II) or functional disorder (type III). Type I mutants show defects in the transcription, processing and translation of mRNA, resulting in null protein expression or a truncated protein that is readily degraded. Type II mutants show a folding disorder in the ER, resulting in impaired trafficking to the plasma membrane. Type III mutants show functional defects such as ligand binding, G-protein coupling and constitutive endocytosis.

reported to cause reduced ligand-binding capacity and maximal AVP-induced cAMP accumulation that is equivalent to wild type (Vmax) without reduced cell surface expression. This indicates that the inability of this mutant to stimulate adenylyl cyclase is indeed caused mainly by its reduced capacity for vasopressin binding (Pan, Wilson, & Gitschier, 1994). In contrast, the R202C V2R mutant has been reported to interfere with the trafficking of the V2 receptor to the cell surface, and thus cause reduced cell surface expression (Tsukaguchi, Matsubara, & Inada, 1995). This indicates that a cysteine replacement in the V2R protein can affect both its ligand-binding capacity and receptor trafficking. In the cases of other mutations in the first and second extracellular loop of V2R, the F105V variant, for example, was reported to show cell surface expression and maximal AVP-induced cAMP formation (Vmax) that is comparable to wild type, but reduced ligand-binding ability (Pasel et al., 2000). The R113W mutation is reported to result in a reduced affinity for AVP and lower cell surface expression (Birnbaumer, Gilbert, & Rosenthal, 1994). The G201D mutation on the other hand results in a decreased cell surface expression but only a minor reduction in ligand-binding affinity and coupling efficiency (Sadeghi, Robertson, Bichet, Innamorati, & Birnbaumer, 1997).

Table 1 Classification of V2 receptor mutations by the complex mechanisms leading to the loss-of-function.

Type	Points of disorder		Mutations in the first and second extracellular loop				DRY/H motif	
			F105V	(R113W)	R181C	G201D	R202C	R137H
Type 1	Protein expression	Null/decreased truncated						
Type II	Localization	ER retention (misfolding)		●	Intact	●	●	●
Type III	Function	Ligand binding	●	●	●	Not mainly	●	Intact
		G-protein coupling				Not mainly		?
		Constitutive endocytosis						●

As shown in Fig. 2, loss-of-function mutations in V2R can be classified into three types. However, most mutations cause a loss-of-function through a combination of mechanisms. Mutations in the first or second extracellular loop of the receptor are assumed to result in disrupted ligand binding but also cause a folding disorder. Conversely, the G201D mutant mainly shows a folding disorder with almost intact ligand-binding and G-protein activation. As an extreme example of this functional complexity, the R137H mutation in the DRY/H motif of V2R results in both beta-arrestin-dependent constitutive endocytosis and a misfolded full-length protein in the ER, resulting in the loss-of-cell surface expression.

Thus, even mutations in the first and second extracellular loop of V2R, which are assumed to be important for ligand binding, may cause a loss-of-function through different and quite complex mechanisms (Fig. 2, Table 1).

It was previously reported that the NDI-causing R137H mutation in the DRY/H motif of the second intracellular loop of V2R failed to stimulate Gs/adenylyl cyclase system even though the binding affinity of this variant for AVP was comparable to the wild-type protein (Rosenthal, Antaramian, Gilbert, & Birnbaumer, 1993). It was subsequently postulated that this mutation might cause a G-protein activation disorder based on its location within the DRY/H motif (Savarese & Fraser, 1992) and it was later found that the R137H variant has lower plasma membrane expression due to both misfolding and beta-arrestin-mediated constitutive endocytosis (Table 1) (Barak, Oakley, Laporte, & Caron, 2001; Kocan et al., 2009; Rochdi et al., 2010; Takahashi et al., 2012). In conclusion, a loss-of-function of V2R through mutation is often due to complex mechanisms that go beyond just the theoretical classification and sub-classification of these variants.

About half of all V2R mutations are missense, 25% are frameshift mutations resulting from nucleotide deletions or insertions, 10% are non-sense mutations, 10% are large deletions, and about the remaining 4% include inframe deletions or insertions, splice-site mutations and multiple or combined mutations (Spanakis et al., 2008). Mutations have now been identified in every domain of V2R, but about twice as many (on a per nucleotide basis) have been reported in transmembrane domains compared with the extracellular or intracellular domains. Ten recurrent mutations (D85N, V88M, R113W, Y128S, R137H, S167L, R181C, R202C, A294P and S315R) have been found in 35 ancestrally independent families, suggesting that they tend to occur sporadically. Consistently, the most frequently found V2R mutations (D85N, V88M, R113W, R137H, S167L, R181C and R202C) arise at potential hot spots (a C-to-T or G-to-A nucleotide substitution occurring at a CpG dinucleotide) (Arthus et al., 2000; Bichet et al., 1994). On the other hand, founder mutations (W71X, L312X) have also now been reported in V2R (Bichet et al., 1993).

3.2 Complete vs partial NDI

Inactivating mutations of V2R are transmitted in a X-linked manner, manifesting as a complete loss-of-function phenotype in most instances. While more than 250 kinds of V2R mutations have been identified to date, a limited number of AVPR2 gene mutations have been reported to result in

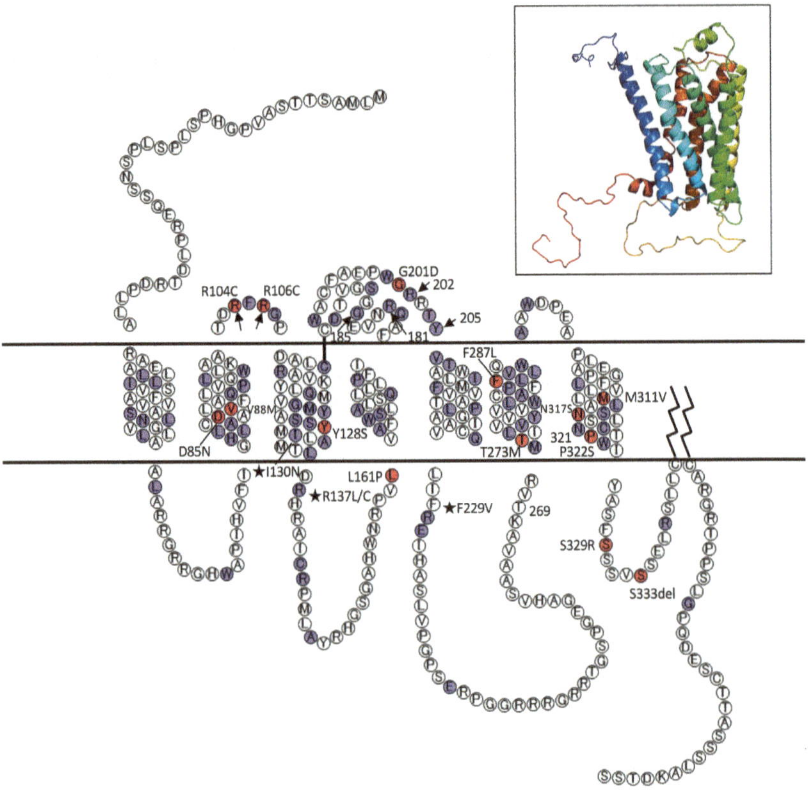

Fig. 3 Disease-causing missense or non-sense mutations of the V2 receptor. Missense or non-sense mutations of V2R that cause NDI are indicated by the blue circles. Red circles with numbers and a substituted amino-acid code denote partial NDI-causing mutations. Arrows with numbers highlight missense mutations substituted by cysteine in the first or second extracellular domain. Stars with numbers and a substituted amino-acid code indicate constitutively active mutations that cause NSIAD. Other types of mutations, such as frameshift deletions or insertions, inframe deletions or insertions, splice-site mutations, large deletions, and complex mutations, are not indicated in this figure. Inset shows the structural model of the V2R.

partial phenotype. To our knowledge, only 13 mutations among them have been tested clinically and confirmed by in vitro analysis (D85N, V88M, R104C, R106C, Y128S, G201D, T273M, F287L, M311V, N317S, P322S, S329R, S333del) (Fig. 3, Table 2) (Ala et al., 1998; Bockenhauer et al., 2009; Chen et al., 2002; Faerch et al., 2008, 2009; Inaba, Hatakeyama, Taniguchi, & Miyamori, 2001; Makita et al., 2016; Neocleous et al., 2012; Pasel et al., 2000; Sadeghi et al., 1997; Sahakitrungruang et al., 2010; Takahashi et al., 2012). D85N, V88M, R104C, R106C are recurrent V2R mutations in families diagnosed with partial NDI.

Table 2 Partial NDI-causing mutations in the V2 receptor and their classification based on functional analysis.

	Type II	Type III
G201D	●	Not mainly
D85N	→	●
F287L, M311V, M317S, P322S	?	●
V88M, R104C, R106C, Y128S, T273M, S329R, S333del	●	●

Reported partial NDI-causing mutations in V2R are classified into type II or III. Most of these mutations likely cause loss-of-function of the receptor via complex mechanisms.

Based on functional analysis, the M311V, M317S and P322S mutations in transmembrane domain VII of V2R have been classified as type III variants that reach the cell surface but with impaired ligand capacity and binding (Neocleous et al., 2012). As mentioned previously, ligand-binding disruption will often happen as a result of mutations in the first and second extracellular domain of V2R. The reason why it may also happen following a transmembrane domain VII mutation was therefore of interest from a functional perspective. Notably in this regard, the M311V variant, in which a Val residue substitutes for Met, typically elicits minor effects on protein folding. The homologous residue in the seventh transmembrane segment of the NK2 tachykinin receptor, M297, participates in binding a peptide agonist (Labrou, Bhogal, Hurrell, & Findlay, 2001), suggesting that V2R residue M311 may play a similar role (Sahakitrungruang et al., 2010). In the case of the P322S mutation in V2R, it has been reported that both a decreased affinity for AVP and reduced maximal activation of adenylyl cyclase might play roles in the resulting partial loss-of-function in comparison with the P322H variant that is completely non-functional (Ala et al., 1998). The proline at position 322 is conserved in the rat V1 and V2 vasopressin receptors, human oxytocin receptor and several other GPCRs. Previously it was suggested that this residue exists in the consensus sequence, NPXXY, in the M3 and angiotensin type I receptors where it plays a key role in their coupling properties by introducing kinks into the transmembrane domain helices (Ala et al., 1998; Hunyady, Bor, Baukal, Balla, & Catt, 1995; Wess, Nanavati, Vogel, & Maggio, 1993). In the case of the P322H variant, it is thought that the hydrogen bond between D85 and the substituted H322 residue may cause P322H to fail to activate G-protein (Ala et al., 1998).

The F287L mutation is located next to a proline at position 286 that is one of the most conserved residues among the GPCRs and has been postulated to be an alpha-helix breaker. The P286R NDI-causing mutation may therefore generate a kink in the transmembrane helix and thus affect the receptor function. The F287L mutation in V2R might also affect the helical conformation of this receptor and alter its binding capability (Chen et al., 2002). While the G201D V2R mutant was reported to show decreased cell surface expression with a minor reduction in ligand-binding affinity and coupling efficiency as mentioned earlier, the D85N variant has reduced ligand-binding affinity and lower adenylyl cyclase activation but no reduction in membrane expression (Ala et al., 1998; Inaba et al., 2001; Sadeghi et al., 1997). Of note also, most V2R mutations (such as V88M, R104C, R106C, Y128S, S333del, and T273M) are associated with a decreased cell surface expression of this receptor with decreased binding affinity, or with a rightward shift in the EC_{50} for AVP-stimulated cAMP accumulation, again suggestive of a mechanistic complexity underlying the resulting phenotypes (Bockenhauer et al., 2009; Makita et al., 2016; Pasel et al., 2000; Sadeghi et al., 1997).

Considering that about 50% of V2R mutations are missense, and about 70% of these cause impaired trafficking, but not defective protein expression, it is interesting that more cases of a partial NDI phenotype have not been reported. We speculate therefore that partial NDI cases might be missed, misdiagnosed or undiagnosed. Indeed, two cases that we previously reported were originally diagnosed with psychogenic polydipsia (Takahashi et al., 2012). However, following a water deprivation test and dDAVP loading test, we found that these patients could concentrate urine in response to AVP and that the principal indication for a possible NDI was the high basal AVP levels in both cases. It is possible that partial NDI will be revealed in further patients with polydipsia and that additional V2R mutations will be found. Careful attention to the basal AVP levels is therefore warranted in such cases. Notably in this regard, the diagnostic criteria for partial NDI are currently obscure. A decision to diagnose complete or partial NDI has generally been based to date on the clinical phenotype such as the daily urine volume, urine osmolality after a water deprivation test or the results of a pitressin/dDAVP loading test. In relation to a third NDI case of ours, a pitressin loading test was positive and a water deprivation test was negative (Makita et al., 2016), suggesting that stronger stimuli may be needed to concentrate urine in some cases. In one prior report, NDI patients were classified retrospectively as having mild, moderate or severe disease (Spanakis et al., 2008). We speculate that some partial NDI cases might be missed even among patients classed with severe disease in which a loading test was not done. Reported NDI cases with some response to loading test (L81F) might have partial NDI from the perspective of an AVP-responsiveness in vitro (Wildin, Cogdell, & Valadez, 1998). Careful examination by loading test may be necessary to diagnose partial NDI.

Phenotype-genotype discordance may also explain why just few cases of partial NDI have been reported. Some mutations causing complete NDI from a clinical point of view produce an AVP-stimulated response at high AVP concentrations in vitro (N317K, N321Y, which have been reported as partial NDI mutants) (Wildin et al., 1998) (R181C) (Sahakitrungruang et al., 2010). This suggests that the level of response to AVP using in vitro assay may not be enough to concentrate urine even in a clinical loading test, or that a genotype-phenotype discordance may reflect the difference between the clinical phenotype and the results of in vitro analysis.

Based on prior analyses that have disclosed the physiological functions of molecules causing rare diseases, we believe that the analysis of V2R mutations causing partial NDI may provide new insights into the physiological operations of this receptor because functional analysis can be performed. Moreover, the analysis of NDI-causing V2R mutations, which show localization disorder, may highlight new possibilities for treatments with pharmacochaperones. This is discussed further below. Hence, for the development of tailored and precision medicines, mutational analysis is desirable in NDI patients even if they show complete NDI.

3.3 Mutation hot spots and phenotype-genotype correlation

Mutations have now been identified in every domain of V2R, but more have been reported in the transmembrane domains compared with the extracellular or intracellular domains of this receptor (Fig. 3). In addition, the existence of frequent mutations may be indicative of hot spots at a CpG dinucleotide. Indeed, the detection of more than half (58%) of the nucleotide substitutions in 26 families could be a consequence of 5-methyl-cytosine deamination at a CpG dinucleotide. Most of the small deletions and insertions in V2R could be attributable to slipped mispairing during DNA replication (Bichet et al., 1994).

It has been reported that mutations affecting ligand-binding domains tend to retain partial signaling in vitro, whereas those that introduce a charge residue in a transmembrane domain are inactivating (Wildin et al., 1998). However, accumulating evidence has disclosed this is not always the case (Fig. 3). Some recurrent mutations (V88M, Y128M) in V2R have been reported to cause both complete and partial NDI, depending on the individual (Arthus et al., 2000; Bichet et al., 1993, 1994; Bockenhauer et al., 2009; Takahashi et al., 2012). Regarding the severity of this disorder, both inter- and intra-familial heterogeneity may exist.

3.4 Pharmacochaperones for loss-of-function mutations of V2R

As referred to earlier, 70% of the NDI-causing mutations produce a protein folding disorder in V2R, resulting in the trapping of the receptor by the quality control system in the ER. It cannot then be transported to the Golgi apparatus or plasma membrane, where the receptor interacts with its ligand. A process to correct this folding disorder would therefore have therapeutic potential for NDI. Ligands which help proteins to fold normally are known as pharmacochaperones and these agents may be candidate NDI drugs. In this section, we discuss the possibility of treating NDI with pharmacochaperones and skip another possible treatment strategy for this disorder by shunting or bypassing the V2R-Gs-cAMP signaling pathway (Bockenhauer & Bichet, 2015).

Several pharmacochaperones have been described to date and classified as antagonist, allosteric or agonist pharmacochaperones (Fig. 4). High-affinity antagonist pharmacochaperones can help mutant receptors to traffic to the plasma membrane, but these agents may not then allow AVP to bind because they are preventing access to the V2R binding site. On the other hand, low-affinity antagonist pharmacochaperone may be readily replaced by AVP, resulting in the activation of mutant receptors. Indeed, it was found that such a pharmacochaperone with low affinity, SR49059 (a non-peptide V1a receptor antagonist), facilitated the concentration of urine in NDI patients (Bernier et al., 2006).

Another possible NDI treatment strategy is the use of agonist pharmacochaperones. A couple of non-peptide agonists have been found to act as pharmacochaperones and enable mutant V2R receptors to mature and traffic to the plasma membrane, but also to activate mutant receptors directly (Jean-Alphonse et al., 2009). In addition to the effect of these molecules as agonist pharmacochaperones, they were found to work as biased agonists that activates cAMP signaling but not activate beta-arrestin/desensitization signaling, resulting in selective activation of downstream signaling. These drugs could thus enable mutant receptors to fold correctly through their pharmacochaperone functions and also operate as an agonist, resulting in cAMP signaling activation. Importantly, they would not recruit beta-arrestin, thus preventing membrane-trafficked mutant receptors from internalization and desensitization.

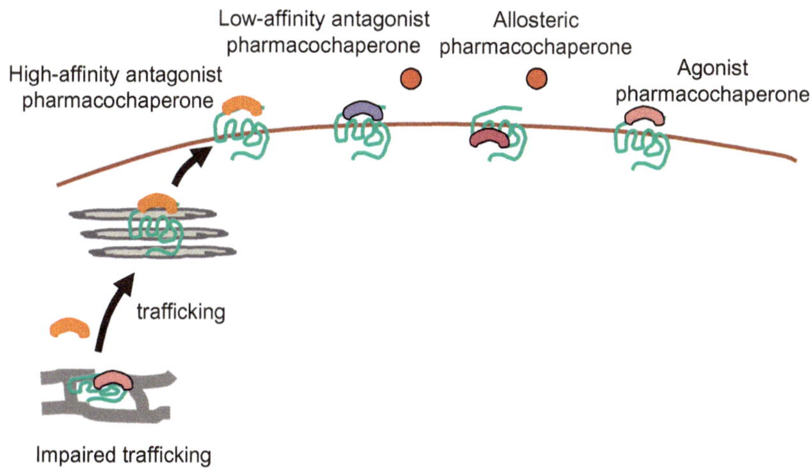

Fig. 4 Therapeutic strategy for NDI that targets V2 receptor loss-of-function mutations causing misfolding. Pharmacochaperones are conceptually the appropriate drugs to target misfolded V2R mutants that cause NDI. These agents are classified as antagonist, allosteric or agonist pharmacochaperones. High-affinity antagonist pharmacochaperones assist mutants to traffic to the plasma membrane but may then block subsequent AVP binding. Low-affinity antagonist pharmacochaperones can be easily replaced by AVP but have relatively weak chaperone activity. The effectiveness of a pharmacochaperone will be inversely related to the eventual accessibility of AVP as the principal agonist of the V2 receptor. Agonist pharmacochaperones can thus correct misfolding, facilitate the trafficking of mutant receptors to the plasma membrane, and activate them on behalf of AVP. Allosteric pharmacochaperones, which would promote V2R trafficking but not disrupt AVP binding, have not been reported yet.

In the field of GPCR signaling, the classical two-state model, in which GPCRs exist in equilibrium between an active and an inactive conformation, has now been replaced by a multi-state model, in which GPCRs are proposed to alter spontaneously between multiple active and inactive conformations, and each ligand is postulated to stabilize a specific conformation of the GPCR, which then leads to a set of specific biological effects. This type of unique ligand behavior has been termed functional selectivity or biased agonism, enabling the development of drugs which only switch on the desired signaling response. Furthermore, it is not limited to orthosteric ligands but also includes allosteric modulators (Makita et al., 2007). In GPCR signaling also, activation is usually followed by desensitization via beta-arrestin. In the field of cardiology, as an example, desensitization was originally described as a mechanism of protection from overstimulation of catecholamine but is now known to cause heart failure (Makita et al., 2013). It is worth noting in this regard that biased ligands which activate a specific and desired signaling pathway have been eagerly sought in multiple clinical fields.

With a focus on partial NDI mutations, because they allow for functional analysis, we found that the antagonist pharmacochaperone tolvaptan can work as an inverse agonist against a wild-type V2R but as a partial agonist against partial NDI-causing mutant receptors by facilitating cAMP accumulation (Takahashi et al., 2012). We found that tolvaptan augments the membrane expression of mutant receptors by acting as a pharmacochaperone and increasing the basal cAMP accumulation level. This finding indicated that certain ligands may change or even reverse the effects of a V2R mutation, depending on the states or systems adopted. This suggests that the effects of an agonist, inverse agonist, or neutral antagonist of a given GPCR ligand are not intrinsically decided by the characteristic of each ligand but by the system adopted. This is referred to as protean agonism after Greek god, Proteus (Kenakin, 2007). Protean agonism may be a unique case of biased agonism as it effects downstream signaling in a manner that is dependent on the receptors but not the ligands.

With respect to NDI treatment strategies, we have also reported on another agonist pharmacochaperone. In contrast to previous findings (Robben et al., 2009), we found that the non-peptide agonist OPC5 enables sustainable signaling by working as an agonist pharmacochaperone for partial NDI mutant receptors (Makita et al., 2016). This agent assists with the maturation of mutant receptors, promotes their trafficking to the plasma membrane, and then activates them as an agonist, which is follow by β-arrestin recruitment and internalization. Overall, OPC5, but not AVP, enables sustainable cAMP accumulation. Our findings suggest that partial agonistic activity is sometimes better (more effective) than strong agonistic activity.

4. Gain-of-function mutations causing nephrogenic syndrome of inappropriate antidiuresis (NSIAD)

4.1 Nephrogenic syndrome of inappropriate antidiuresis

In contrast to NDI, NSIAD is caused by water retention due to a gain-of-function V2R mutation. It was first described in 2005 in two unrelated male infants with severe symptomatic hyponatremia carrying an R137C or R137L mutation in the V2R gene (Feldman et al., 2005). Despite undetectable AVP levels, these patients showed a SIADH-like clinical phenotype, hyponatremia and hypoosmolality with euvolemia or slight hypervolemia. Since this first description, around 30 cases have been reported and several activating mutations have been described although the true prevalence of this may well be underestimated (Carpentier et al., 2012;

Erdelyi et al., 2015) (Fig. 3). Among the gain-of-function V2R mutations thus far identified, R137C has been the most commonly reported.

NSIAD has been described as an X-linked recessive condition in males. Heterozygous females are usually asymptomatic. However, several symptomatic female cases have been reported without skewed X-inactivation. It is understandable that a gain-of-function of V2R in only one allele may be sufficient to cause water retention, depending on its constitutive activity level and intra-familial and inter-familial variabilities (Decaux et al., 2007; Hague et al., 2018; Powlson, Challis, Halsall, Schoenmakers, & Gurnell, 2016). Conversely, in male adult cases with episodic mild hyponatremia, weaker gain-of-function mutations might be disclosed in the future. It may therefore be of value to investigate not only cases of unexplained mild hyponatremia with an undetectable AVP level, but also cases of psychogenic polydipsia showing episodic hyponatremia via a water loading test.

4.2 Mechanism underlying a V2R gain-of-function and its therapeutic implications

Arginine 137 is highly conserved in V2R across species and is located in the DRY/H domain of the second intracellular loop. This region seems to play a central role in the stabilization of the receptor in both its active and inactive forms (Audet & Bouvier, 2012). Indeed, substitution of this arginine for histidine (R137H) has been known to cause a loss-of-function of V2R and resultant NDI. Several reports, including those from our laboratory, have described agonist-independent, arrestin-mediated desensitization and internalization of R137H as well as R137L/C mutants (Barak et al., 2001; Kocan et al., 2009; Rochdi et al., 2010; Takahashi et al., 2012). While most V2R loss-of-function mutations result mainly in a trafficking disorder due to misfolding, the loss-of-function caused by R137H might be caused, at least in part, by constitutive internalization (Barak et al., 2001; Kocan et al., 2009; Rochdi et al., 2010; Takahashi et al., 2012). Structural modeling of V2R has revealed a putative hydrogen bond between the Arg137 and Thr269 residues. It is speculated that loss of this interaction may permit adoption of an activated form of the receptor, independent of AVP binding (Powlson et al., 2016). We speculate that the R137H mutant might be more active and thus not only more internalized but also more unstable/denatured, overall resulting in a loss-of-function because activation may be essentially associated with instability/denaturation and therefore superactivation may be associated with superinstability (Gether et al., 1997; Iiri, Herzmark, Nakamoto, van Dop, & Bourne, 1994).

In contrast to R137H, two other gain-of-function mutations identified in NSIAD patients, F229V and I130N, show constitutive cAMP accumulation without constitutive beta-arrestin recruitment in vitro (Carpentier et al., 2012; Erdelyi et al., 2015). This may implicate biased signaling that prefers activation to desensitization via beta-arrestin in the case of these disease-causing mutants. There is a possibility however that a lower constitutive activity level leads to reduced beta-arrestin recruitment and less internalization, resulting in relatively higher cell surface expression and basal cAMP accumulation. As discussed below, the insensitivity of R137L/C mutants to inverse agonists, contrary to their sensitive effects upon F229V and I130N variants (Carpentier et al., 2012; Decaux et al., 2007), is compatible with the idea that the R137L/C mutations produce more active conformations.

4.3 Therapeutic strategies

In addition to water restriction and osmotic agent urea supplementation, inverse agonists may be logical candidates for NDI therapeutics. Indeed, in vitro studies of tolvaptan and salvaptan have shown that they can inhibit cAMP accumulation by F229V and I130N mutants but not by R137L/C mutants (Carpentier et al., 2012; Decaux et al., 2007). In a clinical setting, a proband carrying the R137L mutation was found to show no response to either of these drugs administrated during a phase III clinical trial (Decaux et al., 2007). Considering the possibility that novel gain-of-function mutations in V2R that cause NSIAD might be disclosed in the future, genetic and functional analysis will continue to be important for designing and researching potential therapeutic options.

Acknowledgment

We thank Masanori Ootaki for support.

References

Ala, Y., Morin, D., Mouillac, B., Sabatier, N., Vargas, R., Cotte, N., et al. (1998). Functional studies of twelve mutant V2 vasopressin receptors related to nephrogenic diabetes insipidus: Molecular basis of a mild clinical phenotype. *Journal of the American Society of Nephrology*, *9*(10), 1861–1872.

Ando, F., Mori, S., Yui, N., Morimoto, T., Nomura, N., Sohara, E., et al. (2018). AKAPs-PKA disruptors increase AQP2 activity independently of vasopressin in a model of nephrogenic diabetes insipidus. *Nature Communications*, *9*(1), 1411. https://doi.org/10.1038/s41467-018-03771-2.

Arthus, M. F., Lonergan, M., Crumley, M. J., Naumova, A. K., Morin, D., De Marco, L. A., et al. (2000). Report of 33 novel AVPR2 mutations and analysis of 117 families with X-linked nephrogenic diabetes insipidus. *Journal of the American Society of Nephrology*, *11*(6), 1044–1054.

Audet, M., & Bouvier, M. (2012). Restructuring G-protein-coupled receptor activation. *Cell, 151*(1), 14–23. https://doi.org/10.1016/j.cell.2012.09.003.

Barak, L. S., Oakley, R. H., Laporte, S. A., & Caron, M. G. (2001). Constitutive arrestin-mediated desensitization of a human vasopressin receptor mutant associated with nephrogenic diabetes insipidus. *Proceedings of the National Academy of Sciences of the United States of America, 98*(1), 93–98. https://doi.org/10.1073/pnas.011303698.

Bernier, V., Morello, J. P., Zarruk, A., Debrand, N., Salahpour, A., Lonergan, M., et al. (2006). Pharmacologic chaperones as a potential treatment for X-linked nephrogenic diabetes insipidus. *Journal of the American Society of Nephrology, 17*(1), 232–243. https://doi.org/10.1681/ASN.2005080854. [pii] ASN.2005080854.

Bichet, D. G., Arthus, M. F., Lonergan, M., Hendy, G. N., Paradis, A. J., Fujiwara, T. M., et al. (1993). X-linked nephrogenic diabetes insipidus mutations in North America and the Hopewell hypothesis. *The Journal of Clinical Investigation, 92*(3), 1262–1268. https://doi.org/10.1172/JCI116698.

Bichet, D. G., Birnbaumer, M., Lonergan, M., Arthus, M. F., Rosenthal, W., Goodyer, P., et al. (1994). Nature and recurrence of AVPR2 mutations in X-linked nephrogenic diabetes insipidus. *American Journal of Human Genetics, 55*(2), 278–286.

Bichet, D. G., & Bockenhauer, D. (2016). Genetic forms of nephrogenic diabetes insipidus (NDI): Vasopressin receptor defect (X-linked) and aquaporin defect (autosomal recessive and dominant). *Best Practice & Research. Clinical Endocrinology & Metabolism, 30*(2), 263–276. https://doi.org/10.1016/j.beem.2016.02.010.

Birnbaumer, M., Gilbert, S., & Rosenthal, W. (1994). An extracellular congenital nephrogenic diabetes insipidus mutation of the vasopressin receptor reduces cell surface expression, affinity for ligand, and coupling to the Gs/adenylyl cyclase system. *Molecular Endocrinology, 8*(7), 886–894. https://doi.org/10.1210/mend.8.7.7984150.

Bockenhauer, D., & Bichet, D. G. (2015). Pathophysiology, diagnosis and management of nephrogenic diabetes insipidus. *Nature Reviews. Nephrology, 11*(10), 576–588. https://doi.org/10.1038/nrneph.2015.89.

Bockenhauer, D., Carpentier, E., Rochdi, D., van't Hoff, W., Breton, B., Bernier, V., et al. (2009). Vasopressin type 2 receptor V88M mutation: Molecular basis of partial and complete nephrogenic diabetes insipidus. *Nephron. Physiology, 114*(1), 1–10. https://doi.org/10.1159/000245059. [pii] 000245059.

Bouley, R., Hasler, U., Lu, H. A., Nunes, P., & Brown, D. (2008). Bypassing vasopressin receptor signaling pathways in nephrogenic diabetes insipidus. *Seminars in Nephrology, 28*(3), 266–278. https://doi.org/10.1016/j.semnephrol.2008.03.010. [pii] S0270-9295(08)00054-5.

Bouley, R., Lu, H. A., Nunes, P., Da Silva, N., McLaughlin, M., Chen, Y., et al. (2011). Calcitonin has a vasopressin-like effect on aquaporin-2 trafficking and urinary concentration. *Journal of the American Society of Nephrology, 22*(1), 59–72. https://doi.org/10.1681/ASN.2009121267. [pii] ASN.2009121267.

Carpentier, E., Greenbaum, L. A., Rochdi, D., Abrol, R., Goddard, W. A., 3rd, Bichet, D. G., et al. (2012). Identification and characterization of an activating F229V substitution in the V2 vasopressin receptor in an infant with NSIAD. *Journal of the American Society of Nephrology, 23*(10), 1635–1640. https://doi.org/10.1681/ASN.2012010077.

Chen, C. H., Chen, W. Y., Liu, H. L., Liu, T. T., Tsou, A. P., Lin, C. Y., et al. (2002). Identification of mutations in the arginine vasopressin receptor 2 gene causing nephrogenic diabetes insipidus in Chinese patients. *Journal of Human Genetics, 47*(2), 66–73. https://doi.org/10.1007/s100380200002.

Decaux, G., Vandergheynst, F., Bouko, Y., Parma, J., Vassart, G., & Vilain, C. (2007). Nephrogenic syndrome of inappropriate antidiuresis in adults: High phenotypic variability in men and women from a large pedigree. *Journal of the American Society of Nephrology, 18*(2), 606–612. https://doi.org/10.1681/ASN.2006090987.

Deen, P. M., Marr, N., Kamsteeg, E. J., & van Balkom, B. W. (2000). Nephrogenic diabetes insipidus. *Current Opinion in Nephrology and Hypertension, 9*(6), 591–595.

Erdelyi, L. S., Mann, W. A., Morris-Rosendahl, D. J., Gross, U., Nagel, M., Varnai, P., et al. (2015). Mutation in the V2 vasopressin receptor gene, AVPR2, causes nephrogenic syndrome of inappropriate diuresis. *Kidney International, 88*(5), 1070–1078. https://doi.org/10.1038/ki.2015.181.

Faerch, M., Christensen, J. H., Corydon, T. J., Kamperis, K., de Zegher, F., Gregersen, N., et al. (2008). Partial nephrogenic diabetes insipidus caused by a novel mutation in the AVPR2 gene. *Clinical Endocrinology, 68*(3), 395–403. https://doi.org/10.1111/j.1365-2265.2007.03054.x. [pii] CEN3054.

Faerch, M., Christensen, J. H., Rittig, S., Johansson, J. O., Gregersen, N., de Zegher, F., et al. (2009). Diverse vasopressin V2 receptor functionality underlying partial congenital nephrogenic diabetes insipidus. *American Journal of Physiology. Renal Physiology, 297*(6), F1518–F1525. https://doi.org/10.1152/ajprenal.00331.2009. [pii] 00331.2009.

Feldman, B. J., Rosenthal, S. M., Vargas, G. A., Fenwick, R. G., Huang, E. A., Matsuda-Abedini, M., et al. (2005). Nephrogenic syndrome of inappropriate antidiuresis. *The New England Journal of Medicine, 352*(18), 1884–1890. https://doi.org/10.1056/NEJMoa042743.

Fujiwara, T. M., Morgan, K., & Bichet, D. G. (1995). Molecular biology of diabetes insipidus. *Annual Review of Medicine, 46*, 331–343. https://doi.org/10.1146/annurev.med.46.1.331.

Gao, M., Cao, R., Du, S., Jia, X., Zheng, S., Huang, S., et al. (2015). Disruption of prostaglandin E2 receptor EP4 impairs urinary concentration via decreasing aquaporin 2 in renal collecting ducts. *Proceedings of the National Academy of Sciences of the United States of America, 112*(27), 8397–8402. https://doi.org/10.1073/pnas.1509565112. [pii] 1509565112.

Gether, U., Ballesteros, J. A., Seifert, R., Sanders-Bush, E., Weinstein, H., & Kobilka, B. K. (1997). Structural instability of a constitutively active G protein-coupled receptor. Agonist-independent activation due to conformational flexibility. *The Journal of Biological Chemistry, 272*(5), 2587–2590.

Hague, J., Casey, R., Bruty, J., Legerton, T., Abbs, S., Oddy, S., et al. (2018). Adult female with symptomatic AVPR2-related nephrogenic syndrome of inappropriate antidiuresis (NSIAD). *Endocrinology, Diabetes & Metabolism Case Reports, 2018*. https://doi.org/10.1530/EDM-17-0139. pii: 17-0139.

Hunyady, L., Bor, M., Baukal, A. J., Balla, T., & Catt, K. J. (1995). A conserved NPLFY sequence contributes to agonist binding and signal transduction but is not an internalization signal for the type 1 angiotensin II receptor. *The Journal of Biological Chemistry, 270*(28), 16602–16609.

Iiri, T., Herzmark, P., Nakamoto, J. M., van Dop, C., & Bourne, H. R. (1994). Rapid GDP release from Gs alpha in patients with gain and loss of endocrine function. *Nature, 371*(6493), 164–168. https://doi.org/10.1038/371164a0.

Inaba, S., Hatakeyama, H., Taniguchi, N., & Miyamori, I. (2001). The property of a novel v2 receptor mutant in a patient with nephrogenic diabetes insipidus. *The Journal of Clinical Endocrinology and Metabolism, 86*(1), 381–385. https://doi.org/10.1210/jcem.86.1.7165.

Jean-Alphonse, F., Perkovska, S., Frantz, M. C., Durroux, T., Mejean, C., Morin, D., et al. (2009). Biased agonist pharmacochaperones of the AVP V2 receptor may treat congenital nephrogenic diabetes insipidus. *Journal of the American Society of Nephrology, 20*(10), 2190–2203. https://doi.org/10.1681/ASN.2008121289. [pii] ASN.2008121289.

Kenakin, T. (2007). Functional selectivity through protean and biased agonism: Who steers the ship? *Molecular Pharmacology, 72*(6), 1393–1401. https://doi.org/10.1124/mol.107.040352. [pii] mol.107.040352.

Kocan, M., See, H. B., Sampaio, N. G., Eidne, K. A., Feldman, B. J., & Pfleger, K. D. (2009). Agonist-independent interactions between beta-arrestins and mutant vasopressin type II receptors associated with nephrogenic syndrome of inappropriate antidiuresis. *Molecular Endocrinology*, *23*(4), 559–571. https://doi.org/10.1210/me.2008-0321.

Kortenoeven, M. L., Trimpert, C., van den Brand, M., Li, Y., Wetzels, J. F., & Deen, P. M. (2012). In mpkCCD cells, long-term regulation of aquaporin-2 by vasopressin occurs independent of protein kinase A and CREB but may involve Epac. *American Journal of Physiology. Renal Physiology*, *302*(11), F1395–F1401. https://doi.org/10.1152/ajprenal.00376.2011. [pii] ajprenal.00376.2011.

Labrou, N. E., Bhogal, N., Hurrell, C. R., & Findlay, J. B. (2001). Interaction of Met297 in the seventh transmembrane segment of the tachykinin NK2 receptor with neurokinin A. *The Journal of Biological Chemistry*, *276*(41), 37944–37949. https://doi.org/10.1074/jbc.M106330200.

Li, J. H., Chou, C. L., Li, B., Gavrilova, O., Eisner, C., Schnermann, J., et al. (2009). A selective EP4 PGE2 receptor agonist alleviates disease in a new mouse model of X-linked nephrogenic diabetes insipidus. *The Journal of Clinical Investigation*, *119*(10), 3115–3126. https://doi.org/10.1172/JCI39680. [pii] 39680.

Li, W., Zhang, Y., Bouley, R., Chen, Y., Matsuzaki, T., Nunes, P., et al. (2011). Simvastatin enhances aquaporin-2 surface expression and urinary concentration in vasopressin-deficient Brattleboro rats through modulation of Rho GTPase. *American Journal of Physiology. Renal Physiology*, *301*(2), F309–F318. https://doi.org/10.1152/ajprenal.00001.2011. [pii] ajprenal.00001.2011.

Lolait, S. J., O'Carroll, A. M., McBride, O. W., Konig, M., Morel, A., & Brownstein, M. J. (1992). Cloning and characterization of a vasopressin V2 receptor and possible link to nephrogenic diabetes insipidus. *Nature*, *357*(6376), 336–339. https://doi.org/10.1038/357336a0.

Makita, N., Kabasawa, Y., Otani, Y., Firman, Sato, J., Hashimoto, M., et al. (2013). Attenuated desensitization of beta-adrenergic receptor by water-soluble N-nitrosamines that induce S-nitrosylation without NO release. *Circulation Research*, *112*(2), 327–334.

Makita, N., Sato, J., Manaka, K., Shoji, Y., Oishi, A., Hashimoto, M., et al. (2007). An acquired hypocalciuric hypercalcemia autoantibody induces allosteric transition among active human Ca-sensing receptor conformations. *Proceedings of the National Academy of Sciences of the United States of America*, *104*(13), 5443–5448. https://doi.org/10.1073/pnas.0701290104. [pii] 0701290104.

Makita, N., Sato, T., Yajima-Shoji, Y., Sato, J., Manaka, K., Eda-Hashimoto, M., et al. (2016). Analysis of the V2 vasopressin receptor (V2R) mutations causing partial nephrogenic diabetes insipidus highlights a sustainable signaling by a non-peptide V2R agonist. *The Journal of Biological Chemistry*, *291*(43), 22460–22471. https://doi.org/10.1074/jbc.M116.733220.

Milano, S., Carmosino, M., Gerbino, A., Svelto, M., & Procino, G. (2017). Hereditary nephrogenic diabetes insipidus: Pathophysiology and possible treatment. An update. *International Journal of Molecular Sciences*, *18*(11), pii: E2385 https://doi.org/10.3390/ijms18112385.

Moeller, H. B., Rittig, S., & Fenton, R. A. (2013). Nephrogenic diabetes insipidus: Essential insights into the molecular background and potential therapies for treatment. *Endocrine Reviews*, *34*(2), 278–301. https://doi.org/10.1210/er.2012-1044. [pii] er.2012-1044.

Neocleous, V., Skordis, N., Shammas, C., Efstathiou, E., Mastroyiannopoulos, N. P., & Phylactou, L. A. (2012). Identification and characterization of a novel X-linked AVPR2 mutation causing partial nephrogenic diabetes insipidus: A case report and review of the literature. *Metabolism*, *61*(7), 922–930. https://doi.org/10.1016/j.metabol.2012.01.005.

Nomura, Y., Onigata, K., Nagashima, T., Yutani, S., Mochizuki, H., Nagashima, K., et al. (1997). Detection of skewed X-inactivation in two female carriers of vasopressin type 2 receptor gene mutation. *The Journal of Clinical Endocrinology and Metabolism, 82*(10), 3434–3437. https://doi.org/10.1210/jcem.82.10.4312.

Olesen, E. T., Rutzler, M. R., Moeller, H. B., Praetorius, H. A., & Fenton, R. A. (2011). Vasopressin-independent targeting of aquaporin-2 by selective E-prostanoid receptor agonists alleviates nephrogenic diabetes insipidus. *Proceedings of the National Academy of Sciences of the United States of America, 108*(31), 12949–12954. https://doi.org/10.1073/pnas.1104691108. [pii] 1104691108.

Pan, Y., Metzenberg, A., Das, S., Jing, B., & Gitschier, J. (1992). Mutations in the V2 vasopressin receptor gene are associated with X-linked nephrogenic diabetes insipidus. *Nature Genetics, 2*(2), 103–106. https://doi.org/10.1038/ng1092-103.

Pan, Y., Wilson, P., & Gitschier, J. (1994). The effect of eight V2 vasopressin receptor mutations on stimulation of adenylyl cyclase and binding to vasopressin. *The Journal of Biological Chemistry, 269*(50), 31933–31937.

Pasel, K., Schulz, A., Timmermann, K., Linnemann, K., Hoeltzenbein, M., Jaaskelainen, J., et al. (2000). Functional characterization of the molecular defects causing nephrogenic diabetes insipidus in eight families. *The Journal of Clinical Endocrinology and Metabolism, 85*(4), 1703–1710. https://doi.org/10.1210/jcem.85.4.6507.

Powlson, A. S., Challis, B. G., Halsall, D. J., Schoenmakers, E., & Gurnell, M. (2016). Nephrogenic syndrome of inappropriate antidiuresis secondary to an activating mutation in the arginine vasopressin receptor AVPR2. *Clinical Endocrinology, 85*(2), 306–312. https://doi.org/10.1111/cen.13011.

Procino, G., Milano, S., Carmosino, M., Barbieri, C., Nicoletti, M. C., Li, J. H., et al. (2014). Combination of secretin and fluvastatin ameliorates the polyuria associated with X-linked nephrogenic diabetes insipidus in mice. *Kidney International, 86*(1), 127–138. https://doi.org/10.1038/ki.2014.10. [pii] ki201410.

Puck, J. M., & Willard, H. F. (1998). X inactivation in females with X-linked disease. *The New England Journal of Medicine, 338*(5), 325–328. https://doi.org/10.1056/NEJM199801293380611.

Robben, J. H., Knoers, N. V., & Deen, P. M. (2005). Characterization of vasopressin V2 receptor mutants in nephrogenic diabetes insipidus in a polarized cell model. *American Journal of Physiology. Renal Physiology, 289*(2), F265–F272. https://doi.org/10.1152/ajprenal.00404.2004.

Robben, J. H., Kortenoeven, M. L., Sze, M., Yae, C., Milligan, G., Oorschot, V. M., et al. (2009). Intracellular activation of vasopressin V2 receptor mutants in nephrogenic diabetes insipidus by non-peptide agonists. *Proceedings of the National Academy of Sciences of the United States of America, 106*(29), 12195–12200. https://doi.org/10.1073/pnas.0900130106. [pii] 0900130106.

Robben, J. H., Sze, M., Knoers, N. V., & Deen, P. M. (2007). Functional rescue of vasopressin V2 receptor mutants in MDCK cells by pharmacochaperones: Relevance to therapy of nephrogenic diabetes insipidus. *American Journal of Physiology. Renal Physiology, 292*(1), F253–F260. https://doi.org/10.1152/ajprenal.00247.2006. [pii] 00247.2006.

Rochdi, M. D., Vargas, G. A., Carpentier, E., Oligny-Longpre, G., Chen, S., Kovoor, A., et al. (2010). Functional characterization of vasopressin type 2 receptor substitutions (R137H/C/L) leading to nephrogenic diabetes insipidus and nephrogenic syndrome of inappropriate antidiuresis: Implications for treatments. *Molecular Pharmacology,* 77(5), 836–845. https://doi.org/10.1124/mol.109.061804.

Rosenthal, W., Antaramian, A., Gilbert, S., & Birnbaumer, M. (1993). Nephrogenic diabetes insipidus. A V2 vasopressin receptor unable to stimulate adenylyl cyclase. *The Journal of Biological Chemistry, 268*(18), 13030–13033.

Rosenthal, W., Seibold, A., Antaramian, A., Lonergan, M., Arthus, M. F., Hendy, G. N., et al. (1992). Molecular identification of the gene responsible for congenital nephrogenic diabetes insipidus. *Nature*, *359*(6392), 233–235. https://doi.org/10.1038/359233a0.

Sadeghi, H., Robertson, G. L., Bichet, D. G., Innamorati, G., & Birnbaumer, M. (1997). Biochemical basis of partial nephrogenic diabetes insipidus phenotypes. *Molecular Endocrinology*, *11*(12), 1806–1813. https://doi.org/10.1210/mend.11.12.0017.

Sahakitrungruang, T., Tee, M. K., Rattanachartnarong, N., Shotelersuk, V., Suphapeetiporn, K., & Miller, W. L. (2010). Functional characterization of vasopressin receptor 2 mutations causing partial and complete congenital nephrogenic diabetes insipidus in Thai families. *Hormone Research in Pædiatrics*, *73*(5), 349–354. https://doi.org/10.1159/000308167. [pii] 000308167.

Satoh, M., Ogikubo, S., & Yoshizawa-Ogasawara, A. (2008). Correlation between clinical phenotypes and X-inactivation patterns in six female carriers with heterozygote vasopressin type 2 receptor gene mutations. *Endocrine Journal*, *55*(2), 277–284.

Savarese, T. M., & Fraser, C. M. (1992). In vitro mutagenesis and the search for structure-function relationships among G protein-coupled receptors. *The Biochemical Journal*, *283*(Pt. 1), 1–19.

Schulz, A., Grosse, R., Schultz, G., Gudermann, T., & Schoneberg, T. (2000). Structural implication for receptor oligomerization from functional reconstitution studies of mutant V2 vasopressin receptors. *The Journal of Biological Chemistry*, *275*(4), 2381–2389.

Spanakis, E., Milord, E., & Gragnoli, C. (2008). AVPR2 variants and mutations in nephrogenic diabetes insipidus: Review and missense mutation significance. *Journal of Cellular Physiology*, *217*(3), 605–617. https://doi.org/10.1002/jcp.21552.

Takahashi, K., Makita, N., Manaka, K., Hisano, M., Akioka, Y., Miura, K., et al. (2012). V2 vasopressin receptor (V2R) mutations in partial nephrogenic diabetes insipidus highlight protean agonism of V2R antagonists. *The Journal of Biological Chemistry*, *287*(3), 2099–2106. https://doi.org/10.1074/jbc.M111.268797.

Tsukaguchi, H., Matsubara, H., & Inada, M. (1995). Expression studies of two vasopressin V2 receptor gene mutations, R202C and 804insG, in nephrogenic diabetes insipidus. *Kidney International*, *48*(2), 554–562.

van den Ouweland, A. M., Dreesen, J. C., Verdijk, M., Knoers, N. V., Monnens, L. A., Rocchi, M., et al. (1992). Mutations in the vasopressin type 2 receptor gene (AVPR2) associated with nephrogenic diabetes insipidus. *Nature Genetics*, *2*(2), 99–102. https://doi.org/10.1038/ng1092-99.

Wess, J., Nanavati, S., Vogel, Z., & Maggio, R. (1993). Functional role of proline and tryptophan residues highly conserved among G protein-coupled receptors studied by mutational analysis of the m3 muscarinic receptor. *The EMBO Journal*, *12*(1), 331–338.

Wildin, R. S., Cogdell, D. E., & Valadez, V. (1998). AVPR2 variants and V2 vasopressin receptor function in nephrogenic diabetes insipidus. *Kidney International*, *54*(6), 1909–1922. https://doi.org/10.1046/j.1523-1755.1998.00214.x.

Yip, K. P. (2006). Epac-mediated Ca(2+) mobilization and exocytosis in inner medullary collecting duct. *American Journal of Physiology. Renal Physiology*, *291*(4), F882–F890. https://doi.org/10.1152/ajprenal.00411.2005. [pii] 00411.2005.

CHAPTER FIVE

Vasopressin inactivation: Role of insulin-regulated aminopeptidase

Don T. Li[a,b], Estifanos N. Habtemichael[a], Jonathan S. Bogan[a,b,*]
[a]Department of Internal Medicine, Section of Endocrinology and Metabolism, Yale University School of Medicine, New Haven, CT, United States
[b]Department of Cell Biology, Yale University School of Medicine, New Haven, CT, United States
*Corresponding author: e-mail address: jonathan.bogan@yale.edu

Contents

Abstract

The physiological importance of vasopressin inactivation has long been appreciated, but the mechanisms and potential pathophysiologic roles of this process remain active subjects of research. Human Placental Leucine Aminopeptidase (P-LAP, encoded by the *LNPEP* gene) is an important determinant of vasopressinase activity during pregnancy and is associated with gestational diabetes insipidus and preeclampsia. Insulin-Regulated Aminopeptidase (IRAP), the rodent homologue of P-LAP, is coregulated with the insulin-responsive glucose transporter, GLUT4, in adipose and muscle cells. Recently, the Tether containing a UBX domain for GLUT4 (TUG) protein was shown to mediate the coordinated regulation of water and glucose homeostasis. TUG sequesters IRAP and GLUT4 intracellularly in the absence of insulin. Insulin and other stimuli cause the proteolytic cleavage of TUG to mobilize these proteins to the cell surface, where IRAP acts to terminate the activity of circulating vasopressin. Intriguingly, genetic variation in *LNPEP* is associated with the vasopressin response and mortality during sepsis, and increased copeptin, a marker of vasopressin secretion, is associated with cardiovascular and metabolic disease. We propose that in the setting of insulin resistance in muscle, increased cell-surface IRAP and accelerated vasopressin degradation cause a compensatory increase in vasopressin secretion. The increased vasopressin concentrations present

Vitamins and Hormones, Volume 113
ISSN 0083-6729
https://doi.org/10.1016/bs.vh.2019.08.017

at the kidneys then contribute to hypertension in the metabolic syndrome. Further analyses of metabolism and of vasopressin and copeptin may yield novel insights into a unified pathophysiologic mechanism linking insulin resistance and hypertension, and potentially other components of the metabolic syndrome, in humans.

1. Introduction: Antidiuretic activity is regulated enzymatically by vasopressin inactivation

Mechanisms underlying the elimination of antidiuretic activity have been of significant interest to the biomedical community since as early as the 1930s, when Heller and colleagues investigated the principles by which this activity is regulated. They found that the "antidiuretic principle" interacted with tissue homogenates and speculated that this binding was important for activity (Heller & Urban, 1935). Further work was done by Dicker and Greenbaum, who incubated vasopressin with tissue slices. This incubation resulted in a partial inactivation of vasopressin's antidiuretic activity, though some tissues appeared to plateau at 20% residual activity (Dicker & Greenbaum, 1954). The incubation of vasopressin with a panel of different tissue homogenates demonstrated that the inactivation processes responsible for regulating vasopressin existed in tissues as diverse as liver, kidney, spleen, and duodenum (Dicker & Greenbaum, 1955, 1956). At the time, these investigators were unable to determine the precise mechanism underlying this behavior; however, by analogy they concluded that the process must involve either the destruction of an intra-peptide disulfide bond, now known to be present in vasopressin, or destruction of the peptide more generally (Dicker & Greenbaum, 1958).

Work then focused on the kinetic properties of arginine-vasopressin (AVP) and injection of vasopressin into animals established that the hormone has a very short half-life. In humans, injected ^{125}I-AVP distributed rapidly into a volume that closely approximated the size of the extracellular fluid space. Metabolic breakdown products of ^{125}I-AVP were detectable within minutes after injection. In these early studies, analysis of product production revealed a mean plasma half-life 24 min (Baumann & Dingman, 1976). However, substantially shorter half-lives were measured in subsequent work. In rats, unlabeled vasopressin was observed to disappear from the circulation with a half-life of 1.6 min (Forsling, Martin, Sturdy, & Burton, 1973). Similar results were obtained in dogs, where studies of ^{125}I-AVP indicated a rapid initial phase, as well as a slower late phase of elimination (Weitzman & Fisher, 1978; Wilson, Weitzman, & Fisher, 1978). Work using ^{3}H-lysine vasopressin showed that it disappeared from the circulation with a half-life of ~1 min in rats (Janaky, Laczi, & Laszlo, 1982). These studies also showed that 1-deamino-8-D-arginine vasopressin (dDAVP) had a substantially longer half-life, consistent with decreased enzymatic degradation. Indeed, Gazis and Sawyer observed that dDAVP is cleared almost entirely through renal excretion, whereas only 6% of infused vasopressin was recovered unchanged in the urine (Gazis & Sawyer, 1978). Other workers observed that the elimination of exogenous vasopressin in rabbits involved both a rapid initial phase and a slower late component; the initial phase had a half-life of 0.9 min and this was the only component observed for endogenous vasopressin (King, Courneya, Tang, Wilson, & Ledsome, 1989). Finally, recent work showed that ^{125}I-AVP had a half-life of 1 min in fasted mice and 1.5 min in insulin-treated wildtype mice, and that this half-life was 3 min in IRAP knockout animals, as discussed below (Wallis, Lankford, & Keller, 2007). Thus, the half-life of vasopressin in the circulation in humans is probably also approximately 1 min. This half-life is substantially shorter than that of another peptide hormone, insulin, which is 4–6 min (Duckworth, Bennett, & Hamel, 1998; Matthews, Rudenski, Burnett, Darling, & Turner, 1985). Importantly, the rapid rate of vasopressin degradation is similar to the average time it takes for blood to transit through the circulatory system, assuming a cardiac output of ~5 L/min and a blood volume of ~5 L. Thus, there may be substantially higher concentrations of vasopressin in arterial blood and at the kidneys, compared to the concentrations present in postcapillary venules of various tissues.

A key step toward understanding the mechanism underlying vasopressin elimination was taken by Burbach and Lebouille, who employed ^{14}C-labeling of vasopressin and studied brain homogenate and membrane fractions. These investigators followed the appearance of proteolytic fragments from the labeled hormone over time to characterize the degradation products and the order in which they were formed (Burbach & Lebouille, 1983). The results showed that peptide fragments were formed sequentially via a proteolytic cleavage process that commenced from the amino terminus onward. This type of behavior suggested the activity of an enzyme with amino-peptidase like behavior, consistent with earlier data (Pliska, Barth, & Thorn, 1971). Importantly, kinetic studies revealed that the cleavage of vasopressin's amino-terminal peptide bond was the rate-limiting step in the degradation pathway (Burbach & Lebouille, 1983). Unlike other aminopeptidases, this enzyme did not require reduction of the disulfide bond involving the amino-terminal cysteine residue. The identification of such a "vasopressinase" then became an important focus of research into vasopressin physiology.

2. Discovery and function of placental leucine aminopeptidase

The discovery of a human vasopressinase was spurred by the observation that the metabolic clearance rate of vasopressin during mid to late pregnancy is markedly increased (up to fourfold) compared with values seen in early pregnancy or in women who are not pregnant. (Gordge, Williams, Huggett, Payne, & Neild, 1995). It was also observed that hormonal disposal rates during pregnancy did not seem to correlate with circulating levels of AVP and striking increases in hormone disposal correlated with increased vasopressinase activity present at these late times. Despite these findings, basal concentrations of AVP were similar throughout pregnancy, indicating that vasopressin production rates must also dramatically increase throughout gestation (Davison, Sheills, Barron, Robinson, & Lindheimer, 1989).

The responsible element for this increased vasopressinase activity was found to be a zinc-dependent aminopeptidase known as placental leucine aminopeptidase (P-LAP, also known as leucyl and cystinyl aminopeptidase, LNPEP, or oxytocinase), which is expressed by placental syncytiotrophoblasts and is also abundant in cardiac and skeletal muscle (Durham & Rewell, 1979; Laustsen et al., 1997; Rogi, Tsujimoto, Nakazato, Mizutani, & Tomoda, 1996; Tsujimoto et al., 1992). This protein is released into maternal serum and its concentration increases throughout gestation to peak at 38 weeks of gestation. These increased circulating levels are associated with an increase of P-LAP mRNA levels in the placenta during third trimester, when compared to the first trimester (Yamahara et al., 2000). In the placenta, expression of P-LAP is regulated by transcription factors including activator protein-2 (AP-2) and Ikaros (Ito et al., 2001, 2002). These factors bind to *cis*-acting elements that are located in the 5′-region of the P-LAP genomic sequence. Data suggest that AP-2 may be the most important transcription factor for the regulation of P-LAP gene expression in human placenta; however, the physiological signals for this regulation are not fully elucidated.

Interestingly, P-LAP appears to have both local and systemic effects during pregnancy. For example, placental blood flow is positively correlated with the abundance of P-LAP in maternal serum (Asada, Mizutani,

Kasugai, Kurauchi, & Tomoda, 1992). This relationship could therefore play a role in the regulation of placental blood flow via P-LAP mediated breakdown of vasoactive peptides such as vasopressin. Furthermore, breakdown of oxytocin by P-LAP might be consistent with the prevention of premature uterine contraction during pregnancy, although P-LAP activity does not significantly decrease until after delivery (Asada et al., 1992). With regards to systemic effects, the increase in P-LAP activity during mid to late pregnancy is usually compensated by a corresponding increase in AVP secretion. When this does not occur, an inadequate increase in AVP synthesis can result in gestational diabetes insipidus (Gordge et al., 1995; Williams et al., 1993). Some cases of gestational diabetes insipidus have been attributed to the increased turnover of AVP through this mechanism (Kondo et al., 2018). These studies imply that both inadequate secretion and increased degradation of vasopressin can cause diabetes insipidus during pregnancy.

Increased serum concentrations of P-LAP have also been reported in patients with preeclampsia or in patients with pregnancy-induced hypertension, when compared with pregnant patients who are normotensive (Durham & Rewell, 1979; Gordge, Robson, Williams, Payne, & Neild, 1994; Spellacy, Usategui-Gomez, & Fernandez-deCastro, 1977). It has therefore been proposed that increased P-LAP may play a pathogenic role in the development of hypertension during pregnancy (Krege & Katz, 1990), although in one study a sharp decrease in P-LAP activity was observed when preeclampsia became severe (Mitsui, Nomura, Itakura, & Mizutani, 2004; Mizutani & Tomoda, 1996). Interestingly, membrane-bound P-LAP is proteolytically cleaved by ADAM12, so that it is released into the circulation in soluble form. The release of soluble P-LAP through cleavage was decreased by the hydroxamic acid-based inhibitors GM6001 and ONO4817 as well as by TNF-alpha protease inhibitor-2; however, whether this regulation is physiologically important is unclear because the same decrease was not observed with tissue inhibitors of metalloproteinase TIMP-1 and TIMP-2 (Ito et al., 2004).

Early arguments for how vasopressin inactivation could be associated with hypertension focused on the idea that the ring structure of vasopressin is critical for its antidiuretic activity, whereas separate C-terminal residues may be more important for its pressor function (Ishikawa, Goldberg, Schrier, Aisenbrey, & Schrier, 1984; Krege & Katz, 1990). This idea was questioned because pharmacological studies of vasopressin structural analogues suggest that the C-terminal residues determine both vasopressor and antidiuretic activity (Manning et al., 1984). Other studies observed a simultaneous loss of both V1 and V2 receptor activation, as well as abrogation of binding to V1a, V1b, and V2 receptors, during the degradation of vasopressin by P-LAP (Gordge et al., 1995; Thibonnier, Coles, Thibonnier, & Shoham, 2002). These results argue that P-LAP is unlikely to result in degradation products that retain significant pressor activity.

An alternative argument for how accelerated vasopressin inactivation may contribute to hypertension in preeclampsia is based on the very short half-life of circulating vasopressin. If vasopressin secretion is increased to compensate for accelerated inactivation, then the kidneys may be exposed to higher than normal vasopressin concentrations, as discussed further below. This would result in sodium and water reabsorption, as well as stimulation of renin release (Bankir, Bichet, & Bouby, 2010; Gonzalez et al., 2016). There is abundant evidence that vasopressin secretion is increased in preeclampsia, based on measurements of the cosecreted protein copeptin; indeed, this increase even precedes the development of overt hypertension (Cornelius, 2014; Jadli et al., 2015; Sandgren et al., 2015; Santillan et al., 2014; Yesil et al., 2017; Yeung et al., 2014). Recently, vasopressin infusion was observed to be sufficient to recapitulate features of preeclampsia in mice (Sandgren et al., 2018). Copeptin also correlates with immunological and proliferative features of preeclampsia (Scroggins et al., 2018; Zhang et al., 2016). Thus, P-LAP likely has both local effects on the placenta and systemic effects on fluid homeostasis, and the notion that it contributes to hypertension during preeclampsia is attractive and merits further study.

3. Discovery and characterization of insulin regulated aminopeptidase (IRAP)

IRAP was identified during work to characterize the major proteins residing in intracellular vesicles containing GLUT4 glucose transporters (Kandror & Pilch, 1994; Kandror, Yu, & Pilch, 1994; Keller, Scott, Mastick, Aebersold, & Lienhard, 1995). The major insulin-responsive glucose transporter, GLUT4 is expressed in adipose and muscle cells. Upon insulin stimulation, it is translocated acutely from an intracellular location to the cell surface to promote glucose uptake. The intracellular location from which GLUT4 is mobilized is the pool of specialized vesicles, termed GLUT4 Storage Vesicles (GSVs; also called Insulin-Regulated Vesicles or IRVs), which is thought to exist as a preformed set of vesicles near the Golgi complex in cells not exposed to insulin (Belman, Habtemichael, & Bogan, 2014; Bogan et al., 2012; Rubin & Bogan, 2009; Xu & Kandror, 2002).

The initial characterization of these vesicles used immunoaffinity techniques to purify GLUT4-containing membranes from a light microsome fraction of adipocytes (Kandror & Pilch, 2011; Mastick, Aebersold, & Lienhard, 1994). IRAP was initially called vp165 or gp160; cloning of the cDNA identified it as an aminopeptidase, so that it was subsequently called Insulin-Regulated (or -Responsive) Aminopeptidase. Subsequent cloning of the human P-LAP gene led to the recognition that IRAP and P-LAP share ~90% sequence identity and are homologous proteins in rodents and humans respectively (Rogi et al., 1996). At the time of this discovery the targets of IRAP aminopeptidase activity were unknown, though the homology suggested that it could target circulating peptides such as vasopressin or oxytocin. IRAP is one of only a few proteins present in GSVs in adipose and muscle cells. Its insulin-stimulated translocation to the cell surface, together with GLUT4, results in vasopressinase activity present at the cell surface of adipocytes (Garza & Birnbaum, 2000; Herbst et al., 1997; Kandror & Pilch, 2011). More broadly, the identification of IRAP was important for the characterization of GSVs and for understanding how insulin regulates glucose uptake (Keller, 2003).

Structurally, IRAP is a type II membrane protein containing an N-terminal 109 residue cytosolic segment, a single transmembrane domain, and 894 residue C-terminal domain that is exposed to either the intraluminal or extracellular environment. Studies have shown that the cytoplasmic tail is mostly responsible for IRAP's subcellular distribution, whereas the central portion of the intraluminal domain contains sites that are essential for IRAP's catalytic function as a zinc-dependent aminopeptidase (Laustsen, Vang, & Kristensen, 2001; Subtil, Lampson, Keller, & McGraw, 2000; Xu, Nan, Fan, Bogan, & Toomre, 2016). However, additional work has shown that the IRAP luminal domain's interaction with sortilin also helps to determine the localization of IRAP (Bogan & Kandror, 2010; Shi & Kandror, 2007). Though the molecular weight of IRAP is theoretically 117,317, the protein is heavily glycosylated such that the molecular mass of the mature protein is 160–165 kDa (Keller, 2003).

IRAP has a wide tissue distribution (Keller et al., 1995; Rogi et al., 1996), but in each tissue that has been studied, IRAP is expressed only in specific cell types. For example, within the placenta only syncytiotrophoblasts express the protein. Other locations that expressed the protein were vascular endothelial cells, as well as gastrointestinal mucosal cells among others (Nagasaka et al., 1997). Most notably, IRAP is strongly expressed in skeletal muscle cells and adipocytes, as well as in the corresponding adipocyte and muscle cell lines, 3T3-L1 adipocytes and L6 myotubes, respectively (Kandror & Pilch, 1994; Ross et al., 1996; Sumitani, Ramlal, Somwar, Keller, & Klip, 1997). These are also primarily the cells in which IRAP has been studied with regard to its subcellular localization and its trafficking itinerary as a companion of GLUT4 (Bogan & Kandror, 2010; Shi & Kandror, 2007). In response to insulin, both IRAP and GLUT4 translocate from GSVs to the cell surface. In the cultured 3T3-L1 adipocyte cell line, this results in an ~eightfold increase of IRAP at the cell surface within 5 min after insulin addition (Ross et al., 1996). In primary cells, this relative increase may be even greater.

3.1 IRAP as a regulator of vesicle trafficking

IRAP was suggested to be more important than GLUT4 itself for the regulated trafficking of GSVs, and this is primarily due to the presence of specific motifs within its cytoplasmic N-terminus (Gross, Farmer, & Pilch, 2004; Hosaka et al., 2005; Jordens, Molle, Xiong, Keller, & McGraw, 2010; Yeh, Sbodio, Tsun, Luo, & Chi, 2007). In particular, IRAP contains an acidic patch at residues 60–71 and a dileucine motif comprising residues 76–77; these residues are contained within a peptide (residues 56–74) identified as critical for IRAP and GLUT4 trafficking (Johnson, Lampson, & McGraw, 2001; Waters et al., 1997; Watson, Hou, & Pessin, 2008). This peptide binds the N-terminal region of the TUG protein, a "Tether, containing a UBX domain, for GLUT4," which traps the GSVs intracellularly by linking them to Golgi matrix proteins through its C-terminus in cells not stimulated with insulin (Belman et al., 2014; Orme & Bogan, 2012; Xu et al., 2011; Yu, Cresswell, Loffler, & Bogan, 2007). In fat and muscle cells, insulin triggers the proteolytic cleavage of TUG at the bond linking residues 164–165 (Belman et al., 2015; Bogan et al., 2012; Habtemichael et al., 2018; Loffler et al., 2013). This liberates the GSVs, so that cargo proteins including IRAP and GLUT4 can be inserted into the plasma membrane, as diagrammed in Fig. 1. In addition, cleavage creates the ubiquitin-like protein modifier, TUGUL (TUG Ubiquitin-Like), which is covalently attached to KIF5B, a kinesin motor that carries GLUT4 from the perinuclear region to the cell surface (Habtemichael et al., 2018; Semiz et al., 2003; Tettamanzi, Yu, Bogan, & Hodsdon, 2006). Because the IRAP peptide binds TUGUL, it participates not only in GSV retention within unstimulated cells, but also in linking the GSVs to the motors to mediate insulin-stimulated translocation to the cell

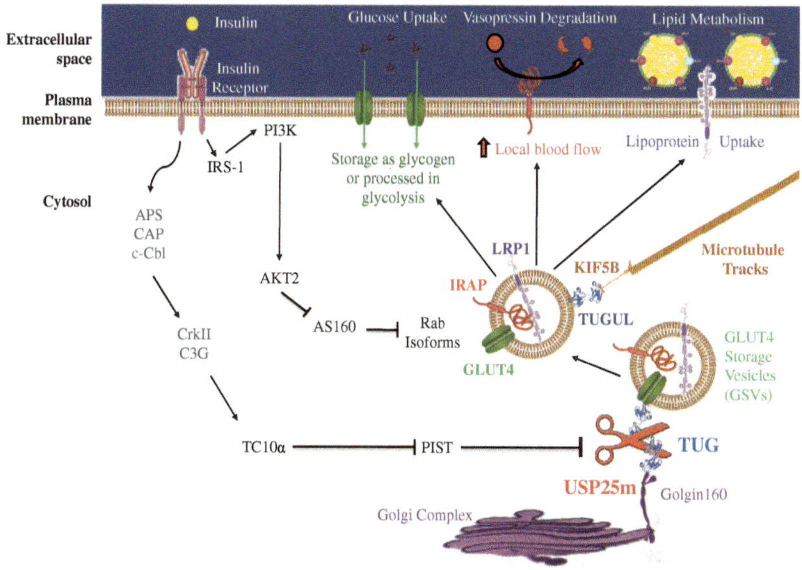

Fig. 1 Coordinated regulation of distinct physiologic effects by insulin-stimulated TUG cleavage. (A) Insulin signals through at least two pathways. How the upstream components of the TC10α pathway function together remains uncertain, and these are shown in gray. Signaling through PIST causes USP25m-mediated cleavage of TUG proteins, and releases GLUT4 Storage Vesicles (GSVs) from an anchoring site comprising Golgin-160 and other proteins linked to the Golgi complex. After cleavage, the TUG N-terminal product (called TUGUL, for TUG Ubiquitin-Like) is covalently attached to KIF5B kinesin motors, so that GSVs are carried to the cell surface on microtubules. The GSV cargos are inserted into the plasma membrane (T-tubules in muscle) and mediate glucose uptake (through GLUT4), vasopressin degradation (through IRAP), and possibly effects on lipoproteins (through LRP1 and sortilin). *Figure adapted from Li, D. T., Habtemichael, E. N., Julca, O., Sales C. I., Westergaard, X. O., DeVries, S. G., et al. (2019). GLUT4 Storage Vesicles: Specialized organelles for regulated trafficking. Yale Journal of Biology and Medicine, 92 (3), 453–470.*

surface. Ubiquitin-like proteolytic processing of TUG causes the insertion of both IRAP and GLUT4 at the cell surface. In skeletal muscle, this process coordinates vasopressin inactivation with glucose uptake (Habtemichael et al., 2015).

Other protein interactions involving IRAP are also important for GSV formation and for the recruitment of the specific cargoes that are enriched in these vesicles. Sortilin is homologous to the yeast vacuolar sorting receptor, Vps10p, and is a transmembrane protein that recruits specific cargoes though its luminal interactions and vesicle budding proteins through its cytosolic domain (Kandror & Pilch, 2011). In muscle and fat cells, sortilin mediates the formation of GSVs and cooperates with retrograde sorting pathways to sequester GLUT4 and IRAP away from the plasma membrane of unstimulated cells (Huang et al., 2013; Lin, Pilch, & Kandror, 1997; Pan, Zaarur, Singh, Morin, & Kandror, 2017; Shi & Kandror, 2005, 2007). According to this model, the major cargoes that are enriched in GSVs are recruited as a unit, through a "mass action" effect involving multiple, low affinity protein interactions (Kandror & Pilch, 2011). Each GSV then contains, on average, 5–6 molecules of GLUT4 and >10 molecules of IRAP (Kupriyanova, Kandror, & Kandror, 2002). Another IRAP-interacting protein important for GLUT4 trafficking is tankyrase, a poly(ADP-ribosyl) transferase that binds an RxxPDG motif in IRAP through its ankyrin repeat domain (Chi & Lodish, 2000; Guo et al., 2012; Sbodio, Lodish, & Chi, 2002; Su, Deshpande, James, & Stockli, 2018; Yeh et al., 2009, 2007). The poly(ADP-ribosylation) activity of tankyrase is likely involved in the regulation of GSVs, possibly through formation of GSVs or maintenance of the GSVs as a distinct, insulin-regulatable pool, but precisely how this occurs is not known. Because IRAP is an essential component of these vesicles, both tankyrase and sortilin are predicted to control IRAP targeting and vasopressin activity, as well as glucose uptake.

3.2 IRAP functions and distribution

The physiologic substrates of IRAP activity were determined by examining peptide processing in IRAP-deficient mice, compared to wildtype controls (Wallis et al., 2007). These studies showed that N-terminal proteolysis of vasopressin by isolated adipocytes or skeletal muscle was abrogated by deletion of IRAP. The sites at which IRAP and P-LAP cleave vasopressin are shown in Fig. 2A. In IRAP knockout mice, the reduction in aminopeptidase activity resulted in a twofold increase in endogenous vasopressin in the plasma, compared to that in wildtype controls. Brain vasopressin peptide and mRNA were reduced >threefold and twofold, respectively, in the IRAP knockout animals, compared to controls. Although the clearance of vasopressin in IRAP-deficient mice suggested that one or more additional peptidolytic pathways may exist, it remains possible that residual clearance resulted from renal excretion in the knockout animals. Regardless, such pathways were unable to compensate for the absence of IRAP-mediated vasopressin inactivation (Wallis et al., 2007). Finally, the specificity of this effect was demonstrated by showing that the processing of circulating oxytocin and angiotensin IV did not depend on IRAP in vivo, implying

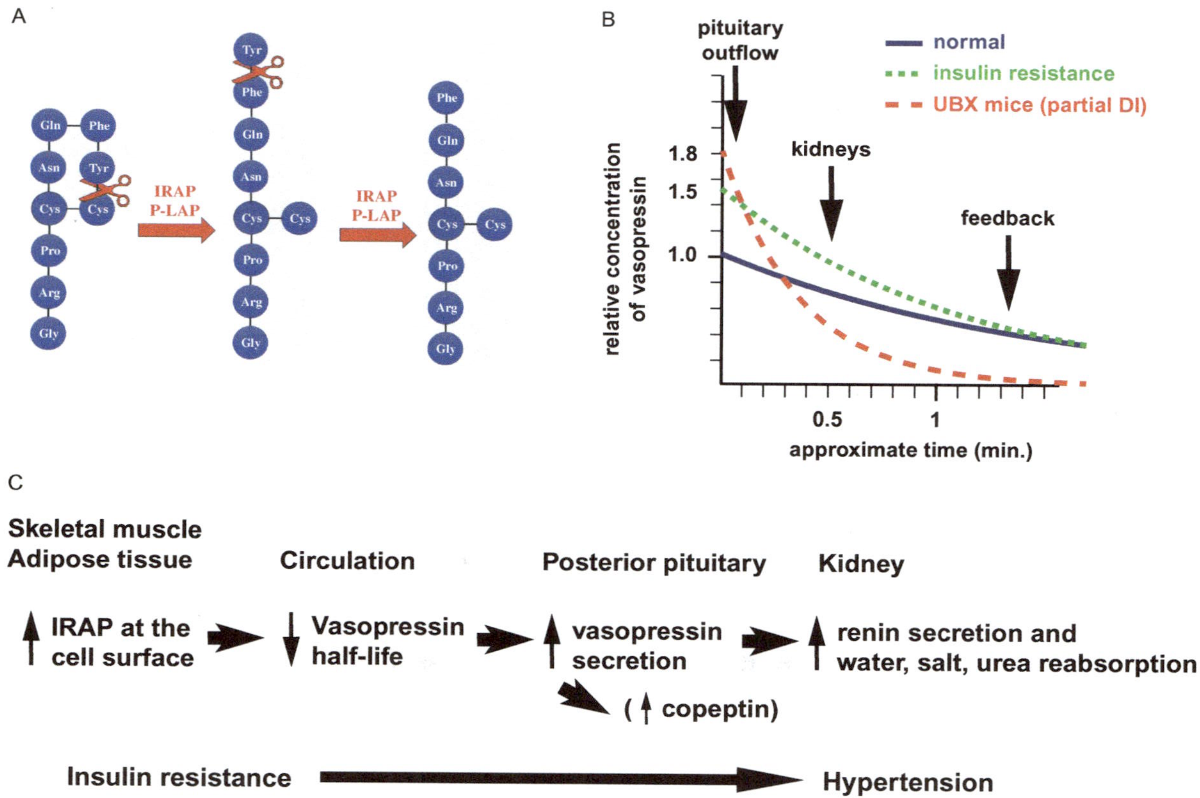

Fig. 2 See legend on next page.

that either the NH_2-terminal cleavage is not rate-limiting step for the degradation of these peptides or that this step is performed by other aminopeptidases (Wallis et al., 2007).

The control of subcellular IRAP distribution and plasma membrane expression may also be important in the regulation of local vasopressin action. For example, within the brain, IRAP was found to be most highly expressed in neurons localized to certain regions such as olfactory domains, hypothalamic nuclei, the hippocampal formation, and in motor and motor associated nuclei (Fernando, Larm, Albiston, & Chai, 2005). Within these neurons, IRAP localized within cell bodies in a vesicular pattern, similar to that observed in fat and muscle cells. This implies that selective translocation of IRAP can modulate the function of the local neuronal environment. Indeed, reduced expression of IRAP in neurons of the paraventricular and suprachiasmatic nuclei are associated with chronic schizophrenia, although the pathophysiologic significance of this finding has not yet been determined (Bernstein et al., 2017).

Another example of how IRAP distribution is controlled is found in the kidney, where immunostaining revealed distinct patterns in the cortex as compared to the collecting duct (Albiston et al., 2011). In the collecting

Fig. 2 Vasopressin processing. (A) The structure of intact vasopressin is shown on the left with the bonds of the cyclic nonapeptide indicated. The cleavage sites used by insulin-regulated aminopeptidase is shown by the scissors and the structure after cleavage of the NH2-terminal peptide bond is shown with sequential cleavage steps. (B) Vasopressin inactivation is modeled as a single exponential with a half-life of 1 min in normal physiology (solid blue curve). To model vasopressin inactivation in the setting of insulin resistance, the half-life is reduced by 25%, and the initial secretion is increased by 50% to compensate for this more rapid degradation (short dashes, green curve). Relative vasopressin concentrations at the kidneys and at a downstream point controlling negative feedback are indicated. The increased concentration at the kidneys may contribute to hypertension. To model physiology in the UBX transgenic mice, the half-life of vasopressin degradation is reduced by 75% (long dashes, red curve). Vasopressin secretion is increased, but is not sufficient to compensate for the accelerated rate of degradation. As a result, vasopressin concentrations at the kidneys are reduced and partial diabetes insipidus was observed in the mice. (C) A model for how insulin resistance in skeletal muscle and adipose tissue can contribute to hypertension. In the setting of insulin resistance, the abundance of IRAP at the cell surface is increased. This results in accelerated vasopressin degradation and in a compensatory increase in vasopressin secretion, which accounts for increased copeptin concentrations. The kidneys receive vasopressin shortly after its release, so that they are exposed to an increased concentration in insulin resistant individuals compared to noninsulin resistant subjects. This results in renin secretion as well as direct effects of vasopressin to cause salt, water, and urea resorption.

duct, IRAP is primarily localized to the luminal surface. Based on these findings, it was proposed that IRAP is important in the generation of free amino acids within the cortex, but plays a localized role in the regulation of vasopressin bioactivity within the collecting duct. Of note, AQP2 protein abundance, which serves as a biomarker of vasopressin activity, was increased twofold in the collecting duct of IRAP knockout mice, compared to wildtype controls (Albiston et al., 2011). This observation supports a role of IRAP to counter vasopressin action in principal cells of the collecting duct.

Importantly, studies of wildtype and IRAP knockout mice demonstrated that insulin stimulated a reduction in the half-life of ^{125}I-vasopressin, from ~1.5 to ~1 min, in wildtype mice, and that in IRAP knockout animals this half-life was ~3 min and was not altered by insulin stimulation (Wallis et al., 2007). These data imply that insulin's effect to increase the turnover of vasopressin was due to effects on IRAP distribution to the cell surface. Furthermore, the specifically of the effect was shown because the clearance of oxytocin was not similarly accelerated (Wallis et al., 2007). Together these data implicate an important role of IRAP on whole-body vasopressin half-life, regulated by IRAP-mediated degradation of circulating vasopressin.

4. TUG mediates coordinated regulation of water and glucose homeostasis

The TUG protein is a critical regulator of insulin-stimulated GSV translocation, which controls the subcellular targeting of GSV cargoes such as IRAP (Bogan, Hendon, McKee, Tsao, & Lodish, 2003; Bogan et al., 2012; Yu et al., 2007). As noted above, intact TUG traps GSVs at the Golgi matrix in unstimulated cells, and insulin stimulates TUG proteolytic cleavage to release IRAP to the cell surface. Insulin also acts through Akt2, AS160/Tbc1D4 and Tbc1D1, and specific Rab proteins to control overall GSV movement (Bogan, 2012). The ability of insulin to regulate GSVs is impaired in insulin-resistant states, such as type 2 diabetes. This impairment can involve Akt2 signaling, yet rodent models of insulin resistance, including high-fat diet-fed rats and mice, also demonstrate impaired TUG cleavage and reduced abundance of the putative TUG protease, Usp25m (Habtemichael et al., 2018). In humans, insulin resistance is characterized by increased abundance of IRAP at the cell surface during fasting in adipocytes and, likely, in skeletal muscle (Garvey et al., 1998; Maianu, Keller, & Garvey, 2001). The implication of this increase in cell surface IRAP is that the rate of vasopressin inactivation may be increased in the setting of insulin resistance, similar to preeclampsia and possibly contributing to hypertension in insulin-resistant individuals, as discussed further below.

To investigate the physiologic effects of TUG cleavage, transgenic $mTUG^{UBX\text{-}Cter}$ mice (here called "UBX mice") were produced (Loffler et al., 2013). In these mice, skeletal muscle-specific, transgenic expression of an artificial TUG fragment (UBX-Cter, residues 377–550) results in degradation of PIST, a mediator of insulin signaling that inhibits TUG cleavage in unstimulated cells. As a result, UBX mice have constitutive proteolytic cleavage of TUG proteins in skeletal muscle during fasting, when insulin concentrations are low. The fasting transgenic mice have a 3.6-fold increase in GLUT4 targeting to T-tubule membranes, increased glucose uptake into muscle, and increased muscle glycogen content, as predicted, compared to controls (Loffler et al., 2013). Whole body fasting glucose turnover was increased by 17%, compared to wildtype control mice. Unexpectedly, metabolic cage studies showed that UBX mice also had a 55% increase in water intake, compared to wildtype controls (Habtemichael et al., 2015). This increased water intake was robust, and was observed on both per weight and a per mouse bases.

Because IRAP is a trafficking partner of GLUT4, it was hypothesized that the increased water intake in UBX mice is due to accelerated degradation of circulating vasopressin. In support of this idea, increased IRAP abundance was observed in T-tubules from transgenic muscles, compared with controls. Water restricted UBX transgenic mice had a dramatic decrease in the abundance of renal AQP2, compared to wildtype controls, indicating reduced vasopressin action. As well, transgenic mice lost more weight during water deprivation, compared to controls, supporting the idea that the mice have a reduced ability to concentrate urine. These results support the idea that increased IRAP-mediated vasopressin inactivation in muscle has a systemic effect, and resulted in a partial diabetes insipidus in the UBX mice. The enormous surface area of the T-tubule system, together with the 3.2-fold increase in IRAP in T-tubule membranes, can contribute substantially to accelerate vasopressin degradation. To study if this was accompanied by a compensatory increase in vasopressin secretion, copeptin concentrations were measured. These were increased in UBX mice, compared to controls, supporting the idea that the reduced vasopressin activity results from accelerated inactivation and not decreased secretion (Habtemichael et al., 2015). Indeed, given that the mice had substantially increased water intake, it seems likely that the demand for vasopressin exceeded the capacity for its secretion. Together, the data show that the termination of vasopressin action is a major physiological effect that results from GSV translocation, and specifically from IRAP at the cell surface in skeletal muscle, which is regulated by TUG cleavage.

From a teleologic standpoint, at least two rationales can be proposed for why glucose uptake and vasopressin inactivation are linked through a common molecular mechanism for GSV translocation (Habtemichael et al., 2015). First, the oxidation of glucose taken up into muscle can produce a substantial volume of water, which must be removed to maintain blood osmolality within a narrow, physiologic range. Second, IRAP translocation may act locally to reduce the pressor effect of vasopressin, and thus to increase blood flow to metabolically active muscles. This effect may act in parallel to nitric oxide-mediated vasorelaxation, and could help to deliver more oxygen and glucose, and to carry away more waste, during periods of increased metabolic activity (Muniyappa & Sowers, 2013; Wagenmakers, Strauss, Shepherd, Keske, & Cocks, 2016).

Of note, the activation of AMP-activated protein kinase, AMPK, results in TUG cleavage and GLUT4 translocation during cardiac ischemia (Quan et al., 2017; Russell et al., 2004; Zaha & Young, 2012). This response has been viewed primarily in terms of fuel substrates for cardiac metabolism. Ischemic myocardium translocates GLUT4 and shifts its substrate preference toward glucose oxidation, from fatty acid oxidation. However, the TUG cleavage response can also be viewed in terms of its effect on IRAP, which would reduce vasopressin activity and may enhance blood flow to ischemic myocardium. A similar mechanism may operate in skeletal muscle, in which AMPK has been implicated in contraction-stimulated glucose uptake (Cartee, 2015; Richter & Hargreaves, 2013). Although though these perspectives focus on local actions, systemic effects of accelerated vasopressin degradation may result because of the large mass of skeletal muscle and the large surface area of the T-tubule system (Fraser, Huang, & Pedersen, 2011). Furthermore, the accelerated degradation of vasopressin may lead to an increased rate of vasopressin secretion, and may account (at least in part) for increased copeptin observed in insulin resistant states (discussed below) and after ischemia (Afzali et al., 2013; Katan et al., 2009; Kehl et al., 2012; Staub et al., 2009).

5. Implications of vasopressin degradation: Copeptin and the metabolic syndrome

Because TUG cleavage and GSV translocation integrate the activities of GLUT4 and IRAP, dysregulation of this pathway may account for the association of hypertension and insulin resistance in the metabolic syndrome (Czech, 2017; Enhorning et al., 2011). GSVs also contain cargoes (e.g., LRP1, sortilin) that bind apolipoproteins (e.g., ApoE, ApoA-V), and it is possible that their dysregulation contributes to altered lipid metabolism as well (Habtemichael et al., 2015). We propose that dysregulation of these vesicles can help to account for the association of copeptin with insulin resistance in muscle, such as that characteristic of type 2 diabetes, obesity, and polycystic ovary syndrome (Abbasi et al., 2012; Asferg, Andersen, Linneberg, Goetze, & Jeppesen, 2014; Canivell et al., 2018; Enhorning et al., 2013, 2010; Karbek et al., 2014; Rothermel et al., 2016; Roussel et al., 2016; Saleem et al., 2009; Taskin, Bulbul, Adali, Hismiogullari, & Inceboz, 2015; Wannamethee et al., 2015). If so, then copeptin may correlate specifically with postprandial insulin resistance, and not with fasting measures of insulin resistance (e.g., HOMA-IR) that result primarily from effects in the liver.

Impaired GLUT4 regulation results in insulin resistance and impaired IRAP translocation may contribute to altered vasopressin dynamics in insulin-resistant individuals. The impaired regulation of these proteins can reflect a combination of increased abundance at the cell surface during the basal, unstimulated state, as well as decreased abundance at the cell surface in the presence of insulin. A highly significant, 4.3-fold increase in cell surface abundance of IRAP was observed in unstimulated adipocytes obtained from individuals with type 2 diabetes, as compared to healthy controls (Maianu et al., 2001). A similar trend was observed in skeletal muscle where increased IRAP is present in membrane fractions containing sarcolemma and T-tubule markers in patients with insulin resistance or diabetes, compared with healthy controls, in both basal and insulin-stimulated states (Garvey et al., 1998). The interesting implication of this data is that increased copeptin concentrations in insulin-resistant individuals may result, at least in part, from accelerated vasopressin degradation triggering a compensatory increase in vasopressin secretion.

For this compensatory increase in vasopressin secretion to contribute to hypertension, two factors are required. First, the half-life of circulating vasopressin would have to be short, so that the kidneys are exposed to higher concentrations even if venous vasopressin concentrations are unaffected. As discussed above, the half-life of vasopressin is probably ~1 min, similar to the time it takes for blood to circulate, so that this condition is fulfilled. Second, the feedback signals that regulate vasopressin secretion, directly or indirectly, must lie further downstream, and not simply at the level of renal perfusion. Vasopressin secretion is subject to several layers of regulation, in addition to effects mediated by plasma osmolality and sodium concentration. For example, vasopressin may downregulate its own synthesis in the hypothalamus, and baroreflex and anticipatory signals are implicated in vasopressin secretion (Bankir, Bichet, & Morgenthaler, 2017; Bichet, 2018; Choe et al., 2015; Han et al., 2015; Milutinovic-Smiljanic, Sarenac, Lozic-Djuric, Murphy, & Japundzic-Zigon, 2013; Rehbein et al., 1986). If these negative feedback signals are mitigated by vasopressin degradation, then a compensatory response to increase vasopressin secretion would result. These scenarios for normal physiology and insulin resistance are diagrammed in Fig. 2B. In addition, the likely scenario in UBX mice is plotted. Here, vasopressin secretion is unable to fully compensate for the accelerated degradation resulting from translocated IRAP, so that partial diabetes insipidus results.

The idea that increased vasopressin concentrations can act on the kidney to contribute to hypertension has received substantial support. Mechanisms that are involved in this response include signaling through both V1a and V2 receptors to cause sodium and water resorption, as well as renin secretion from both the collecting duct and from macula densa (Bankir et al., 2010; Koshimizu et al., 2012; Mironova, Bugaj, Roos, Kohan, & Stockand, 2012; Mironova et al., 2015). More broadly, the model we propose, diagrammed in Fig. 2C, may help to substantiate mechanisms underlying the association of copeptin with cardiovascular mortality and outcomes in patients with insulin resistance and diabetes (Asferg et al., 2014; Bjornstad et al., 2016). These studies demonstrate that copeptin may be important in a range of disease and can serve as a surrogate marker for systemic—or, perhaps more properly, renal—activity of vasopressin (Schiel, Perenthaler, Steveling, & Stein, 2016; Tasevska, Enhorning, Persson, Nilsson, & Melander, 2016). The strong relationship between various medical comorbidities and the level of circulating copeptin may therefore be a reflection on the systemic effects of local IRAP redistribution (Zhao, Wu, Zhou, & Wang, 2017; Zhu et al., 2016). This potential physiological effect of vasopressin degradation deserves to be further explored.

One important study that demonstrates the feasibility of this approach, as well as its potential importance was work that tested the association between single nucleotide polymorphisms (SNPs) in genes related to the vasopressin pathway and outcomes in patients with septic shock (Nakada et al., 2011). These investigators found that SNPs in IRAP were significantly associated with 28-day mortality in both their initial derivation and subsequent validation cohorts. Specifically, the TT variation had a 51.0% mortality rate as compared to 34.5% in patients with other variations. Furthermore, they found that the TT genotype was associated with a significant increase in the plasma vasopressin clearance and accounted for 80% of the variance of serum sodium concentrations in cardiac surgical patients. Together, the data suggest that vasopressin degradation maybe vitally important to survival and outcomes in patients with septic shock (Nakada et al., 2011).

6. Conclusion

Over the last century, a growing body of work has increasingly suggested that vasopressin inactivation is important for the regulation of both cardiovascular and metabolic pathways. For example, the observation that vasopressin turnover is dramatically increased during late pregnancy led to the discovery of PLAP-mediated inactivation, which was subsequently associated with both preeclampsia and gestational diabetes insipidus. Which of these opposing phenotypes results may depend on whether the capacity for vasopressin secretion is exceeded. This scenario may be similar to the distinction we propose between hypertension associated with insulin resistance and partial diabetes insipidus in the UBX transgenic mice.

The characterization of IRAP as a trafficking partner of GLUT4 and an important marker of GSV identity opened a novel line of inquiry into the implications of vasopressin inactivation. Specifically, these findings suggested that coordination of distinct metabolic and physiologic phenotypes could be achieved through regulation of common storage vesicles. Insight into the molecular mechanism underlying this coordinated activity was provided by the elucidation of TUG-mediated vesicle sequestration and insulin-stimulated vesicular trafficking.

The potential physiologic rationales for this metabolic coordination include clearance of the free water generated by glucose oxidation, recruitment of increased blood flow to metabolically active tissues, and potentially the systemic elimination of water and metabolites. Whatever the rationale, studies involving IRAP mutations have shown that this process is important in survival rates after sepsis. Additionally, measurement of copeptin levels, which may also serve as a surrogate for vasopressin inactivation given that vasopressin secretion and inactivation are likely connected, have shown that this process is implicated in both diabetes and cardiovascular outcomes.

Together, the body of work surrounding vasopressin inactivation implies that this physiologic process is intimately involved in metabolic homeostasis.

Further elucidation of the molecular mechanisms involved, as well as the pathophysiological implications of perturbed vasopressin homeostasis, will yield important novel insights into human health.

Acknowledgments

The authors acknowledge support from the National Institutes of Health Grant R01 DK092661 (to J.S.B.) and Grant F30 DK115037 (to D.T.L.) by American Diabetes Association Grant 1-17-IBS-40 (to J.S.B.).

References

Abbasi, A., Corpeleijn, E., Meijer, E., Postmus, D., Gansevoort, R. T., Gans, R. O., et al. (2012). Sex differences in the association between plasma copeptin and incident type 2 diabetes: The Prevention of Renal and Vascular Endstage Disease (PREVEND) study. *Diabetologia, 55*(7), 1963–1970. https://doi.org/10.1007/s00125-012-2545-x.

Afzali, D., Erren, M., Pavenstadt, H. J., Vollert, J. O., Hertel, S., Waltenberger, J., et al. (2013). Impact of copeptin on diagnosis, risk stratification, and intermediate-term prognosis of acute coronary syndromes. *Clinical Research in Cardiology, 102*(10), 755–763. https://doi.org/10.1007/s00392-013-0583-0.

Albiston, A. L., Yeatman, H. R., Pham, V., Fuller, S. J., Diwakarla, S., Fernando, R. N., et al. (2011). Distinct distribution of GLUT4 and insulin regulated aminopeptidase in the mouse kidney. *Regulatory Peptides, 166*(1–3), 83–89. https://doi.org/10.1016/j.regpep.2010.09.003.

Asada, Y., Mizutani, S., Kasugai, M., Kurauchi, O., & Tomoda, Y. (1992). Relationship between the changes in placental blood flow resistance assessed by Doppler technique and maternal serum placental aminopeptidases, which degrade vaso-active peptides, in pre-eclampsia. *Hormone and Metabolic Research, 24*(7), 342–344. https://doi.org/10.1055/s-2007-1003329.

Asferg, C. L., Andersen, U. B., Linneberg, A., Goetze, J. P., & Jeppesen, J. L. (2014). Copeptin, a surrogate marker for arginine vasopressin secretion, is associated with higher glucose and insulin concentrations but not higher blood pressure in obese men. *Diabetic Medicine, 31*(6), 728–732. https://doi.org/10.1111/dme.12411.

Bankir, L., Bichet, D. G., & Bouby, N. (2010). Vasopressin V2 receptors, ENaC, and sodium reabsorption: A risk factor for hypertension? *American Journal of Physiology. Renal Physiology, 299*(5), F917–F928. https://doi.org/10.1152/ajprenal.00413.2010.

Bankir, L., Bichet, D. G., & Morgenthaler, N. G. (2017). Vasopressin: Physiology, assessment and osmosensation. *Journal of Internal Medicine, 282*(4), 284–297. https://doi.org/10.1111/joim.12645.

Baumann, G., & Dingman, J. F. (1976). Distribution, blood transport, and degradation of antidiuretic hormone in man. *The Journal of Clinical Investigation, 57*(5), 1109–1116.

Belman, J. P., Bian, R. R., Habtemichael, E. N., Li, D. T., Jurczak, M. J., Alcazar-Roman, A., et al. (2015). Acetylation of TUG protein promotes the accumulation of GLUT4 glucose transporters in an insulin-responsive intracellular compartment. *The Journal of Biological Chemistry, 290*(7), 4447–4463. https://doi.org/10.1074/jbc.M114.603977.

Belman, J. P., Habtemichael, E. N., & Bogan, J. S. (2014). A proteolytic pathway that controls glucose uptake in fat and muscle. *Reviews in Endocrine & Metabolic Disorders, 15*(1), 55–66. https://doi.org/10.1007/s11154-013-9276-2.

Bernstein, H. G., Muller, S., Dobrowolny, H., Wolke, C., Lendeckel, U., Bukowska, A., et al. (2017). Insulin-regulated aminopeptidase immunoreactivity is abundantly present in human hypothalamus and posterior pituitary gland, with reduced expression in paraventricular and suprachiasmatic neurons in chronic schizophrenia. *European Archives of Psychiatry and Clinical Neuroscience, 267*(5), 427–443. https://doi.org/10.1007/s00406-016-0757-7.

Bichet, D. G. (2018). Vasopressin and the regulation of thirst. *Annals of Nutrition & Metabolism, 72*(Suppl. 2), 3–7. https://doi.org/10.1159/000488233.

Bjornstad, P., Maahs, D. M., Jensen, T., Lanaspa, M. A., Johnson, R. J., Rewers, M., et al. (2016). Elevated copeptin is associated with atherosclerosis and diabetic kidney disease in adults with type 1 diabetes. *Journal of Diabetes and its Complications, 30*(6), 1093–1096. https://doi.org/10.1016/j.jdiacomp.2016.04.012.

Bogan, J. S. (2012). Regulation of glucose transporter translocation in health and diabetes. *Annual Review of Biochemistry, 81*, 507–532. https://doi.org/10.1146/annurev-biochem-060109-094246.

Bogan, J. S., Hendon, N., McKee, A. E., Tsao, T. S., & Lodish, H. F. (2003). Functional cloning of TUG as a regulator of GLUT4 glucose transporter trafficking. *Nature, 425*(6959), 727–733. https://doi.org/10.1038/nature01989.

Bogan, J. S., & Kandror, K. V. (2010). Biogenesis and regulation of insulin-responsive vesicles containing GLUT4. *Current Opinion in Cell Biology, 22*(4), 506–512. https://doi.org/10.1016/j.ceb.2010.03.012.

Bogan, J. S., Rubin, B. R., Yu, C., Loffler, M. G., Orme, C. M., Belman, J. P., et al. (2012). Endoproteolytic cleavage of TUG protein regulates GLUT4 glucose transporter translocation. *The Journal of Biological Chemistry, 287*(28), 23932–23947. https://doi.org/10.1074/jbc.M112.339457.

Burbach, J. P., & Lebouille, J. L. (1983). Proteolytic conversion of arginine-vasopressin and oxytocin by brain synaptic membranes. Characterization of formed peptides and mechanisms of proteolysis. *The Journal of Biological Chemistry, 258*(3), 1487–1494.

Canivell, S., Mohaupt, M., Ackermann, D., Pruijm, M., Guessous, I., Ehret, G., et al. (2018). Copeptin and insulin resistance: Effect modification by age and 11 beta-HSD2 activity in a population-based study. *Journal of Endocrinological Investigation, 41*(7), 799–808. https://doi.org/10.1007/s40618-017-0807-7.

Cartee, G. D. (2015). Roles of TBC1D1 and TBC1D4 in insulin- and exercise-stimulated glucose transport of skeletal muscle. *Diabetologia, 58*(1), 19–30. https://doi.org/10.1007/s00125-014-3395-5.

Chi, N. W., & Lodish, H. F. (2000). Tankyrase is a golgi-associated mitogen-activated protein kinase substrate that interacts with IRAP in GLUT4 vesicles. *The Journal of Biological Chemistry, 275*(49), 38437–38444. https://doi.org/10.1074/jbc.M007635200.

Choe, K. Y., Han, S. Y., Gaub, P., Shell, B., Voisin, D. L., Knapp, B. A., et al. (2015). High salt intake increases blood pressure via BDNF-mediated downregulation of KCC2 and impaired baroreflex inhibition of vasopressin neurons. *Neuron, 85*(3), 549–560. https://doi.org/10.1016/j.neuron.2014.12.048.

Cornelius, D. C. (2014). Copeptin: A new biomarker that is specific for preeclampsia? *Hypertension, 64*(6), 1189–1191. https://doi.org/10.1161/HYPERTENSIONAHA.114.04255.

Czech, M. P. (2017). Insulin action and resistance in obesity and type 2 diabetes. *Nature Medicine, 23*(7), 804–814. https://doi.org/10.1038/nm.4350.

Davison, J. M., Sheills, E. A., Barron, W. M., Robinson, A. G., & Lindheimer, M. D. (1989). Changes in the metabolic clearance of vasopressin and in plasma vasopressinase throughout human pregnancy. *The Journal of Clinical Investigation, 83*(4), 1313–1318.

Dicker, S. E., & Greenbaum, A. L. (1954). The degree of inactivation of the antidiuretic activity of vasopressin by the kidneys and the liver of rats. *The Journal of Physiology, 126*(1), 116–123.

Dicker, S. E., & Greenbaum, A. L. (1955). Preliminary study of the mechanism of inactivation of vasopressin by kidney homogenates. *The Journal of Physiology, 127*(2), 39–40P.

Dicker, S. E., & Greenbaum, A. L. (1956). Inactivation of the antidiuretic activity of vasopressin by tissue homogenates. *The Journal of Physiology*, *132*(1), 199–212.
Dicker, S. E., & Greenbaum, A. L. (1958). The destruction of the antidiuretic activity of vasopressin by—SH active compounds. *The Journal of Physiology*, *141*(1), 107–116.
Duckworth, W. C., Bennett, R. G., & Hamel, F. G. (1998). Insulin degradation: Progress and potential. *Endocrine Reviews*, *19*(5), 608–624. https://doi.org/10.1210/edrv.19.5.0349.
Durham, B. H., & Rewell, R. E. (1979). Serum cystyl aminopeptidase activity in the 36th week of pregnancy. *Journal of Clinical Pathology*, *32*(4), 318–320.
Enhorning, S., Bankir, L., Bouby, N., Struck, J., Hedblad, B., Persson, M., et al. (2013). Copeptin, a marker of vasopressin, in abdominal obesity, diabetes and microalbuminuria: The prospective Malmo Diet and Cancer Study cardiovascular cohort. *International Journal of Obesity*, *37*(4), 598–603. https://doi.org/10.1038/ijo.2012.88.
Enhorning, S., Struck, J., Wirfalt, E., Hedblad, B., Morgenthaler, N. G., & Melander, O. (2011). Plasma copeptin, a unifying factor behind the metabolic syndrome. *The Journal of Clinical Endocrinology and Metabolism*, *96*(7), E1065–E1072. https://doi.org/10.1210/jc.2010-2981.
Enhorning, S., Wang, T. J., Nilsson, P. M., Almgren, P., Hedblad, B., Berglund, G., et al. (2010). Plasma copeptin and the risk of diabetes mellitus. *Circulation*, *121*(19), 2102–2108. https://doi.org/10.1161/CIRCULATIONAHA.109.909663.
Fernando, R. N., Larm, J., Albiston, A. L., & Chai, S. Y. (2005). Distribution and cellular localization of insulin-regulated aminopeptidase in the rat central nervous system. *The Journal of Comparative Neurology*, *487*(4), 372–390. https://doi.org/10.1002/cne.20585.
Forsling, M. L., Martin, M. J., Sturdy, J. C., & Burton, A. M. (1973). Observations on the release and clearance of neurophysin and the neurohypophysial hormones in the rat. *The Journal of Endocrinology*, *57*(2), 307–315.
Fraser, J. A., Huang, C. L., & Pedersen, T. H. (2011). Relationships between resting conductances, excitability, and t-system ionic homeostasis in skeletal muscle. *The Journal of General Physiology*, *138*(1), 95–116. https://doi.org/10.1085/jgp.201110617.
Garvey, W. T., Maianu, L., Zhu, J. H., Brechtel-Hook, G., Wallace, P., & Baron, A. D. (1998). Evidence for defects in the trafficking and translocation of GLUT4 glucose transporters in skeletal muscle as a cause of human insulin resistance. *The Journal of Clinical Investigation*, *101*(11), 2377–2386. https://doi.org/10.1172/jci1557.
Garza, L. A., & Birnbaum, M. J. (2000). Insulin-responsive aminopeptidase trafficking in 3T3-L1 adipocytes. *The Journal of Biological Chemistry*, *275*(4), 2560–2567.
Gazis, D., & Sawyer, W. H. (1978). Elimination of infused arginine-vasopressin and its long-acting deaminated analogue in rats. *The Journal of Endocrinology*, *78*(2), 179–186.
Gonzalez, A. A., Cifuentes-Araneda, F., Ibaceta-Gonzalez, C., Gonzalez-Vergara, A., Zamora, L., Henriquez, R., et al. (2016). Vasopressin/V2 receptor stimulates renin synthesis in the collecting duct. *American Journal of Physiology Renal Physiology*, *310*(4), F284–F293. https://doi.org/10.1152/ajprenal.00360.2015.
Gordge, M. P., Robson, S. C., Williams, D. J., Payne, N. N., & Neild, G. H. (1994). Serum vasopressinase and platelet responses to arginine vasopressin in normal pregnancy, pregnancy-induced hypertension and pre-eclampsia. *Platelets*, *5*(2), 90–95. https://doi.org/10.3109/09537109409005518.
Gordge, M. P., Williams, D. J., Huggett, N. J., Payne, N. N., & Neild, G. H. (1995). Loss of biological activity of arginine vasopressin during its degradation by vasopressinase from pregnancy serum. *Clinical Endocrinology*, *42*(1), 51–58.
Gross, D. N., Farmer, S. R., & Pilch, P. F. (2004). Glut4 storage vesicles without Glut4: Transcriptional regulation of insulin-dependent vesicular traffic. *Molecular and Cellular Biology*, *24*(16), 7151–7162. https://doi.org/10.1128/mcb.24.16.7151-7162.2004.

Guo, H. L., Zhang, C., Liu, Q., Li, Q., Lian, G., Wu, D., et al. (2012). The Axin/TNKS complex interacts with KIF3A and is required for insulin-stimulated GLUT4 translocation. *Cell Research*, *22*(8), 1246–1257. https://doi.org/10.1038/cr.2012.52.

Habtemichael, E. N., Alcazar-Roman, A., Rubin, B. R., Grossi, L. R., Belman, J. P., Julca, O., et al. (2015). Coordinated regulation of vasopressin inactivation and glucose uptake by action of TUG protein in muscle. *The Journal of Biological Chemistry*, *290*(23), 14454–14461. https://doi.org/10.1074/jbc.C115.639203.

Habtemichael, E. N., Li, D. T., Alcazar-Roman, A., Westergaard, X. O., Li, M., Petersen, M. C., et al. (2018). Usp25m protease regulates ubiquitin-like processing of TUG proteins to control GLUT4 glucose transporter translocation in adipocytes. *The Journal of Biological Chemistry*, *293*(27), 10466–10486. https://doi.org/10.1074/jbc.RA118.003021.

Han, S. Y., Bouwer, G. T., Seymour, A. J., Korpal, A. K., Schwenke, D. O., & Brown, C. H. (2015). Induction of hypertension blunts baroreflex inhibition of vasopressin neurons in the rat. *The European Journal of Neuroscience*, *42*(9), 2690–2698. https://doi.org/10.1111/ejn.13062.

Heller, H., & Urban, F. F. (1935). The fate of the antidiuretic principle of postpituitary extracts in vivo and in vitro. *The Journal of Physiology*, *85*(4), 502–518.

Herbst, J. J., Ross, S. A., Scott, H. M., Bobin, S. A., Morris, N. J., Lienhard, G. E., et al. (1997). Insulin stimulates cell surface aminopeptidase activity toward vasopressin in adipocytes. *The American Journal of Physiology*, *272*(4 Pt. 1), E600–E606. https://doi.org/10.1152/ajpendo.1997.272.4.E600.

Hosaka, T., Brooks, C. C., Presman, E., Kim, S. K., Zhang, Z., Breen, M., et al. (2005). p115 Interacts with the GLUT4 vesicle protein, IRAP, and plays a critical role in insulin-stimulated GLUT4 translocation. *Molecular Biology of the Cell*, *16*(6), 2882–2890. https://doi.org/10.1091/mbc.E05-01-0072.

Huang, G., Buckler-Pena, D., Nauta, T., Singh, M., Asmar, A., Shi, J., et al. (2013). Insulin responsiveness of glucose transporter 4 in 3T3-L1 cells depends on the presence of sortilin. *Molecular Biology of the Cell*, *24*(19), 3115–3122. https://doi.org/10.1091/mbc.E12-10-0765.

Ishikawa, S., Goldberg, J. P., Schrier, D. M., Aisenbrey, G. A., & Schrier, R. W. (1984). Interrelationship between subpressor effects of vasopressin and other vasoactive hormones in the rat. *Mineral and Electrolyte Metabolism*, *10*(3), 184–189.

Ito, N., Nomura, S., Iwase, A., Ito, T., Kikkawa, F., Tsujimoto, M., et al. (2004). ADAMs, a disintegrin and metalloproteinases, mediate shedding of oxytocinase. *Biochemical and Biophysical Research Communications*, *314*(4), 1008–1013.

Ito, T., Nomura, S., Okada, M., Katsumata, Y., Iwase, A., Kikkawa, F., et al. (2001). Transcriptional regulation of human placental leucine aminopeptidase/oxytocinase gene. *Molecular Human Reproduction*, 7(9), 887–894.

Ito, T., Nomura, S., Okada, M., Katsumata, Y., Kikkawa, F., Rogi, T., et al. (2002). Ap-2 and Ikaros regulate transcription of human placental leucine aminopeptidase/oxytocinase gene. *Biochemical and Biophysical Research Communications*, *290*(3), 1048–1053. https://doi.org/10.1006/bbrc.2001.6325.

Jadli, A., Sharma, N., Damania, K., Satoskar, P., Bansal, V., Ghosh, K., et al. (2015). Promising prognostic markers of preeclampsia: New avenues in waiting. *Thrombosis Research*, *136*(2), 189–195. https://doi.org/10.1016/j.thromres.2015.05.011.

Janaky, T., Laczi, F., & Laszlo, F. A. (1982). Biological half-lives and organ distribution of tritiated 8-lysine-vasopressin and 1-deamino-8-D-arginine-vasopressin in Brattleboro rats. *Annals of the New York Academy of Sciences*, *394*, 116–127.

Johnson, A. O., Lampson, M. A., & McGraw, T. E. (2001). A di-leucine sequence and a cluster of acidic amino acids are required for dynamic retention in the endosomal recycling compartment of fibroblasts. *Molecular Biology of the Cell*, *12*(2), 367–381.

Jordens, I., Molle, D., Xiong, W., Keller, S. R., & McGraw, T. E. (2010). Insulin-regulated aminopeptidase is a key regulator of GLUT4 trafficking by controlling the sorting of GLUT4 from endosomes to specialized insulin-regulated vesicles. *Molecular Biology of the Cell, 21*(12), 2034–2044. https://doi.org/10.1091/mbc.E10-02-0158.

Kandror, K. V., & Pilch, P. F. (1994). gp160, a tissue-specific marker for insulin-activated glucose transport. *Proceedings of the National Academy of Sciences of the United States of America, 91*(17), 8017–8021.

Kandror, K. V., & Pilch, P. F. (2011). The sugar is sIRVed: Sorting Glut4 and its fellow travelers. *Traffic, 12*(6), 665–671. https://doi.org/10.1111/j.1600-0854.2011.01175.x.

Kandror, K. V., Yu, L., & Pilch, P. F. (1994). The major protein of GLUT4-containing vesicles, gp160, has aminopeptidase activity. *The Journal of Biological Chemistry, 269*(49), 30777–30780.

Karbek, B., Ozbek, M., Karakose, M., Topaloglu, O., Bozkurt, N. C., Cakir, E., et al. (2014). Copeptin, a surrogate marker for arginine vasopressin, is associated with cardiovascular risk in patients with polycystic ovary syndrome. *Journal of Ovarian Research, 7*, 31. https://doi.org/10.1186/1757-2215-7-31.

Katan, M., Fluri, F., Morgenthaler, N. G., Schuetz, P., Zweifel, C., Bingisser, R., et al. (2009). Copeptin: A novel, independent prognostic marker in patients with ischemic stroke. *Annals of Neurology, 66*(6), 799–808. https://doi.org/10.1002/ana.21783.

Kehl, D. W., Iqbal, N., Fard, A., Kipper, B. A., De La Parra Landa, A., & Maisel, A. S. (2012). Biomarkers in acute myocardial injury. *Translational Research, 159*(4), 252–264. https://doi.org/10.1016/j.trsl.2011.11.002.

Keller, S. R. (2003). The insulin-regulated aminopeptidase: A companion and regulator of GLUT4. *Frontiers in Bioscience, 8*, s410–s420.

Keller, S. R., Scott, H. M., Mastick, C. C., Aebersold, R., & Lienhard, G. E. (1995). Cloning and characterization of a novel insulin-regulated membrane aminopeptidase from Glut4 vesicles. *The Journal of Biological Chemistry, 270*(40), 23612–23618.

King, K. A., Courneya, C. A., Tang, C., Wilson, N., & Ledsome, J. R. (1989). Pharmacokinetics of vasopressin and atrial natriuretic peptide in anesthetized rabbits. *Endocrinology, 124*(1), 77–83. https://doi.org/10.1210/endo-124-1-77.

Kondo, T., Nakamura, M., Kitano, S., Kawashima, J., Matsumura, T., Ohba, T., et al. (2018). The clinical course and pathophysiological investigation of adolescent gestational diabetes insipidus: A case report. *BMC Endocrine Disorders, 18*(1), 4. https://doi.org/10.1186/s12902-018-0234-6.

Koshimizu, T. A., Nakamura, K., Egashira, N., Hiroyama, M., Nonoguchi, H., & Tanoue, A. (2012). Vasopressin V1a and V1b receptors: From molecules to physiological systems. *Physiological Reviews, 92*(4), 1813–1864. https://doi.org/10.1152/physrev.00035.2011.

Krege, J. H., & Katz, V. L. (1990). A proposed relationship between vasopressinase altered vasopressin and preeclampsia. *Medical Hypotheses, 31*(4), 283–287.

Kupriyanova, T. A., Kandror, V., & Kandror, K. V. (2002). Isolation and characterization of the two major intracellular Glut4 storage compartments. *The Journal of Biological Chemistry, 277*(11), 9133–9138. https://doi.org/10.1074/jbc.M106999200.

Laustsen, P. G., Rasmussen, T. E., Petersen, K., Pedraza-Diaz, S., Moestrup, S. K., Gliemann, J., et al. (1997). The complete amino acid sequence of human placental oxytocinase. *Biochimica et Biophysica Acta, 1352*(1), 1–7.

Laustsen, P. G., Vang, S., & Kristensen, T. (2001). Mutational analysis of the active site of human insulin-regulated aminopeptidase. *European Journal of Biochemistry, 268*(1), 98–104.

Lin, B. Z., Pilch, P. F., & Kandror, K. V. (1997). Sortilin is a major protein component of Glut4-containing vesicles. *The Journal of Biological Chemistry, 272*(39), 24145–24147.

Loffler, M. G., Birkenfeld, A. L., Philbrick, K. M., Belman, J. P., Habtemichael, E. N., Booth, C. J., et al. (2013). Enhanced fasting glucose turnover in mice with disrupted action of TUG protein in skeletal muscle. *The Journal of Biological Chemistry, 288*(28), 20135–20150. https://doi.org/10.1074/jbc.M113.458075.

Maianu, L., Keller, S. R., & Garvey, W. T. (2001). Adipocytes exhibit abnormal subcellular distribution and translocation of vesicles containing glucose transporter 4 and insulin-regulated aminopeptidase in type 2 diabetes mellitus: Implications regarding defects in vesicle trafficking. *The Journal of Clinical Endocrinology and Metabolism, 86*(11), 5450–5456. https://doi.org/10.1210/jcem.86.11.8053.

Manning, M., Olma, A., Klis, W., Kolodziejczyk, A., Nawrocka, E., Misicka, A., et al. (1984). Carboxy terminus of vasopressin required for activity but not binding. *Nature, 308*(5960), 652–653.

Mastick, C. C., Aebersold, R., & Lienhard, G. E. (1994). Characterization of a major protein in GLUT4 vesicles. Concentration in the vesicles and insulin-stimulated translocation to the plasma membrane. *The Journal of Biological Chemistry, 269*(8), 6089–6092.

Matthews, D. R., Rudenski, A. S., Burnett, M. A., Darling, P., & Turner, R. C. (1985). The half-life of endogenous insulin and C-peptide in man assessed by somatostatin suppression. *Clinical Endocrinology, 23*(1), 71–79.

Milutinovic-Smiljanic, S., Sarenac, O., Lozic-Djuric, M., Murphy, D., & Japundzic-Zigon, N. (2013). Evidence for involvement of central vasopressin V1b and V2 receptors in stress-induced baroreflex desensitization. *British Journal of Pharmacology, 169*(4), 900–908. https://doi.org/10.1111/bph.12161.

Mironova, E., Bugaj, V., Roos, K. P., Kohan, D. E., & Stockand, J. D. (2012). Aldosterone-independent regulation of the epithelial Na+ channel (ENaC) by vasopressin in adrenalectomized mice. *Proceedings of the National Academy of Sciences of the United States of America, 109*(25), 10095–10100. https://doi.org/10.1073/pnas.1201978109.

Mironova, E., Chen, Y., Pao, A. C., Roos, K. P., Kohan, D. E., Bugaj, V., et al. (2015). Activation of ENaC by AVP contributes to the urinary concentrating mechanism and dilution of plasma. *American Journal of Physiology. Renal Physiology, 308*(3), F237–F243. https://doi.org/10.1152/ajprenal.00246.2014.

Mitsui, T., Nomura, S., Itakura, A., & Mizutani, S. (2004). Role of aminopeptidases in the blood pressure regulation. *Biological & Pharmaceutical Bulletin, 27*(6), 768–771.

Mizutani, S., & Tomoda, Y. (1996). Effects of placental proteases on maternal and fetal blood pressure in normal pregnancy and preeclampsia. *American Journal of Hypertension, 9*(6), 591–597.

Muniyappa, R., & Sowers, J. R. (2013). Role of insulin resistance in endothelial dysfunction. *Reviews in Endocrine & Metabolic Disorders, 14*(1), 5–12. https://doi.org/10.1007/s11154-012-9229-1.

Nagasaka, T., Nomura, S., Okamura, M., Tsujimoto, M., Nakazato, H., Oiso, Y., et al. (1997). Immunohistochemical localization of placental leucine aminopeptidase/oxytocinase in normal human placental, fetal and adult tissues. *Reproduction, Fertility, and Development, 9*(8), 747–753.

Nakada, T. A., Russell, J. A., Wellman, H., Boyd, J. H., Nakada, E., Thain, K. R., et al. (2011). Leucyl/cystinyl aminopeptidase gene variants in septic shock. *Chest, 139*(5), 1042–1049. https://doi.org/10.1378/chest.10-2517.

Orme, C. M., & Bogan, J. S. (2012). The ubiquitin regulatory X (UBX) domain-containing protein TUG regulates the p97 ATPase and resides at the endoplasmic reticulum-golgi intermediate compartment. *The Journal of Biological Chemistry, 287*(9), 6679–6692. https://doi.org/10.1074/jbc.M111.284232.

Pan, X., Zaarur, N., Singh, M., Morin, P., & Kandror, K. V. (2017). Sortilin and retromer mediate retrograde transport of Glut4 in 3T3-L1 adipocytes. *Molecular Biology of the Cell, 28*(12), 1667–1675. https://doi.org/10.1091/mbc.E16-11-0777.

Pliska, V., Barth, T., & Thorn, N. A. (1971). Some metabolites of specifically tritium labelled lysine-vasopressin: Identification by thin-layer chromatography and determination by the liquid scintillation technique. *Acta Endocrinologica, 67*(1), 1–11.

Quan, N., Sun, W., Wang, L., Chen, X., Bogan, J. S., Zhou, X., et al. (2017). Sestrin2 prevents age-related intolerance to ischemia and reperfusion injury by modulating substrate metabolism. *The FASEB Journal, 31*(9), 4153–4167. https://doi.org/10.1096/fj.201700063R.

Rehbein, M., Hillers, M., Mohr, E., Ivell, R., Morley, S., Schmale, H., et al. (1986). The neurohypophyseal hormones vasopressin and oxytocin. Precursor structure, synthesis and regulation. *Biological Chemistry Hoppe-Seyler, 367*(8), 695–704.

Richter, E. A., & Hargreaves, M. (2013). Exercise, GLUT4, and skeletal muscle glucose uptake. *Physiological Reviews, 93*(3), 993–1017. https://doi.org/10.1152/physrev.00038.2012.

Rogi, T., Tsujimoto, M., Nakazato, H., Mizutani, S., & Tomoda, Y. (1996). Human placental leucine aminopeptidase/oxytocinase. A new member of type II membrane-spanning zinc metallopeptidase family. *The Journal of Biological Chemistry, 271*(1), 56–61.

Ross, S. A., Scott, H. M., Morris, N. J., Leung, W. Y., Mao, F., Lienhard, G. E., et al. (1996). Characterization of the insulin-regulated membrane aminopeptidase in 3T3-L1 adipocytes. *The Journal of Biological Chemistry, 271*(6), 3328–3332.

Rothermel, J., Kulle, A., Holterhus, P. M., Toschke, C., Lass, N., & Reinehr, T. (2016). Copeptin in obese children and adolescents: Relationships to body mass index, cortisol and gender. *Clinical Endocrinology, 85*(6), 868–873. https://doi.org/10.1111/cen.13235.

Roussel, R., El Boustany, R., Bouby, N., Potier, L., Fumeron, F., Mohammedi, K., et al. (2016). Plasma copeptin, AVP gene variants, and incidence of type 2 diabetes in a cohort from the community. *The Journal of Clinical Endocrinology and Metabolism, 101*(6), 2432–2439. https://doi.org/10.1210/jc.2016-1113.

Rubin, B. R., & Bogan, J. S. (2009). Intracellular retention and insulin-stimulated mobilization of GLUT4 glucose transporters. *Vitamins and Hormones, 80*, 155–192. https://doi.org/10.1016/s0083-6729(08)00607-9.

Russell, R. R., 3rd, Li, J., Coven, D. L., Pypaert, M., Zechner, C., Palmeri, M., et al. (2004). AMP-activated protein kinase mediates ischemic glucose uptake and prevents postischemic cardiac dysfunction, apoptosis, and injury. *The Journal of Clinical Investigation, 114*(4), 495–503. https://doi.org/10.1172/JCI19297.

Saleem, U., Khaleghi, M., Morgenthaler, N. G., Bergmann, A., Struck, J., Mosley, T. H., Jr., et al. (2009). Plasma carboxy-terminal provasopressin (copeptin): A novel marker of insulin resistance and metabolic syndrome. *The Journal of Clinical Endocrinology and Metabolism, 94*(7), 2558–2564. https://doi.org/10.1210/jc.2008-2278.

Sandgren, J. A., Deng, G., Linggonegoro, D. W., Scroggins, S. M., Perschbacher, K. J., Nair, A. R., et al. (2018). Arginine vasopressin infusion is sufficient to model clinical features of preeclampsia in mice. *JCI Insight, 3*(19). https://doi.org/10.1172/jci.insight.99403. pii: 99403.

Sandgren, J. A., Scroggins, S. M., Santillan, D. A., Devor, E. J., Gibson-Corley, K. N., Pierce, G. L., et al. (2015). Vasopressin: The missing link for preeclampsia? *American Journal of Physiology. Regulatory, Integrative and Comparative Physiology, 309*(9), R1062–R1064. https://doi.org/10.1152/ajpregu.00073.2015.

Santillan, M. K., Santillan, D. A., Scroggins, S. M., Min, J. Y., Sandgren, J. A., Pearson, N. A., et al. (2014). Vasopressin in preeclampsia: A novel very early human pregnancy biomarker and clinically relevant mouse model. *Hypertension, 64*(4), 852–859. https://doi.org/10.1161/HYPERTENSIONAHA.114.03848.

Sbodio, J. I., Lodish, H. F., & Chi, N. W. (2002). Tankyrase-2 oligomerizes with tankyrase-1 and binds to both TRF1 (telomere-repeat-binding factor 1) and IRAP (insulin-responsive aminopeptidase). *The Biochemical Journal, 361*(Pt. 3), 451–459.

Schiel, R., Perenthaler, T. J., Steveling, A., & Stein, G. (2016). Plasma copeptin in children and adolescents with type 1 diabetes mellitus in comparison to healthy controls. *Diabetes Research and Clinical Practice, 118*, 156–161. https://doi.org/10.1016/j.diabres.2016.06.019.

Scroggins, S. M., Santillan, D. A., Lund, J. M., Sandgren, J. A., Krotz, L. K., Hamilton, W. S., et al. (2018). Elevated vasopressin in pregnant mice induces T-helper subset alterations consistent with human preeclampsia. *Clinical Science (London, England)*, *132*(3), 419–436. https://doi.org/10.1042/CS20171059.

Semiz, S., Park, J. G., Nicoloro, S. M., Furcinitti, P., Zhang, C., Chawla, A., et al. (2003). Conventional kinesin KIF5B mediates insulin-stimulated GLUT4 movements on microtubules. *The EMBO Journal*, *22*(10), 2387–2399. https://doi.org/10.1093/emboj/cdg237.

Shi, J., & Kandror, K. V. (2005). Sortilin is essential and sufficient for the formation of Glut4 storage vesicles in 3T3-L1 adipocytes. *Developmental Cell*, *9*(1), 99–108. https://doi.org/10.1016/j.devcel.2005.04.004.

Shi, J., & Kandror, K. V. (2007). The luminal Vps10p domain of sortilin plays the predominant role in targeting to insulin-responsive Glut4-containing vesicles. *The Journal of Biological Chemistry*, *282*(12), 9008–9016. https://doi.org/10.1074/jbc.M608971200.

Spellacy, W. N., Usategui-Gomez, M., & Fernandez-deCastro, A. (1977). Plasma human placental lactogen, oxytocinase, and placental phosphatase in normal and toxemic pregnancies. *American Journal of Obstetrics and Gynecology*, *127*(1), 10–16.

Staub, D., Morgenthaler, N. G., Buser, C., Breidthardt, T., Potocki, M., Noveanu, M., et al. (2009). Use of copeptin in the detection of myocardial ischemia. *Clinica Chimica Acta*, *399*(1–2), 69–73. https://doi.org/10.1016/j.cca.2008.09.016.

Su, Z., Deshpande, V., James, D. E., & Stockli, J. (2018). Tankyrase modulates insulin sensitivity in skeletal muscle cells by regulating the stability of GLUT4 vesicle proteins. *The Journal of Biological Chemistry*, *293*(22), 8578–8587. https://doi.org/10.1074/jbc.RA117.001058.

Subtil, A., Lampson, M. A., Keller, S. R., & McGraw, T. E. (2000). Characterization of the insulin-regulated endocytic recycling mechanism in 3T3-L1 adipocytes using a novel reporter molecule. *The Journal of Biological Chemistry*, *275*(7), 4787–4795.

Sumitani, S., Ramlal, T., Somwar, R., Keller, S. R., & Klip, A. (1997). Insulin regulation and selective segregation with glucose transporter-4 of the membrane aminopeptidase vp165 in rat skeletal muscle cells. *Endocrinology*, *138*(3), 1029–1034. https://doi.org/10.1210/endo.138.3.5010.

Tasevska, I., Enhorning, S., Persson, M., Nilsson, P. M., & Melander, O. (2016). Copeptin predicts coronary artery disease cardiovascular and total mortality. *Heart*, *102*(2), 127–132. https://doi.org/10.1136/heartjnl-2015-308183.

Taskin, M. I., Bulbul, E., Adali, E., Hismiogullari, A. A., & Inceboz, U. (2015). Circulating levels of obestatin and copeptin in obese and nonobese women with polycystic ovary syndrome. *European Journal of Obstetrics, Gynecology, and Reproductive Biology*, *189*, 19–23. https://doi.org/10.1016/j.ejogrb.2015.03.006.

Tettamanzi, M. C., Yu, C., Bogan, J. S., & Hodsdon, M. E. (2006). Solution structure and backbone dynamics of an N-terminal ubiquitin-like domain in the GLUT4-regulating protein, TUG. *Protein Science*, *15*(3), 498–508. https://doi.org/10.1110/ps.051901806.

Thibonnier, M., Coles, P., Thibonnier, A., & Shoham, M. (2002). Molecular pharmacology and modeling of vasopressin receptors. *Progress in Brain Research*, *139*, 179–196.

Tsujimoto, M., Mizutani, S., Adachi, H., Kimura, M., Nakazato, H., & Tomoda, Y. (1992). Identification of human placental leucine aminopeptidase as oxytocinase. *Archives of Biochemistry and Biophysics*, *292*(2), 388–392.

Wagenmakers, A. J., Strauss, J. A., Shepherd, S. O., Keske, M. A., & Cocks, M. (2016). Increased muscle blood supply and transendothelial nutrient and insulin transport induced by food intake and exercise: Effect of obesity and ageing. *The Journal of Physiology*, *594*(8), 2207–2222. https://doi.org/10.1113/jphysiol.2014.284513.

Wallis, M. G., Lankford, M. F., & Keller, S. R. (2007). Vasopressin is a physiological substrate for the insulin-regulated aminopeptidase IRAP. *American Journal of Physiology.*

Endocrinology and Metabolism, *293*(4), E1092–E1102. https://doi.org/10.1152/ajpendo.00440.2007.
Wannamethee, S. G., Welsh, P., Papacosta, O., Lennon, L., Whincup, P. H., & Sattar, N. (2015). Copeptin, insulin resistance, and risk of incident diabetes in older men. *The Journal of Clinical Endocrinology and Metabolism*, *100*(9), 3332–3339. https://doi.org/10.1210/JC.2015-2362.
Waters, S. B., D'Auria, M., Martin, S. S., Nguyen, C., Kozma, L. M., & Luskey, K. L. (1997). The amino terminus of insulin-responsive aminopeptidase causes Glut4 translocation in 3T3-L1 adipocytes. *The Journal of Biological Chemistry*, *272*(37), 23323–23327.
Watson, R. T., Hou, J. C., & Pessin, J. E. (2008). Recycling of IRAP from the plasma membrane back to the insulin-responsive compartment requires the Q-SNARE syntaxin 6 but not the GGA clathrin adaptors. *Journal of Cell Science*, *121*(Pt 8), 1243–1251. https://doi.org/10.1242/jcs.017517.
Weitzman, R. E., & Fisher, D. A. (1978). Arginine vasopressin metabolism in dogs. I. Evidence for a receptor-mediated mechanism. *The American Journal of Physiology*, *235*(6), E591–E597. https://doi.org/10.1152/ajpendo.1978.235.6.E591.
Williams, D. J., Metcalfe, K. A., Skingle, L., Stock, A. I., Beedham, T., & Monson, J. P. (1993). Pathophysiology of transient cranial diabetes insipidus during pregnancy. *Clinical Endocrinology*, *38*(6), 595–600.
Wilson, K. C., Weitzman, R. E., & Fisher, D. A. (1978). Arginine vasopressin metabolism in dogs. II. Modeling and system analysis. *The American Journal of Physiology*, *235*(6), E598–E605. https://doi.org/10.1152/ajpendo.1978.235.6.E598.
Xu, Z., & Kandror, K. V. (2002). Translocation of small preformed vesicles is responsible for the insulin activation of glucose transport in adipose cells. Evidence from the in vitro reconstitution assay. *The Journal of Biological Chemistry*, *277*(50), 47972–47975. https://doi.org/10.1074/jbc.C200486200.
Xu, Y., Nan, D., Fan, J., Bogan, J. S., & Toomre, D. (2016). Optogenetic activation reveals distinct roles of PIP3 and Akt in adipocyte insulin action. *Journal of Cell Science*, *129*(10), 2085–2095. https://doi.org/10.1242/jcs.174805.
Xu, Y., Rubin, B. R., Orme, C. M., Karpikov, A., Yu, C., Bogan, J. S., et al. (2011). Dual-mode of insulin action controls GLUT4 vesicle exocytosis. *The Journal of Cell Biology*, *193*(4), 643–653. https://doi.org/10.1083/jcb.201008135.
Yamahara, N., Nomura, S., Suzuki, T., Itakura, A., Ito, M., Okamoto, T., et al. (2000). Placental leucine aminopeptidase/oxytocinase in maternal serum and placenta during normal pregnancy. *Life Sciences*, *66*(15), 1401–1410.
Yeh, T. Y., Beiswenger, K. K., Li, P., Bolin, K. E., Lee, R. M., Tsao, T. S., et al. (2009). Hypermetabolism, hyperphagia, and reduced adiposity in tankyrase-deficient mice. *Diabetes*, *58*(11), 2476–2485. https://doi.org/10.2337/db08-1781.
Yeh, T. Y., Sbodio, J. I., Tsun, Z. Y., Luo, B., & Chi, N. W. (2007). Insulin-stimulated exocytosis of GLUT4 is enhanced by IRAP and its partner tankyrase. *The Biochemical Journal*, *402*(2), 279–290. https://doi.org/10.1042/bj20060793.
Yesil, A., Kanawati, A., Helvacioglu, C., Kaya, C., Ozgun, C. G., & Cengiz, H. (2017). Identification of patients at risk for preeclampsia with the use of uterine artery Doppler velocimetry and copeptin. *The Journal of Maternal-Fetal & Neonatal Medicine*, *30*(22), 2763–2768. https://doi.org/10.1080/14767058.2016.1262841.
Yeung, E. H., Liu, A., Mills, J. L., Zhang, C., Mannisto, T., Lu, Z., et al. (2014). Increased levels of copeptin before clinical diagnosis of preeclampsia. *Hypertension*, *64*(6), 1362–1367. https://doi.org/10.1161/HYPERTENSIONAHA.114.03762.
Yu, C., Cresswell, J., Loffler, M. G., & Bogan, J. S. (2007). The glucose transporter 4-regulating protein TUG is essential for highly insulin-responsive glucose uptake in 3T3-L1 adipocytes. *The Journal of Biological Chemistry*, *282*(10), 7710–7722. https://doi.org/10.1074/jbc.M610824200.

Zaha, V. G., & Young, L. H. (2012). AMP-activated protein kinase regulation and biological actions in the heart. *Circulation Research, 111*(6), 800–814. https://doi.org/10.1161/CIRCRESAHA.111.255505.

Zhang, L., Wang, X., Cao, H., Chen, Y., Chen, X., Zhao, X., et al. (2016). Vasopressin V1A receptor mediates cell proliferation through GRK2-EGFR-ERK1/2 pathway in A7r5 cells. *European Journal of Pharmacology, 792,* 15–25. https://doi.org/10.1016/j.ejphar.2016.10.023.

Zhao, Q., Wu, X. X., Zhou, J., & Wang, X. (2017). Elevated plasma levels of copeptin linked to diabetic retinopathy in type 2 diabetes. *Molecular and Cellular Endocrinology, 442,* 106–112. https://doi.org/10.1016/j.mce.2016.12.003.

Zhu, F. X., Wu, H. L., Tu, K. S., Chen, J. X., Zhang, M., & Shi, C. (2016). Serum levels of copeptin are associated with type 2 diabetes and diabetic complications in Chinese population. *Journal of Diabetes and its Complications, 30*(8), 1566–1570. https://doi.org/10.1016/j.jdiacomp.2016.07.017.

CHAPTER SIX

Molecular aspects of aquaporins

Suchun Li[a], Chunling Li[a,*], Weidong Wang[a,b,*]

[a]Department of Pathophysiology, Zhongshan School of Medicine, Sun Yat-sen University, Guangzhou, China
[b]Department of Nephrology, The Seventh Affiliated Hospital, Sun Yat-sen University, Shenzhen, China
[*]Corresponding authors: e-mail address: lichl3@mail.sysu.edu.n; wangwd6@mail.sysu.edu.cn

Contents

Abstract

Water can diffuse freely through the lipid bilayer at a limited rate or by water channel proteins in a rapid speed. Aquaporins (AQPs) are a family of membrane water channel proteins that mainly function as regulators of intracellular and intercellular water flow. They are distributed wildly in specific cell types in multiple organs and tissues. Since the first AQP was identified, 13 AQPs have been characterized in mammals. Structural analysis shows that AQPs are homotetramers with each AQP monomer containing six

Vitamins and Hormones, Volume 113
ISSN 0083-6729
https://doi.org/10.1016/bs.vh.2019.08.019

transmembrane α-helices, two half helices and five connecting loops with two conserved asparagine–proline–alanine (NPA) motifs embedding into the plasma membrane. AQPs are demonstrated to selectively transport water but also some other small molecules. The cellular functions of aquaporins are regulated mainly by posttranslational modifications, e.g., phosphorylation, ubiquitination, glycosylation, subcellular distribution, degradation, and protein interactions. Aquaporins, in particular, AQP2 plays an important role in some disease conditions such as water loss and gain. Insights into the molecular mechanisms regulating AQPs and its clinical significance are proving to be fundamental for development of novel therapeutic targets or reliable diagnostic and prognostic biomarkers. In this chapter, we summarize the molecular aspects of aquaporins which include the isoforms, crystal structure and the molecular mechanisms underlying the regulation of AQPs, with most focus on arginine vasopressin-regulated AQP2.

1. Discovery of the first aquaporin

Water is the major component of cells in all forms of life and water homeostasis and energy balance are essential for survival and adaptation of living cells. Water can diffuse freely through the lipid bilayer at a limited rate, but it was noted that water permeability in red blood cells was much higher than expected for spontaneous diffusion. This observation indicated the existence of a water channel protein, and in fact, prompted the search for specialized proteins that may exist to facilitate water fluxing across membranes.

The membrane water channel was identified by Agre and colleagues around 1988–1992 (Denker, Smith, Kuhajda, & Agre, 1988; Smith & Agre, 1991), which fundamentally provides evidence that movement of water across membrane can be facilitated. Water transport is subject to refined control by a specific gateway, a protein channel inserted in hydrophobic bilayer membrane. Compared to simple diffusion through lipid bilayer, lower activation energy for transport is a typical feature of water channel-mediated diffusion. It is well established now that water channels facilitate the permeation of water and small solutes across membranes. The driving force for water movement is the chemical and physical potential of water between both sides of the membrane (e.g., osmotic and/or hydrostatic pressure gradients) (Carbrey & Agre, 2009).

Agre and his colleagues had initially purified by chance a novel protein from the red blood cell membrane, with a nonglycosylated component of 28 kDa and a glycosylated component migrating as a diffuse band of

35–60 kDa (Denker et al., 1988). The 28 kDa membrane-bound species were named CHannel-forming Integral Protein of 28 kDa (CHIP28). The CHIP28 protein is an abundant integral membrane protein in mammalian red blood cells and renal proximal tubules and belongs to a family of membrane proteins with unknown functions. Later on, the amino acid sequence of CHIP 28 was identified (Smith & Agre, 1991) and cDNA was subsequently cloned (Preston & Agre, 1991). Expression of CHIP28 in *Xenopus laevis* oocytes was later shown to induce rapid water diffusion down osmotic gradients across the plasma membrane, leading to the identification of the first membrane water channels (Preston, Carroll, Guggino, & Agre, 1992). The role of CHIP 28 protein as a functional water channel was confirmed through further studies on the reconstitution of purified protein in liposomes (Zeidel, Ambudkar, Smith, & Agre, 1992). Shortly after establishing the water pore function of CHIP28, homologous proteins in mammals, bacteria and plants were found to function as water channels, and the term "aquaporin, AQP" was coined for members of this protein family (Agre, Sasaki, & Chrispeels, 1993). Accordingly, CHIP28 was renamed to aquaporin 1, abbreviated as AQP1.

2. Classification and isoforms of mammalian aquaporins

Since the discovery of AQP1, several hundred AQP homologs have been identified from all kingdoms of life and at least 13 AQPs (AQP0–AQP12) have been discovered in mammals, nine of them are expressed in the kidney. At least based on their permeability and structural similarity, mammalian AQPs are divided into three subfamilies: (a) classical aquaporins, which are water-selective aquaporins though some aquaporins transport urea and other solutes as well (AQP0, AQP1, AQP2, AQP4, AQP5, AQP6, and AQP8); (b) aquaglyceroporins (GLPs), permeable to water, glycerol, urea, and other solutes (AQP3, AQP7, AQP9, and AQP10); (c) subcellular aquaporins or superaquaporins (AQP11 and AQP12), which have low homology with the other AQPs. The distribution and permeability of mammalian AQPs are shown in Table 1.

2.1 Classical aquaporins

The first subfamily is water selective or specific aquaporins, also named as "orthodox," "classical" aquaporins, including AQP0, AQP1, AQP2, AQP4, AQP5, AQP6 and AQP8. This subfamily of AQPs has been extensively studied, which helps us define regulation of AQP expression in the body and

Table 1 Mammalian AQPs and their distributions.

	Transport	Distributions
Aquaporins		
AQP0	Water (Chandy, Zampighi, Kreman, & Hall, 1997)	eye
AQP1	Water (Denker et al., 1988)	brain, eye, kidney, heart, lung, gastrointestinal tract, salivary gland, liver, ovary, testis, muscle, erythrocytes, spleen
AQP2	Water (Christensen et al., 1998)	kidney, ear, ductus deferens
AQP4	Water (Ishibashi, Hara, & Kondo, 2009)	Brain, kidney, salivary gland, heart, gastrointestinal tract, muscle
AQP5	Water (Ishibashi et al., 2009)	Salivary gland, lung, gastrointestinal tract, ovary, eye, kidney
AQP6	Water, urea, glycerol, anion (Holm, Klaerke, & Zeuthen, 2004)	Brain, kidney
AQP8	Water, urea, ammonia (Soria et al., 2010)	Testis, liver, pancreas, ovary, lung, kidney
Aquaglyceroporins		
AQP3	Water, urea, glycerol, ammonia (Ishibashi et al., 2009)	Kidney, heart, ovary, eye, salivary gland, gastrointestinal tract, Respiratory tract, brain, erythrocyte, fat
AQP7	Water, urea, glycerol, ammonia (Ishibashi et al., 2009)	Testis, heart, kidney, ovary, fat
AQP9	Water, urea, glycerol (Ishibashi et al., 2009)	Liver, spleen, testis, ovary, leukocyte
AQP10	Water, urea, glycerol (Ishibashi et al., 2009)	Gastrointestinal tract
Superaquaporins		
AQP11	Water? (Gorelick, Praetorius, Tsunenari, Nielsen, & Agre, 2006)	Testis, heart, kidney, ovary, muscle, gastrointestinal tract, leukocytes, liver, brain
AQP12	Unknown (Ishibashi, Tanaka, & Morishita, 2014)	Pancreas

their potential roles in physiological and pathophysiological conditions. The inclusion of AQP6 and AQP8 into this category is, on the other hand, primarily based on sequence homology, since AQP6 is also permeated by anions (Yasui, Hazama, et al., 1999), and AQP8 perhaps by both water and urea (Ishibashi, Kuwahara, Kageyama, et al., 1997).

AQP0, first discovered as the 28kDa Major Intrinsic Protein of Lens (MIP), is the most abundant protein in the lens fiber cells, accounting for approximately 50% of the membrane protein content (Alcala, Lieska, & Maisel, 1975). It plays an important role in maintaining the homeostasis and transparency of lens (Clemens et al., 2013). The water permeability of AQP0 was found 40 times lower than AQP1 (Chandy et al., 1997).

AQP1 was the first aquaporin to be characterized in human erythrocytes (Denker et al., 1988; Preston et al., 1992). AQP1 is expressed nearly ubiquitously in the body (Table 1), where it deeply involved in water transport and fluid secretion. AQP1 also plays an important role in angiogenesis, cell migration and cell growth (Nico & Ribatti, 2010), e.g., reduction of AQP1 expression blocked angiogenesis and slowed the progression of tumors (Bin & Shi-Peng, 2011). Moreover, AQP1 is the first AQP that was found to function as a gas channel enhancing CO_2 and NH_3 permeability (Geyer, Musa-Aziz, Qin, & Boron, 2013; Nakhoul, Davis, Romero, & Boron, 1998; Prasad, Coury, Finn, & Zeidel, 1998). AQP1 was also reported to function as a cGMP-gated cation channel (Anthony et al., 2000).

AQP2 is an arginine vasopressin (AVP)-regulated aquaporin which is expressed mainly in the kidney, and AQP2 mutation causes nephrogenic diabetes insipidus (NDI) in human (Deen et al., 1994). Currently, 65 mutations of the AQP2 gene have been described as causes of autosomal NDI, most of which show a recessive inheritance (Milano, Carmosino, Gerbino, Svelto, & Procino, 2017). AQP2 is known to transport only water but not any other small molecules. AQP2 is expressed in principal cells of the collecting ducts in the kidney (Fushimi et al., 1993) and is abundant both in the apical plasma membrane and subapical vesicles (Nielsen, DiGiovanni, Christensen, Knepper, & Harris, 1993). Translocation of AQP2 from intracellular compartments to the apical membrane is dependent on the binding of vasopressin to its V2 receptor located in the basolateral plasma membrane (Marples, Knepper, Christensen, & Nielsen, 1995; Nielsen et al., 1993), by which vasopressin increases the water permeability and water is therefore reabsorbed in the collecting ducts (Fig. 3). AQP2 was found in the endolymphatic sac in the ear in humans, suggesting an active role in the homeostasis of the endolymph (Taguchi et al., 2007). AQP2 was also expressed in the rete testis, efferent ducts, and epididymis of dogs, although its functional role in the reproductive tract is still unknown (Domeniconi, Orsi, Justulin, Leme Beu, & Felisbino, 2008). Nowadays, regulation of AQP2 protein and gene expression, posttranscriptional modification, and intracellular trafficking are very well studied and will be discussed in the following sections of this chapter.

AQP4 is highly expressed at glia cells in the brain, where it regulates water balance. AQP4 null mice appeared normal, although detailed studies revealed they had impaired sight, hearing, and olfactory abilities (Lu, Zhang, Zador, & Verkman, 2008). AQP4 is also expressed at skeletal myocytes and gastric parietal cells. However, AQP4 null mice suffered no abnormalities in these organs (Li, Patil, & Verkman, 2002; Moore, Ma, Yang, & Verkman, 2000; Nielsen, King, Christensen, & Agre, 1997). A very mild urine concentration defect was observed in AQP4 null mice as AQP4 is expressed at the basolateral membrane of the principal cells mainly in the medullary collecting ducts (van Hoek, Ma, Yang, Verkman, & Brown, 2000).

AQP5 is expressed in glandular epithelia, alveolar epithelium and secretory glands where it is involved in the generation of saliva, tears and pulmonary secretions (Song, Sonawane, & Verkman, 2002; Song & Verkman, 2001). AQP5 is also found in type-B intercalated cells in the kidney connecting tubules and collecting ducts where AQP5 colocalizes with the apical transporter pendrin (Procino et al., 2011). However, AQP5 knockout mice develop grossly normally (Ma et al., 1999).

AQP6 is localized in the H^+-ATPase bearing intercalated cells of renal collecting duct (Ohshiro et al., 2001). Interestingly, proven through transmission electron microscopy, AQP6 is located not only on the cell surface but also in intracellular vesicles of the intercalated cells (Yasui, Hazama, et al., 1999). Although the water permeability of AQP6 is markedly low basally, when exposed to low pH, the water permeability of AQP6 is rapidly activated (Yasui, Kwon, Knepper, Nielsen, & Agre, 1999). Moreover, AQP6 also transports urea, glycerol, and nitrate (Holm et al., 2004; Ikeda et al., 2002).

AQP8 is expressed in many organs and tissues, especially the pancreas, liver, salivary glands, kidneys, testes and lung. Several studies showed that AQP8 transports water and ammonia (Liu, Nagase, Huang, Calamita, & Agre, 2006; Saparov, Liu, Agre, & Pohl, 2007). As a member of classical aquaporins, AQP8 is unique in that its genome has six exons rather than four (Koyama et al., 1998; Michalek, 2016). AQP8 is a water channel first found in intracellular domains of the proximal tubule and the collecting duct cells

(Elkjaer et al., 2001). In the kidney, mitochondrial AQP8 was demonstrated play an important role in mitochondrial ammonia transport and the adaptive response of proximal tubule to acidosis (Molinas, Trumper, & Marinelli, 2012; Soria et al., 2010).

2.2 Aquaglyceroporins

The second subfamily is aquaglyceroporins (GLPs) that are permeable to water and other small uncharged molecules (ammonia, urea, in particular glycerol). The aquaglyceroporins, including AQP3, AQP7, AQP9 and AQP10, can be distinguished from classical aquaporins based on amino acid sequence alignments (Borgnia, Nielsen, Engel, & Agre, 1999). Both AQP3 and AQP7 are found to express in the kidney. AQP3 is the first mammalian aquaglyceroporin to be cloned, and it is permeable to glycerol and water (Echevarria, Windhager, Tate, & Frindt, 1994; Yang & Verkman, 1997). AQP7, AQP9, and AQP10 transport water, glycerol, and urea when expressed in Xenopus oocytes (Ishibashi, Kuwahara, Gu, et al., 1997; Ishibashi, Morinaga, Kuwahara, Sasaki, & Imai, 2002; Tsukaguchi, Weremowicz, Morton, & Hediger, 1999). AQP9 is also permeable to a wide range of other solutes in oocytes (Tsukaguchi et al., 1999). Most aquaglyceroporins which transport glycerol and urea are less understood yet.

AQP3 is widely expressed in the kidney, skin, heart, eye, digestive system and respiratory system. In most cases, this protein is expressed in epithelial cells or on the lumen surface. AQP3 is permeable to water, glycerol, and urea. AQP3 is the most abundant skin aquaglyceroporin, where AQP3 facilitates water and glycerol transport, playing an important role in hydration of mammalian skin epidermis, proliferation and differentiation of keratinocytes (Bollag, Xie, Zheng, & Zhong, 2007; Boury-Jamot et al., 2006). In the kidney, AQP3 is localized in the basolateral plasma membranes of cortical and outer medullary collecting duct principal cells, where it mediates the basolateral exit of water (Ishibashi, Sasaki, Fushimi, et al., 1997). Upon long-term vasopressin stimulation, expression of both apical AQP2 and basolateral AQP3 are increased, which raises trans-epithelial water transport capacity of collecting duct principal cells. However, unlike AQP2, there is no evidence for shuttling of AQP3 as no sub-basolateral AQP3-containing vesicles have been demonstrated (Ecelbarger et al., 1995). Emerging evidence demonstrated that AQP3 is regulated on short-term basis likely via cAMP-PKA (cyclic adenosine monophosphate-protein kinase A pathway (Hua et al., 2015; Jourdain et al., 2014;

Marlar et al., 2014). In the kidney the increased basolateral diffusion of AQP3 induced by elevated intracellular cAMP likely altered AQP3 interactions with other proteins or lipids in the plasma membrane, which may be a physiological adaptation to the increased water flow mediated by apical AQP2 (Marlar et al., 2014). AQP3 was recently shown to transport H_2O_2 through the plasma membrane (Almasalmeh, Krenc, Wu, & Beitz, 2014; Bienert & Chaumont, 2014), which likely play an important role in initiating intracellular signalings in cell migration (Hara-Chikuma et al., 2012), inflammation (Hara-Chikuma et al., 2015), and cancer progression (Hara-Chikuma, Watanabe, & Satooka, 2016; Marlar, Jensen, Login, & Nejsum, 2017; Satooka & Hara-Chikuma, 2016).

AQP7 is permeable to water and glycerol and is expressed in the brush border of the proximal tubule S3 segments of the kidney (Nejsum et al., 2000), where AQP7 is colocalized with AQP1. AQP7 null mice have reduced water permeability in the proximal tubule brush border membrane (Sohara et al., 2005), whereas they do not exhibit a urinary concentrating defect or water-balance abnormality.

2.3 Superaquaporins

The third subfamily, called "superaquaporins" or "unorthodox aquaporins," has low conserved amino acid sequences around the Asn-Pro-Ala (NPA) boxes unclassifiable to the first two subfamilies. Mammalian AQP11 and AQP12 are the only two members in this subfamily. The structure and function of AQP11 and AQP12 are currently poorly understood.

AQP11 has a conventional N-terminal NPA signature motif and a unique amino acid sequence pattern that includes an Asn-Pro-Cys (NPC) motif, which appears essential for full expression of molecular function (Ikeda et al., 2011). AQP11 is expressed widely in the testis, thymus, kidney, intestine, and liver. In the kidney, AQP11 is intracellularly expressed in the proximal tubular epithelial cells. Although detailed subcellular localization of AQP11 remains unknown. AQP11 has been observed colocalized with markers of the endoplasmic reticulum in transiently transfected cells (Morishita et al., 2005) and in the kidney from HA-tagged AQP11-transgenic mice (Inoue et al., 2014). Functional studies of AQP11 are also controversial. AQP11 expressed in Xenopus oocytes did not transport water (Gorelick et al., 2006), whereas AQP11 reconstituted into liposomes had a high water channel activity (Yakata et al., 2007). Interestingly, AQP11 knockout mice were fatal with polycystic kidneys (Morishita et al., 2005) which are similar to human polycystic kidney diseases (PKD).

AQP12 is more closely related to AQP11 than to other aquaporins. With regard to the signature motifs, the first NPA motif of AQP12 is substituted by an Asn-Pro-Thr (NPT) motif and the C-terminal NPA motif is conserved (Calvanese, Pellegrini-Calace, & Oliva, 2013; Itoh et al., 2005). No evidence so far shows renal localization of AQP12 in animals.

3. Structural characterization of mammalian aquaporins

Aquaporins facilitate water flow across cellular membrane along osmotic gradients, while excluding the passage of ions and protons. The high-resolution structures provided greater insights into the atomic-level mechanisms of water and solute conduction and of proton and/or ion exclusion of AQPs.

Crystal structures have established that AQPs are homotetramers. Each AQP monomer is about 30 kDa and holds a functional pore (where water passing through) created by six transmembrane α-helices (1–6), two half helices (B and E) and five connecting loops (A–E) with two conserved asparagine–proline–alanine (NPA) motifs embedding into the plasma membrane. The N- and C-terminal domains are located in the cytoplasm (Fig. 1A). In the membrane, four AQP monomers self-assemble into a tetrameric biological unit (Fig. 1B), by this way an additional tetrameric pore is formed, which is mostly hydrophobic. The fifth pore is thought to be responsible for the translocation of dissolved gasses and ionic species (Huber, Tsujita, & Nakada, 2012). It may transport urea, glycerol, H_2O_2, NH_3, CO_2, etc. passing through plasma membrane.

The six transmembrane α-helices surround a single, narrow aqueous pore (Sui, Han, Lee, Walian, & Jap, 2001) (Fig. 1A and B), where two half helices (B and E) (NPA signature motif) fold into the channel from opposite sides of the membrane. This arrangement determines substrates selectivity. The positions of loops B and E are stabilized through ion pairs and hydrogen bonds with neighboring transmembrane helices. This pore has electrostatic interactions which renders the AQP outer cone walls in a hydrophobic state, causing repulsion (Sachdeva & Singh, 2014).

The extracellular vestibule of AQP monomer contains a constriction region made up of aromatic and arginine residues (known as the ar/R constriction or the selectivity filter) (Fig. 1C), which is important for preventing proton conduction. The ar/R constriction makes up the narrowest region of the pore, and has a diameter of ~2.8 Å in human AQP1, ~1.5 Å in human AQP4 and ~3.4 Å for the less selective

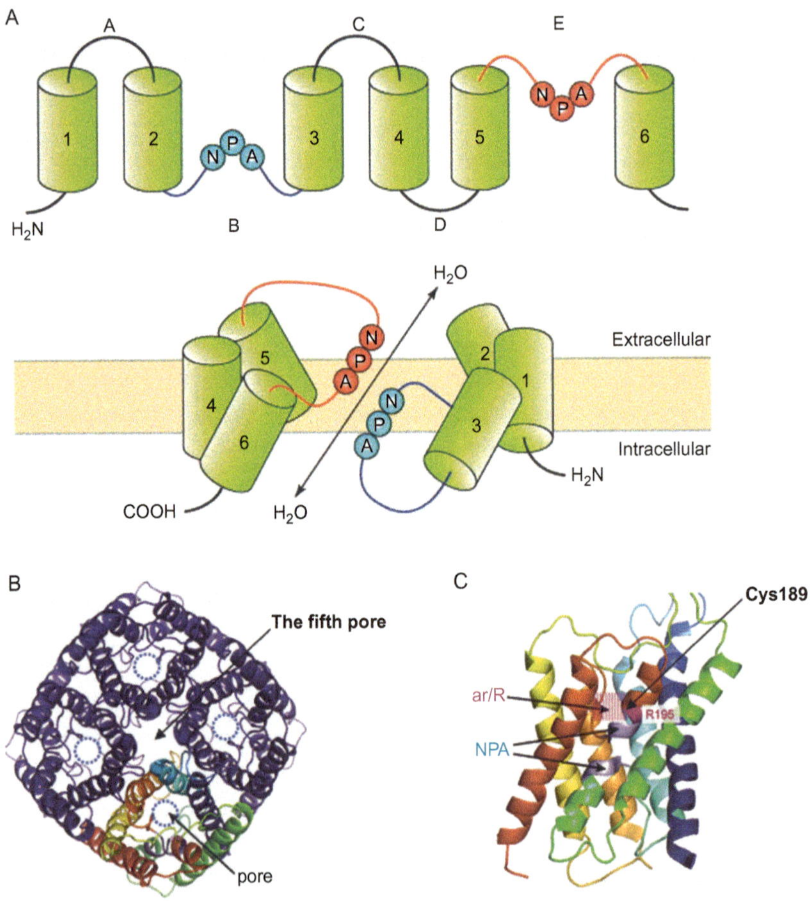

Fig. 1 Structural characteristics of AQP protein (A) The AQP monomer is formed by six transmembrane α-helices (1–6), two half helices (B and E) and five connecting loops (A–E) with two conserved asparagine–proline–alanine (NPA) motifs embedding into the plasma membrane. (B) The tetramer of AQP viewed from extracellular side. Each AQP monomer has an individual pore and the fifth pore is formed in the center (arrow). (C) The aromatic and arginine residues (ar/R), located close to the extracellular entrance of the pore, is the narrowest region in the AQP pores. The Arg195 and NPA motifs are shown in magenta and light blue, respectively. The residue Cys189 is the mercury binding site that known to inhibit AQP. *Panel A: Reproduced with permission from Zeuthen, T. (2001). How water molecules pass through aquaporins.* Trends in Biochemical Sciences, *26(2), 77–79; Panel B and C: Reproduced with permission from Tani, K., & Fujiyoshi, Y. (2014). Water channel structures analysed by electron crystallography.* Biochimica et Biophysica Acta, 1840*(5), 1605–1613. doi:10.1016/j.bbagen.2013.10.007.*

aquaglyceroporins (Fu et al., 2000; Ho et al., 2009; Sui et al., 2001). The narrow interior pore region of AQPs presents a potential challenge in the identification of small molecules that target the aqueous pore region.

The central pore formed by two short helices B and E together with the tightly stacked NPA motifs is very constrict in each AQP monomer. The constriction lined with hydrophobic amino acid residues. The diameter of the constriction is so small (approximately 3 Å) that only one water molecule can pass through the constriction, driven by hydrogen-bond formation with the Asn residues, which lowers the energy barrier for the water molecule entering the constriction (Benga, 2012) (Fig. 2). Molecular dynamics simulation of aquaporins revealed that the move pattern of water molecules through the channel is very characteristic. The water molecules are forced to pass through the channel in a single line manner (Sui et al., 2001) and orientate themselves in the local electrical field formed by the atoms of the channel. Upon entering the channel from the extracellular mouth, the water molecules face with their oxygen atoms down the channel. At the level of the central constriction they reverse orientation, facing with the oxygen atoms up (Benga, 2012; Murata et al., 2000) (Fig. 2).

Some aquaporins possess a cysteine residue in the E loop near the pore, which is responsible for functional sensitivity to mercury. The residue Cys189 in AQP1 is the site for mercurial binding, thus inhibiting water transport through AQP1 (Jung, Preston, Smith, Guggino, & Agre, 1994; Preston, Jung, Guggino, & Agre, 1993). When mercury interacts with Cys189, the pore is collapsed by conformational changes at the ar/R region (Fig. 1C), and thus water movements through AQP1 are inhibited. The inhibition mechanisms were elucidated later by molecular dynamics simulations (Hirano et al., 2010).

4. Protein modification of AQP2

Posttranslational modifications (PTMs) are usually highly dynamic processes that alter the properties of the target proteins. Through changes in protein conformation, these PTMs have been shown to modulate the localization, stability, activity, and interacting partners of their substrate proteins, thus playing pivotal roles in intracellular signaling, protein maturation and folding. The cellular functions of many eukaryotic membrane proteins, including the AVP-regulated AQP2, are regulated by posttranslational modifications.

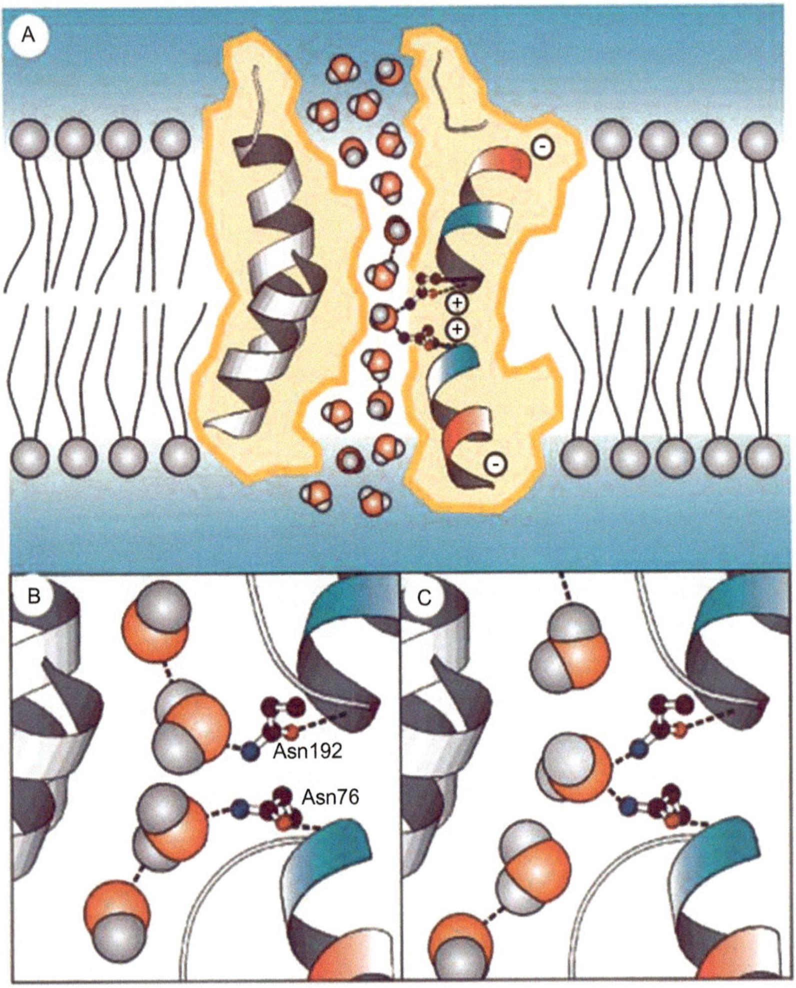

Fig. 2 Schematic representations explaining the mechanism for blocking proton permeation of AQP1. (A) A diagram illustrating how partial charges from the helix dipoles restrict the orientation of the water molecules passing through the constriction of the pore. (B) and (C) A diagram illustrating hydrogen bonding of a water molecule with Asn 76 and/or Asn 192, which extend their amido groups into the constriction of the pore. *Reproduced with permission from Murata, K., Mitsuoka, K., Hirai, T., Walz, T., Agre, P., Heymann, J. B., et al. (2000). Structural determinants of water permeation through aquaporin-1.* Nature, 407*(6804), 599–605. doi:10.1038/35036519.*

AQP2 is expressed in the principal cells of the kidney collecting ducts and its apical or intracellular distribution is mainly regulated by AVP. Stimulation with AVP results in a predominant translocation of AQP2 to apical membrane localization from subapical compartments (Fig. 3).

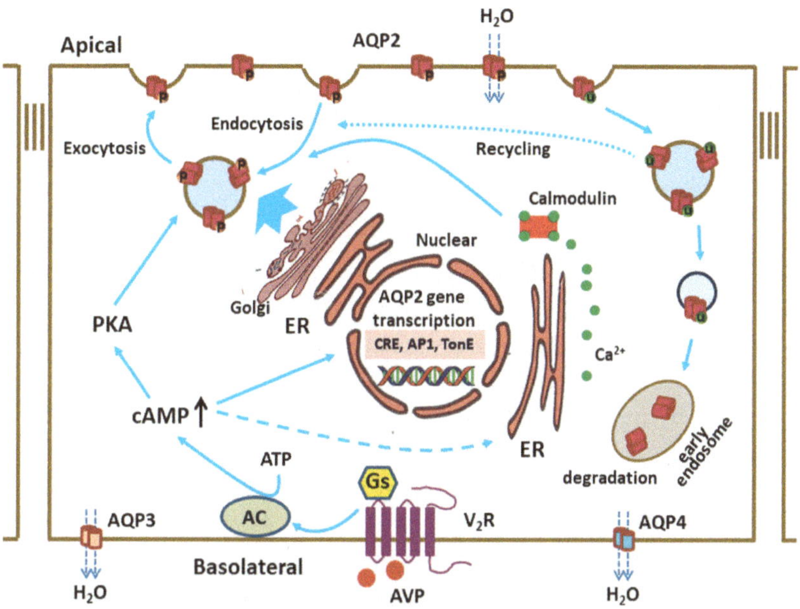

Fig. 3 Arginine vasopressin-mediated AQP2 regulation in the kidney collecting duct principal cells. AVP binds to its V2 receptor (V2R) at the basolateral membrane and activates Gs protein-mediated signaling cascade. Coupling with Gs protein, V2R activates adenylyl cyclases (AC) and increases intracellular cAMP, which subsequently activates protein kinase A (PKA) and phosphorylation of AQP2, resulting in the translocation of AQP2 bearing vesicles to the apical plasma membrane (trafficking and exocytosis). AVP stimulation also results in increased intracellular Ca^{2+} levels via Ca^{2+} release from calmodulin-dependent ryanodine-sensitive intracellular stores, which induces apical membrane expression of AQP2. AVP also induces phosphorylation of a cAMP-response element binding protein (CREB), which binds to cyclic AMP-responsive element (CRE) and AP-1 at the AQP2 gene promoter region and thus increases AQP2 gene transcription. Once the water balance is restored, AVP levels drop and AQP2 is removed from the apical membrane (endocytosis). After internalization, AQP2 can either be recycled to the apical membrane upon restimulation or follow the route of degradation via multivesicular bodies and lysosomes. Driven by the transcellular osmotic gradient, water enters principal cells through AQP2 and passes through basolateral plasma membrane via AQP3 and AQP4 to the blood. ER, endoplasmic reticulum. *Reproduced with permission from Li, C., & Wang, W. (2017). Molecular biology of aquaporins.* Advances in Experimental Medicine and Biology, 969, *1–34. doi:10.1007/978-94-024-1057-0_1.*

This renders the apical membrane highly permeable to water and is a key event in formation of concentrated urine and thus regulating body water balance. Recent publications demonstrated that phosphorylation and ubiquitination are major mechanisms that regulate AQP2 activity, subcellular sorting and distribution, degradation, and protein interactions.

Thus, in this section, we mainly focus on PTMs of AVP-mediated regulation of AQP2 with major emphasis on phosphorylation and ubiquitination.

4.1 Phosphorylation

Protein phosphorylation and dephosphorylation are common dynamic posttranslational processes often involved in regulation of protein function and cellular distribution. Phosphorylation is one of well-studied PTMs in the regulation of channel gating and membrane accumulation of AQP2. However, it should be emphasized that although dephosphorylation is likely a key event in the modulation of AQP2, this process is less well described in the literature.

AQP2 contains numerous putative phosphorylation sites for various protein kinases, e.g., PKA, PKG, PKC, and casein kinase II (Brown, 2003) based upon bioinformatic analysis. However, beside sequence recognition, a protein kinase must fulfill several criteria to play a role in modulating channel function: (1) the kinase must be present in the same cell type as the target protein; (2) the kinase must colocalize with the target protein; and (3) the kinase must be active under particular conditions, e.g., AVP stimulation. Phosphorylation of AQP2 at multiple sites regulates its translocation to the apical membrane from intracellular vesicles. Phosphorylation of the serine at position 256 (S256) in the C-terminal tail of AQP2 was the first identified and the best characterized phosphorylation site of AQP2. Large scale phospho-proteomic analysis later demonstrated that beside of S256 phosphorylation site, the polyphosphorylated sites of AQP2 contains S261, S264, and S269 (Fushimi, Sasaki, & Marumo, 1997; Hoffert et al., 2008; Hoffert, Pisitkun, Wang, Shen, & Knepper, 2006).

The targeting of AQP2 to the apical plasma membrane of collecting duct principal cells is mainly regulated by the actions of AVP on the V2R. The V2R binds to the stimulatory guanine nucleotide-binding protein alpha-subunit (Gsα), resulting in activation of adenylyl cyclases (AC) and increased intracellular cAMP. Elevation of cAMP and activation of protein kinase A (PKA) can phosphorylated AQP2 on Ser256, which is pivotal for AVP-mediated AQP2 membrane targeting (Knepper, Kwon, & Nielsen, 2015; Li & Wang, 2017; Nielsen et al., 2002) (Fig. 3). S256-AQP2 is a target for PKA-induced phosphorylation, although other kinases may also target this residue (Fushimi et al., 1997; Katsura, Gustafson, Ausiello, & Brown, 1997), however, PKA alone cannot phosphorylate the downstream sites at S261, S264, and S269 (Hoffert et al., 2008). S256 is a substrate for Golgi casein kinase 2, and phosphorylation by this kinase is required for transition of AQP2 through the Golgi (Procino et al., 2003).

Phosphorylated AQP2 at S256 (pS256-AQP2) is detected in both intracellular vesicles and the apical plasma membrane in the collecting duct, where its abundance is increased in response to AVP treatment (Christensen, Zelenina, Aperia, & Nielsen, 2000). pS256-AQP2 was able to be detected in nonstimulated vasopressin-deficient Brattleboro rats, indicating that S256 is highly phosphorylated even in low levels of circulating AVP (Christensen et al., 2000). Phosphorylation of AQP2 at S256 seems important for apical targeting of AQP2 and thus water transport. Vasopressin or forskolin treatment failed to induce translocation of AQP2 when AQP2-S256A mutant was transfected to LLC-PK1 cells (Fushimi et al., 1997; Katsura et al., 1997). Importantly, congenital progressive hydronephrosis, polyuria, and urinary concentrating defect were observed in a mouse strain with an amino acid substitution at S256 for preventing phosphorylation. As expected, vasopressin treatment was not effective in these mice (McDill, Li, Kovach, Ding, & Chen, 2006).

The abundance of phosphorylated AQP2 at S264 (pS264-AQP2) was found to increase in both the apical and basolateral plasma membrane of principal cells after acute specific V2R agonist dDAVP treatment (Fenton et al., 2008). Phosphorylated S269-AQP2 (pS269-AQP2) is not observed in any intracellular compartments but detected exclusively in the apical plasma membrane of the principal cells of the collecting duct (Moeller, Knepper, & Fenton, 2009), indicating a role of pS269-AQP2 in plasma membrane targeting. Phosphorylated S261-AQP2 (pS261-AQP2), on the other hand, is predominantly localized within the cell in compartments different from the endoplasmic reticulum, Golgi apparatus, and lysosomes (Hoffert et al., 2007).

In IMCD tubule suspensions, dDAVP or exogenous cAMP increases phosphorylation of AQP2 at S256, S264, and S269, which remain high as long as the agonist is present (Hoffert et al., 2008). Phosphorylation of AQP2 at S256 occurs initially and rapidly reaches the highest magnitude, whereas maximal phosphorylation at the other sites (S264 and S269) occurs relatively slow. S256 phosphorylation appears required and strongly facilitates phosphorylation of S264 and S269, as the S264 and S269 phosphoforms of AQP2 are not observed in cells expressing the S256 mutated form of AQP2 or in kidney sections from a mouse model with a mutation of S256 (Hoffert et al., 2008). In fact, S256 phosphorylation alone is necessary and sufficient for membrane accumulation of AQP2 induced by AVP (or cAMP), independent of the phosphorylation state of any other sites in the C terminus, e.g., S264 or S269 (Arthur et al., 2015). These observations strongly suggest that S256 phosphorylation is a priming event for phosphorylation of S264 and S269 and plays a critical role in intracellular translocation of AQP2. Earlier studies revealed that the increased monophosphorylation of AQP2 at S256 with vasopressin stimulation in rat IMCD coincided with decreased phosphorylation of AQP2 at S261, which was associated with intracellular vesicle distribution, suggesting that phosphorylation of S256 and S261 may inversely regulate AQP2 trafficking (Hoffert et al., 2007, 2006). dDAVP stimulation results in decreased phosphorylation at S261, which is associated with reduced stability of AQP2 (Nedvetsky et al., 2010). In fact phosphorylation of S261 on AQP2 occurs after ubiquitin-mediated endocytosis, suggesting that this phosphorylation at S261 does not induce AQP2 ubiquitination itself, but likely stabilizes ubiquitinated AQP2 and intracellular localization of AQP2 (Tamma, Robben, Trimpert, Boone, & Deen, 2011) (Fig. 4).

PKG, an AGC kinase family member (like PKA), is proposed to modulate AQP2 trafficking; however, its role remains controversial. The agonist of PKG, cGMP, has been shown to induce translocation of AQP2 to the plasma membrane in AQP2-transfected LLC-PK1 cells and in isolated kidney slices (Bouley et al., 2000). Atrial natriuretic peptide (ANP), L-arginine, or cGMP phosphodiesterase type 5 (PDE5) inhibitors sildenafil citrate increases intracellular cGMP levels, resulting in accumulation of AQP2 in plasma membrane (Boone, Kortenoeven, Robben, & Deen, 2010; Bouley et al., 2000; Wang, Li, Nejsum, et al., 2006). In contrast, one study showed that ANP and NO (nitric oxide) signaling deceased S256-AQP2 phosphorylation and reduced AQP2 in the plasma membrane, antagonizing vasopressin-mediated water permeability in inner medullary collecting duct cells (Klokkers et al., 2009). Further studies are warranted to clarify the role of PKG-cGMP signaling pathway in promoting AQP2 trafficking and expression.

Some other kinases other than PKA/PKG appear to regulate AQP2 phosphorylation and intracellular trafficking, e.g., phosphatidylinositol-3-Kinase/Akt, Ca^{2+}/calmodulin-dependent protein kinase II, GSK3, cyclin-dependent kinases. This has been nicely reviewed recently (Jung & Kwon, 2016; Vukicevic, Schulz, Faust, & Klussmann, 2016). For example, AKT (also known as PKB) was recently shown to mediate vasopressin-stimulated AQP2 membrane accumulation (Kim et al., 2011).

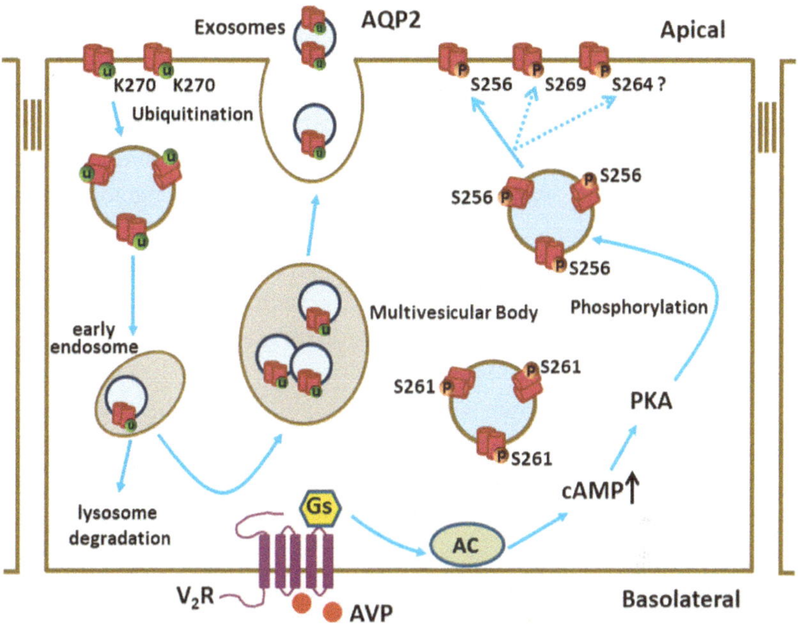

Fig. 4 Phosphorylation and ubiquitination of AQP2 determines its intracellular localization. AVP or forskolin induces phosphorylation of AQP2 monomers at S256, followed by increased phosphorylation at S269 and S264 and reduced phosphorylation at S261, resulting in redistribution of AQP2 from intracellular vesicles to the apical membrane. After AVP withdrawal, AQP2 is ubiquitinated at K270, which induces AQP2 internalization and lysosomal degradation, or sorted to the multivesicular body (MVB) and released in exosomes into the urine via exocytosis. *Reproduced with permission from Li, C., & Wang, W. (2017). Molecular biology of aquaporins.* Advances in Experimental Medicine and Biology, 969, *1–34. doi:10.1007/978-94-024-1057-0_1.*

However, it should be noted that PKA is still the principal kinase for AQP2 trafficking and body fluid homeostasis. This conception was supported by a recent finding that a dominant negative PKA mutation in the kidney elicits a diabetes insipidus phenotype (Gilbert, Yang, Su, & McKnight, 2015). The V2R-cAMP-PKA signaling pathway in AQP2 regulation has been widely accepted, however, evidence supporting other intracellular signaling mechanisms coupling V2R stimulation to AQP2 membrane targeting are emerging, especially roles of local mediators in modulating the effect of AVP (Olesen & Fenton, 2017).

4.2 Ubiquitination

There are two major protein degradative pathways to function in mammalian cells, which are the ubiquitin proteasome pathway and lysosomal proteolysis pathway. Ubiquitination is a more complex PTM whereby ubiquitin (Ub) are covalently attached to a target protein by a process involving a hierarchical set of three enzymes (Gao & Karin, 2005). Ubiquitin labels protein through a conjugation system comprising E1 activation, E2 conjugation, and E3 ligation enzymes. Attachment of ubiquitin can affect proteins in numerous ways, e.g., promote endocytosis, degradation, and at the molecular level affect binding of interaction partners. Ubiquitination can promote protein internalization via endocytotic pathway, followed by their degradation in lysosomes (Gunnarson, Zelenina, & Aperia, 2004). Protein ubiquitination is reversed by deubiquitinating enzymes (DUBs), which is essential for cellular homeostasis.

AQP1 is the first known aquaporin that is able to be ubiquitinated and degraded by the proteasome. Exposure to hypertonic medium induced a decrease of AQP1 ubiquitination and markedly increased stability of AQP1 protein, thereby contributing to overall protein abundance (Leitch, Agre, & King, 2001).

Although three putative potential attachment sites (cytosolic lysine residues) for AQP2 ubiquitination were found at positions 228, 238, and 270, site mutation study revealed that AQP2 is ubiquitinated only at the COOH-terminal K270 with one K63-linked chain of mainly two to three ubiquitin moieties (Kamsteeg et al., 2006). The ubiquitination of AQP2 at the plasma membrane enhances endocytosis, sorting to intraendosomal vesicles and degradation (Kamsteeg et al., 2006). Five common isoforms of E3 ligases (UBR4, UHRF1, NEDD 4, BRE1B, and Cullin-5) are putatively associated with dDAVP-induced AQP2 regulation (Lee et al., 2011). Cullin-5 (CUL5), for example, a member of the cullin gene family of scaffold proteins of the E3 complex and a vasopressin-activated calcium-mobilizing receptor, was upregulated during dDAVP withdrawal in IMCD cells, which was associated with increased AQP2 among the ubiquitinated proteins in intracellular vesicles fractions. This finding suggests that CUL5 may play a role in the attachment of Ub to AQP2, resulting in ubiquitination and internalization of AQP2, and reduction of AQP2 abundance after dDAVP withdrawal, presumably via lysosomal and/or proteosomal degradation (Lee et al., 2011). By using Bayes' theorem to integrate information from multiple large-scale proteomic and transcriptomic datasets, a recent study identified the E3 ligases most likely to interact with AQP2 as (in order of probability): NEDD4 and NEDD4L (tied for first), AMFR, STUB1, ITCH, ZFPL1 among 377 E3 ubiquitin ligases examined (Medvar, Raghuram, Pisitkun, Sarkar, & Knepper, 2016). A later study confirmed the role of NEDD4/NEDD4L in mediating ubiquitination and degradation of AQP2 and also found a link protein between AQP2 and NEDD4/NEDD4L (Trimpert et al., 2017).

Phosphorylation and ubiquitination are highly dynamic in the cell and they likely act in concert and could greatly promote fine regulation of protein function (Hunter, 2007). Phosphorylation of AQP2 at S256 and dephosphorylation at S261 cause its translocation from intracellular vesicles to the apical membrane, whereas ubiquitination of AQP2 at K270 induces its internalization and lysosomal degradation, or released in exosomes into the urine via exocytosis (Fig. 4). The subcellular distribution of AQP2 is finely tuned by phosphorylation at S256, S264, S269 and/or intracellular S261 phosphorylation as well as K270 ubiquitination, by this way AQP2 is dynamically regulated.

The potential interplay between polyubiquitylation and polyphosphorylation of AQP2 was examined (Moeller, Aroankins, Slengerik-Hansen, Pisitkun, & Fenton, 2014; Tamma et al., 2011). dDAVP or forskolin induces phosphorylation of AQP2 monomers at S256, followed by increased phosphorylation at S269 and S264 and reduced phosphorylation at S261, resulting in steady-state redistribution of AQP2 from vesicles to the apical membrane, whereas increased AQP2 ubiquitination induced endocytosis and steady-state redistribution of AQP2 to the intracellular vesicles (Fig. 4). Phosphorylation often occurs as a priming event for ubiquitination and ubiquitination can regulate protein phosphorylation by regulation of kinase activities (Hunter, 2007). AQP2 phosphorylation was demonstrated to be able to override dominant endocytic signal of K63-linked polyubiquitylation. In polarized epithelial cells and kidney tissue, distribution of AQP2 on the plasma membrane is regulated by phosphorylation at S256 and S269. The rate of AQP2 endocytosis was reduced by prolonging phosphorylation specifically at S269. AQP2 phosphorylation at S269 and ubiquitylation at K270 can occur in parallel, with increased S269 phosphorylation and decreased AQP2 endocytosis occurring when K270 polyubiquitylation levels are maximal (Moeller et al., 2014). The study suggests that site-specific phosphorylation can counteract polyubiquitylation and determine its final localization.

4.3 Glutathionylation

Dysregulation of redox homeostasis and production of excessive free radicals may contribute to the pathogenesis of kidney diseases. Excessive reactive oxygen species (ROS) produced in the kidney may affect the expression and the activity of many transporters and channels, including aquaporins. As an important PTM, S-glutathionylation exerts protection of cysteine residues against irreversible oxidation during redox imbalance. Glutathione, one of major cellular antioxidant molecules, is continuously converted into the reduced form of glutathione. In mpkCCD cells, vasopressin stimulated translation of seven glutathione S-transferase (GST) proteins functioning to conjugate the tripeptide glutathione to substrates, e.g., cysteine, likely indicating the involvement of redox into vasopressin-activated signal transduction pathway (Sandoval et al., 2013). Topological analysis of AQP2 suggests that Cys75 and Cys79 on cytosolic B-loop might be targets of S-glutathionylation (Tamma et al., 2014). The S-glutathionylation of AQP2 is tightly modulated by intracellular ROS content both in kidney tissue and in HEK cells stably transfected AQP2. Specifically, an oxidant inducer caused a significant increase in AQP2 S-glutathionylation secondary to an increase in ROS content, indicating that this redox sensitive PTM is linked to the redox condition of the tissue, however, whether S-glutathionylation affects the localization and the activity of AQP2 is not reported (Tamma et al., 2014). Interestingly, a very recent study elegantly demonstrated that hydrogen sulfide (H2S), a gas molecule with reductive feature, upregulates renal AQP2 protein expression and promotes urine concentration (Luo et al., 2019); however, whether glutathionylation of AQP2 is affected by H_2S was not investigated.

4.4 Glycosylation

The process of attaching both O- and N-glycans to extracellular domains of membrane proteins can be important for the apical location of these proteins. AQPs contain N-linked glycosylation consensus sites in their extracellular loops, some of which are not efficiently recognized during protein synthesis by oligosaccharyltransferase, generating a mixture of glycosylated and nonglycosylated species. N-glycosylation is not believed to be important in the transport function of aquaporins. The site of N-glycosylation in AQP1 is Asn42, which lies in a potential N-glycosylation consensus sequence. The early study using site-directed mutant of Asn42 showed that the non-glycosylation of AQP1 failed to affect water permeability in oocytes (Preston et al., 1993).

A role of glycosylation of AQP2 seems complex. The glycosylated form of AQP2 has a shorter half-life than the nonglycosylated form (Sandoval et al., 2013), indicating that N-linked glycosylation is not necessary for the stability of AQP2. However, the glycosylation seems important for cell surface expression of AQP2 (Hendriks et al., 2004) but not for routing because inhibition of glycosylation does not prevent delivery of AQP2 to the plasma membrane in response to increased cAMP (Baumgarten, Van De Pol, Wetzels, Van Os, & Deen, 1998). In addition, glycosylation appears not essential for tetramerization of AQP2 in the endoplasmic reticulum, as part of tetrameric complexes contain one or more nonglycosylated AQP2 molecules (Hendriks et al., 2004), whereas addition of a single *N*-linked oligosaccharide moiety can partially compensate for ER folding defects of AQP2 induced by disease-related mutations (Buck, Eledge, & Skach, 2004).

4.5 Acetylation of AQP protein

N-terminal acetylation has been proposed to be a determinant of protein stability. AQP2 was one of the proteins identified with N-terminal acetylation (Sandoval et al., 2013). The lysine acetylome and phosphoproteome in the rat native inner medullary collecting duct was recently determined (Hyndman et al., 2018). In the vasopressin-regulated water reabsorption pathway, eight proteins were acetylated, including the novel identification of the basolateral water channel, AQP3, acetylated at K282 (Hyndman et al., 2018), while the significance of AQP3 acetylation is unclear. A reduction in histone H4 acetylation level of AQP5 induced by TNF-α caused a decrease of AQP5 expression and the fluid secretion from salivary acinar cells (Yamamura et al., 2012). The broad-spectrum histone deacetylase (HDAC) inhibitor induced AQP3 mRNA and protein expression in a dose- and time-dependent manner in normal keratinocytes, indicating a potential regulation of AQP3 expression by HDAC (Yamamura et al., 2012). Modifications of histone deacetylase activities are also found to play a key role in AQP1 regulation (Huysseune et al., 2009). However, these studies about acetylation mainly focus on histones and their roles in modulating chromatin structure and gene transcription, which more belong to epigenetic regulation but not PTMs.

4.6 PTMs of AQP2 by other molecules

While arginine vasopressin is the key hormone that regulate AQP2 expression, many other molecules/hormones can also influence AQP2 membrane accumulation and cellular expression, e.g., prostaglandins (Olesen, Rutzler, Moeller, Praetorius, & Fenton, 2011; Tamma, Wiesner, et al., 2003; Zelenina et al., 2000), bradykinin (Tamma, Carmosino, Svelto, & Valenti, 2005), dopamine (Boone, Kortenoeven, Robben, Tamma, & Deen, 2011), arginine/NO (Bouley et al., 2000), hydrogen sulfide (Luo et al., 2019), atrial natriuretic peptide (ANP) (Klokkers et al., 2009; Wang, Li, Nejsum, et al., 2006), oxytocin (Li et al., 2008), angiotensin II (Li et al., 2011). However, PTMs of AQP2 by these molecules is largely unknown and further investigations are definitely warranted.

5. Molecular machinery of AQP2 trafficking

AQPs are important regulators of transcellular water flow that expressed in a wide range of tissues and organs in mammals. As membrane integral proteins, AQPs are synthesized and cotranslationally inserted in the endoplasmic reticulum (ER) membrane and travel across the secretory pathway to reach their target membrane. The transport of water mediating by AQPs is regulated either by opening or closing the channels (gating), a conformational change, or by altering the density of AQPs in particular membrane. These regulations of AQPs are often associated with certain physiological or pathophysiological conditions.

AQP2 is expressed in the kidney connecting tubule and collecting duct, where it plays an important role in body water balance. In general, AQP2 is regulated through two mechanisms. In short-term regulation, trafficking between intracellular vesicles and plasma membrane is associated with post-translational modification of AQP2 (see Section 4) and vesicle-transporting systems dependent on cell signaling; in long-term regulation, changes of protein abundance are regulated by transcription and mRNA/protein stability (see Section 6).

Arginine vasopressin is the key hormone to regulate AQP2 trafficking and expression. Upon vasopressin stimulation, AQP2 translocates from intracellular storage vesicles to the apical plasma membrane, rendering the cell water permeability. It should be noted that AQP2 is also found in the basolateral membrane in some conditions, such as aldosterone treatment (de Seigneux et al., 2007; Nielsen et al., 2006). Intracellular trafficking and absolute abundance of AQP2 protein in the kidney collecting duct principal cells are finely regulated, which are orchestrated by multiple intracellular signaling molecules directing *exo*- and endocytic trafficking, by several cytoskeletal proteins as the tracks for the correct trafficking, by proteins controlling cAMP signaling, and by transcriptional factors directing translation of AQP2 protein (Jung & Kwon, 2016). In this section, we mainly discuss the molecular mechanisms by which AQP2-bearing vesicles are transported to the apical membrane.

5.1 Gating

Gating of AQPs is a rapid way to control the membrane permeability of water flow into and out of the cell, which has been reported for plant, yeast and mammalian AQPs. Changes of pH and phosphorylation are two common signals through which water pores are gated in plants and mammals (Gonen, Sliz, Kistler, Cheng, & Walz, 2004; Nemeth-Cahalan, Kalman, & Hall, 2004; Tornroth-Horsefield et al., 2006; Tournaire-Roux et al., 2003). Phosphorylation-mediated gating has been reported for mammalian AQP4 (Zelenina, Zelenin, Bondar, Brismar, & Aperia, 2002), while gating of mammalian AQP0 is via pH and calcium (Gonen et al., 2004; Nemeth-Cahalan & Hall, 2000). In mammals, the mechanism of AQP regulation by pH is not as clear as for plants. A pinching mechanism for AQP0 is also suggested, in which small conformational changes due to protonation pinch the constriction region, limiting the passage of molecules (Tornroth-Horsefield, Hedfalk, Fischer, Lindkvist-Petersson, & Neutze, 2010). Although mechanisms of AQP2 trafficking are well studied, the view on gating of AQP2 via phosphorylation is still debatable (Eto, Noda, Horikawa, Uchida, & Sasaki, 2010; Moeller, Knepper, et al., 2009). A comprehensive analysis of water permeability of AQP2 in oocytes failed to provide any evidence that mutation of phosphorylation site of AQP2 alters the water transport function of AQP2 (Moeller, MacAulay, Knepper, & Fenton, 2009), thus not in favor of supporting that phosphorylation mediated gating of AQP2. In fact, the predominant regulatory mechanism of mammalian aquaporins is trafficking.

5.2 Trafficking of AQP2

The regulation of trafficking of AQP2-containing membrane vesicles to and from the apical plasma membrane of principal cells is a rapid and short-term regulation, which occurs within several minutes. The trafficking of AQP2 is regulated at the transcriptional and/or translational level, and also involves shuttle of AQP2 between intracellular storage vesicles and the target membrane. PTMs, especially phosphorylation, are one of important mechanisms regulating redistribution of AQP2 in the collecting duct principal cells, as discussed before. Here, we summarize the most important and recent studies on the cellular transport machinery mediating (1) trafficking of AQP2 to the apical plasma membrane, (2) AQP2 vesicles targeting to the apical plasma membrane, (3) docking and fusion of AQP2-bearing vesicles with the apical plasma membrane (exocytosis), (4) removal of AQP2 from the apical plasma membrane (endocytosis) (Fig. 5).

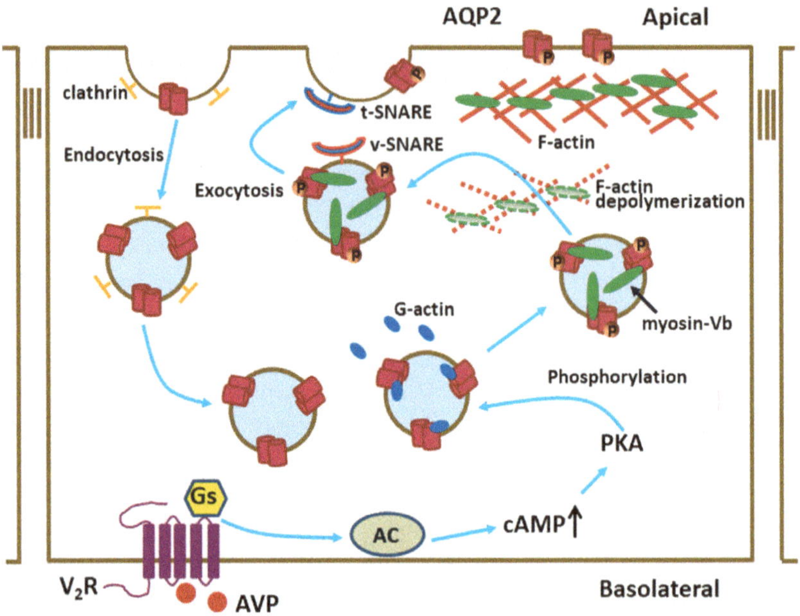

Fig. 5 Exocytosis and endocytosis of AQP2 in the collecting duct principal cells. Docking and fusing of AQP2-bearing vesicles (exocytosis) is mediated by SNARE mechanisms, involving vesicle v-SNAREs and target membrane t-SNAREs. PKA-induced phosphorylation of AQP2 at S256 reduced the direct binding of AQP2 to G-actin, but increased the affinity of AQP2 to myosin-Vb, one of the central regulators in apical trafficking. This interaction results in a reduced binding of myosin-Vb to F-actin, resulting in F-actin destabilization that allows translocation of AQP2 vesicles to the plasma membrane. Removal of AQP2 from the apical plasma membrane (endocytosis) is a way by which AQP2 is recycled. AQP2 accumulates in clathrin-coated pits and is internalized via a clathrin-mediated process. SNARE: soluble N-ethylmaleimide-sensitive factor attachment protein receptors. *Reproduced with permission from Li, C., & Wang, W. (2017). Molecular biology of aquaporins.* Advances in Experimental Medicine and Biology, 969, *1–34. doi:10.1007/978-94-024-1057-0_1.*

5.2.1 The cAMP-mediated effect of vasopressin on AQP2 trafficking

In the absence of vasopressin, AQP2 is localized mainly in subapical vesicles of the collecting duct principal cells. Stimulating by high osmolality or a reduction of effective circulating blood volume, vasopressin, secreted from the posterior pituitary, binds to its Gs-coupled type-2 receptor (V2R) located on the basolateral membrane of the renal collecting duct principal cells. This interaction initiates a complex intracellular signaling cascade resulting in activation of AC, which increases intracellular cyclic adenosine monophosphate (cAMP) levels, activation of PKA, and phosphorylation of AQP2. AC6 is the principal AC mediating these responses, as only

AC6-deficient mice have urine concentration defects (Kittikulsuth, Stuart, & Kohan, 2014; Kittikulsuth, Stuart, Van Hoek, et al., 2014; Rieg et al., 2010; Roos, Strait, Raphael, Blount, & Kohan, 2012). The subsequent activation of PKA leads to phosphorylation of AQP2 at serine 256, which is required for AQP2 translocation from intracellular storage vesicles to the apical membrane (Fig. 3).

As PKA and many of its substrates are present throughout the cell, localization of PKA to specific sites is necessary for a timely and spatially effective activation of PKA. This process is mediated by A-kinase anchoring proteins (AKAPs). AKAPs, as scaffolding proteins, tether PKA in close proximity of its substrates (Langeberg & Scott, 2015). It is demonstrated that AQP2-bearing vesicles contained PKA activity (Lande, Jo, Zeidel, Somers, & Harris, 1996) and the anchoring of PKA to AKAPs was a prerequisite for the AQP2 shuttle (Henn et al., 2005; Klussmann, Maric, Wiesner, Beyermann, & Rosenthal, 1999). It was reported that several splice variant of AKAP18, AKAP18delta, and AKAP220 were involved in the trafficking of AQP2 (Henn et al., 2004; McSorley et al., 2006; Okutsu et al., 2008). Very recent data demonstrates an involvement of AKAP-Lbc-mediated RhoA activation in the control of AQP2 trafficking (Schrade et al., 2018).

5.2.2 Role of calcium in AQP2 trafficking

In addition to increasing cAMP levels in collecting duct principal cells, AVP stimulation also results in rapid increased intracellular Ca^{2+} levels. Studies have provided evidences for the involvement of intracellular Ca^{2+} mobilization in vasopressin-mediated AQP2 trafficking, though the mechanisms for this is not completely clear. Ryanodine and calmodulin inhibitors were shown to block vasopressin-stimulated translocation of AQP2 to the plasma membrane (Chou et al., 2000; Levine, Kachadorian, Levin, & Schlondorff, 1981) (Fig. 3). It is reported that calcium is necessary for AQP2 insertion into the plasma membrane since inhibition of the AVP-induced calcium response prevents AQP2 trafficking but does not affect intracellular cAMP levels (Chou et al., 2000). Evidence suggests that elevated intracellular calcium induced by rosiglitazone facilitated AQP2 exposure at the apical plasma membrane, thus increasing collecting duct water permeability in MCD4 renal cells (Procino et al., 2015). However, it is also reported that a very low calcium level is required for the trafficking of AQP2 and intracellular Ca^{2+} chelator did not reduce forskolin-elicited permeability increment (Lorenz et al., 2003). Besides intracellular Ca^{2+} levels, extracellular Ca^{2+} is also known to be involved in AQP2 regulation (Procino et al., 2004; Valenti et al., 2002, 2000). In fact, both PKA and the calcium signaling pathways coordinately modulate AQP2 activity. Exchange protein directly activated by cAMP (Epac) is a key molecule that enhances calcium signaling in response to cAMP and calcium signaling (Ando & Uchida, 2018). Activation of Epac appears to be involved in long-term regulation of AQP2 by dDAVP, but a role of calcium in this regulation was not investigated in a recent in vitro study (Kortenoeven et al., 2012).

5.2.3 AQP2 vesicles targeting to the apical plasma membrane of principal cells

While diffusion is the principal mechanism of molecular movement on a small distance, movement over greater distances requires the participant of the microtubules and actin cytoskeletons (Agutter & Wheatley, 2000; Fuchs & Karakesisoglou, 2001). The translocation of AQP2 bearing vesicles to and from the apical membrane is believed to require organized dynamics of the cytoskeleton and its associated proteins. The microtubule-associated motor protein dynein and dynactin are important in vasopressin-regulated trafficking of AQP2 bearing vesicles. The actin cytoskeleton which anchors the AQP2-bearing vesicles in the unstimulated cell is also essential for the trafficking of AQP2-bearing vesicles.

Dynamic interactions between AQP2 and the actin cytoskeleton are critical for initiating AQP2 apical targeting. Vasopressin or forskolin has been shown to depolymerize F-actin locally, resulting in the fusion of AQP2-carrying vesicles with the apical membrane (Lei et al., 2018; Simon, Gao, Franki, & Hays, 1993; Yui, Lu, Bouley, & Brown, 2012). Recent findings showed that the integrin-linked kinase (ILK), a protein in cytoskeleton regulation, is important in orchestrating dynamic cytoskeletal architecture during recycling of AQP2. In vitro analysis showed that vasopressin-induced AQP2 membrane accumulation was prevented after ILK inhibition, which was associated with increased F-actin polymerization (Mamuya et al., 2016). Consistent with this, ILK conditional-knockdown mice developed polyuria along with decreased AQP2 expression (Cano-Penalver et al., 2014). It is reported that RhoA, a small GTP-binding protein that participates in polymerization of actin, and SPA-1, a GTPase-activating enzyme for Rap1 that is involved in assembly of F-actin, are two proteins required for the trafficking of AQP2 bearing vesicles (Noda et al., 2004; Tamma, Klussmann, et al., 2003). Indeed, AQP2 itself can directly modulate the local actin cytoskeleton depolymerization and subsequent exocytosis. PKA-induced phosphorylation of AQP2 at S256 reduced the direct binding of AQP2 to G-actin, but increased the affinity of AQP2 to myosin-Vb, one of the central regulators in apical trafficking. This interaction causes a reduced binding of myosin-Vb to F-actin, resulting in F-actin destabilization that allows translocation of AQP2 vesicles to the plasma membrane (Noda et al., 2008; Noda, Horikawa, Katayama, & Sasaki, 2005) (Fig. 5). Recently, AKAP220, the A-kinase anchoring protein 220, is demonstrated to control vasopressin-independent aspects of AQP2 trafficking at apical membranes of principal cells via RhoA-mediated modulation of actin-barrier dynamics (Okutsu et al., 2008; Whiting et al., 2016). Deletion of AKAP220 disrupted the apical actin network of the organoids, and resulted in impaired actin-barrier formation and over-accumulation of AQP2 and RhoA at the apical surface of collecting duct cells in the knockout mice, indicating an important role of AKAP220 in regualting AQP2 trafficking through apical actin barrier in the kidney (Whiting et al., 2016).

5.2.4 Docking and fusion of AQP2 bearing vesicles with the apical membrane (exocytosis)

Fusion of AQP2-bearing vesicles is the last step of AQP2 translocated to the apical membrane. Docking and fusing of AQP2-bearing vesicles is mediated by SNARE (Soluble N-ethylmaleimide-sensitive factor attachment protein receptors) mechanisms, involving vesicle v-SNAREs (soluble NSF attachment protein receptors) and target membrane t-SNAREs (Fig. 5). v-SNAREs such as vesicle-associated membrane polypeptide, VAMP-2, VAMP-3, and VAMP-8 are present in the collecting duct principal cells and colocalize with AQP2 in the same vesicles (Franki, Macaluso, Gao, & Hays, 1995; Jo, Harris, Amendt-Raduege, Majewski, & Hammond, 1995; Nielsen et al., 1995; Wang et al., 2010). Tetanus toxin, by cleaving the SNARE protein VAMP-2, abolishes vasopressin-induced AQP2 translocation to the plasma membrane, suggesting an involvement of v-SNARES in AQP2 docking (Gouraud et al., 2002). t-SNAREs, such as syntaxin-4, syntaxin 3, SNAP23, and SNAP25, are observed in the apical membrane of principal cells (Inoue et al., 1998; Mandon, Chou, Nielsen, & Knepper, 1996). Knockdown of Munc18, a protein-inhibiting SNARE mediated membrane fusion, increased AQP2 membrane accumulation, whereas, knockdown of VAMP-2, VAMP-3, syntaxin 3, and SNAP23 inhibited AQP2 fusion at the apical membrane (Procino et al., 2008). The SNARE-associated protein Snapin, which links AQP2 and the t-SNARE complex, can aid AQP2 trafficking from storage vesicles to the apical plasma membrane by association with syntaxin-3 and SNAP23 (Mistry et al., 2009). It is noted that many other proteins (e.g., annexin-2, GTPase, AKT substrate) are also involved in AQP2 trafficking and exocytosis, although their precise roles and how they interact with AQP2 (or AQP2-bearing vesicles) remains to be fully established (Moeller & Fenton, 2012).

5.2.5 *Removal of AQP2 from the apical plasma membrane (endocytosis)*

Removal of AQP2 from the apical plasma membrane (endocytosis) is a way by which AQP2 is recycled. AQP2 accumulates in clathrin-coated pits and is internalized via a clathrin-mediated process (Brown, Weyer, & Orci, 1988; Sun et al., 2002) (Fig. 5). This process depends on a GTPase, dynamin, which is involved in the formation and pinching off of clathrin-coated pits. Heat shock protein 70 (hsp70) and heat shock cognate 70 (hsc70), two AQP2 direct interactors, are involved in uncoating of clathrin-coated vesicles and believed to regulate AQP2 endocytosis (Lu et al., 2007). Mimicking phosphorylation of AQP2 at S256 and S269 decreased their interaction with clathrin, hsp/hsc70 and dynamin, thus induced a decreased rate of AQP2 endocytosis (Lu et al., 2007; Moeller, Praetorius, Rutzler, & Fenton, 2010). Thus, phosphorylation of AQP2 may alter the efficiency of pit maturation and clathrin-coated vesicle, and modulate quantity of AQP2 in clathrin-coated pits and internalization (Fenton, Pedersen, & Moeller, 2013). Caveolin-1, a major protein in membrane microdomain caveolae, was internalized together with AQP2 after forskolin removal in MDCK cells, indicating caveolin-1 or caveolae-dependent pathway in AQP2 internalization (Aoki et al., 2012).

After internalization, AQP2 can either be recycled to the apical membrane upon restimulation or follow the route of degradation via multivesicular bodies and lysosomes (Fig. 3). Some AQP2 transferred to multivesicular body is excreted into the urine as exosomes (Oshikawa, Sonoda, & Ikeda, 2016) (Fig. 4). Regulated endocytosis of AQP2 contributes significantly to final plasma membrane levels of AQP2. Inhibition of endocytosis and subsequent proteosomal and/or lysosomal degradation of AQP2 can increase the amount of AQP2 at the apical membrane, an alternative way to increase the water permeability of collecting ducts.

6. AQP2 protein abundance by vasopressin and others

6.1 Vasopressin-mediated transcriptional regulation of *Aqp2*

In addition to the short-term regulation, vasopressin-mediated long-term regulation of AQP2, which occurs over a period of hours to days, is also essential for water reabsorption and urine concentration in the kidney. The long-term regulation of AQP2 is presented by changing AQP2 protein abundance, largely due to transcriptional regulation of the AQP2 gene (*Aqp2)* in the kidney collecting duct principal cells. Regulation of AQP2 gene expression is mediated via differential DNA binding of transcription factors and co-regulators, as well as via chromatin modification.

It is well known that the 5′-flanking region of *Aqp2* has a cyclic AMP-responsive element (CRE), a TATA box, two GATA consensus sequences, an AP-1 site, an AP-2 site, and three *E*-boxes. Both the CRE and AP1 site are essential for the *Aqp2* promoter activation (Yasui, Zelenin, Celsi, & Aperia, 1997). The activation of PKA stimulated by AVP, causes phosphorylation of a cAMP-response element binding protein (CREB), which binds to the DNA and increases transcription of the *Aqp2*. However, a recent study on accessibility and binding sites for CREB and C/EBPβ (CCAAT/enhancer binding protein-β) in mpkCCD cells, by genome-wide mapping of DNA, identified two enhancer regions pertinent to transcriptional regulation of the *Aqp2* and showed C/EBPβ (but not CREB) binding, indicating that any roles for CREB in the regulation of AQP2 gene transcription is likely to be indirect (Jung, Raghuram, Lee, & Knepper, 2018).

Since AVP can increase intracellular Ca^{2+} levels, it should be noted that calmodulin-activated kinase II and calmodulin-activated kinase IV other than PKA can phosphorylate CREB (Sun, Enslen, Myung, & Maurer, 1994), indicating potential regulation of Ca^{2+} signaling pathways in the transcription of *Aqp2*.

Recent studies quantified vasopressin-induced changes in the nuclear proteome of mpkCCD cells by large-scale proteomics profiling and demonstrated a number of transcription factors, such as JunB, Elf3, and Hmbox1 that have putative binding sites in the 5′-flanking region of the gene coding for the AQP2 (Schenk et al., 2012).

6.2 Other regulators involved in transcriptional regulation of *Aqp2*

6.2.1 Transcriptional factors

AQP2 protein abundance by transcriptional regulation is likely to be combinatorial and integrates the effects of multiple transcriptional regulators. A systems-level analysis of cell-specific *Aqp2* expression in renal collecting duct revealed many transcriptional regulators and transcriptional regulators binding elements that were involved in the transcription of the *Aqp2*. The transcriptional regulators that bind to ETS (Ets-like factors), HOX (homeobox-binding factors), RXR (retinoid X receptor family), CREB, and GATA (GATA-binding factors) of the *Aqp2* are likely to be involved in cell-specific regulation of *Aqp2* expression (Yu et al., 2009), providing further insights into the transcription regulation of the *Aqp2*. Several other transcription factors or molecules, such as NFAT family (TonEBP and NFATc) (Hasler et al., 2006), NF-κB (Hasler et al., 2008), GATA2 (Yu et al., 2014), microRNAs, nuclear receptors etc. have also been reported to involve in this transcriptional regulation. These recent developments reveal that there is far more to transcriptional regulation of AQP2 gene than the classical transcription factor cAMP responsive element-binding protein.

Hypertonicity affects transcription of many genes through the interaction between the tonicity-responsive enhancer (TonE) and its transcription factor TonEBP. TonEBP knockout mice show downregulated protein expression of AQP2, confirming the role of TonE/TonEBP in AQP2 transcription (Hasler et al., 2006). Nuclear factor of activated T cells 5 (NFAT5) is an osmoregulatory transcription factor that directly upregulates the expression of AQP2 in response to calcium and osmotic stress (Li et al., 2007). Stimulation by lipopolysaccharide (LPS) or hypertonicity increased NF-κB activity which negatively regulates AQP2 transcription via directly binding to κB binding sites in the *Aqp2* promoter (Hasler et al., 2008) and likely antagonizes the effects of CRE and TonEBP (Hasler et al., 2008). GATA2, a member of the GATA transcription factor family, was predominantly expressed in the collecting ducts of mouse kidney (Yu et al., 2014). Renal tubular cell-specific deletion of GATA2 induced significantly decreased AQP2 expression and polyuria, indicating an important role of GATA2 as a direct transactivator of AQP2 (Yu et al., 2014).

6.2.2 MicroRNAs

MicroRNAs (miRNAs) are a class of super-regulatory molecules, which post-transcriptionally control the expression of multiple sets of genes that are usually functionally related (Xie et al., 2005). MicroRNAs may act as part of a ribonucleoprotein (RNP) complex to the complementary RNA, which provide a sequence-specific binding component to allow the RNP to act on a particular target (Mohr & Mott, 2015). These findings suggest that by regulating the transcription of target genes, miRNAs may be involved in maintaining water and electrolyte homeostasis in the body by coordinating pivotal biological functions. Recently, the latest study demonstrated that mice with microRNA-132 silence caused acute diuresis, reduced renal total and the apical plasma membrane expression of AQP2 and abrogated increase in AVP levels. This study showed that miRNA-132 play an essential role in AVP production and/or release via a transcriptional repressor methyl CpG-binding protein 2 (MECP2) binding directly to the enhancer region of AVP gene (Murgatroyd et al., 2009), thereby affecting water reabsorption in the kidney (Bijkerk et al., 2018). However, blocking miRNA-132 actually resulted in increased AQP2 levels in vitro, suggesting the possibility of AVP-independent regulation of AQP2 by miRNA-132, which requires further studies (Bijkerk et al., 2018). Another study demonstrated that overexpression of microRNA-32 or microRNA-137, which was not identified as vasopressin-responsive micro-RNAs, decreased dDAVP-regulated AQP2 mRNA and protein levels in mpkCCDc14 cells, indicating that the interaction of miRNAs with the AQP2 regulatory pathway is likely vasopressin-independent (Kim, Jung, Lee, & Kwon, 2015).

6.2.3 Nuclear receptors

AVP-independent increases in AQP2 protein abundance induced by the ligand-activated transcription factor Farnesoid X receptor (FXR) was recently shown to contribute to the regulation of urine volume (Zhang et al., 2014). Activation of FXR dramatically increased AQP2 expression, while FXR knockout mice exhibited a polyuria phenotype with decreased AQP2 expression. Luciferase reporter and Chip assays revealed a putative FXR binding site (FXRE) in AQP2 gene promoter region, suggesting a direct targeting of FXR to the AQP2 gene. Other nuclear receptors such as peroxisome proliferator-activated receptor-γ (PPAR-γ) (Tiwari, Blasi, Heyen, McHarg, & Ecelbarger, 2008; Zhou et al., 2015), Estrogen Receptor Alpha (ERα) (Cheema et al., 2015), glucocorticoid receptor (Chen et al., 2005; Saito et al., 2000), and liver X receptors (Lu et al., 2016; Su et al., 2017) may also capable to regulate renal water balance and urine concentration, which were summarized in a recent review (Zhang, Wang, & Guan, 2016).

7. Pathophysiology of vasopressin-regulated AQP2

Tight regulation of water homeostasis is essential for most physiological processes in mammals. Urinary excretion of water and salt is finely regulated in the kidney to allow a rapid adaptation to water intakes and losses so that the blood osmolality (approximately 300 mOsm/kg·H_2O)

and electrolyte concentration in intra- and extracellular fluids are stably maintained. Increases in plasma osmolality or decreases of the blood volume stimulate release of AVP from posterior pituitary into bloodstream, which promotes AQP2 protein intracellular trafficking and increases AQP2 protein abundance, thus, water is reabsorbed and urine is concentrated in the kidney. Several clinically important water-balance disorders are associated with dysregulation of AQP2, which is supported by a variety of evidence from animal models of diseases and clinical conditions.

7.1 AQP2 in conditions with loss of body water

Diabetes insipidus (DI) is a clinical condition characterized by polyuria, polydipsia and hypotonic urine. The classifications for DI mainly include central DI (CDI), nephrogenic DI (NDI), and others (Kortenoeven & Fenton, 2014; Moeller, Fuglsang, & Fenton, 2016). NDI is further classified as congenital and acquired NDI. Congenital NDI is a rare inherited disorder and presents from birth, due to loss of function mutations in the genes encoding the V2R (AVPR2 gene) or AQP2 (AQP2 gene) (Moeller et al., 2016). Mutation in the AQP2 gene contributes about 10% of NDI cases and more than 50 mutations are currently reported (Moeller et al., 2016; Moeller, Rittig, & Fenton, 2013). These gene mutations may cause dysfunctional AQP2 by different mechanisms, including mis-sorting or misfold of AQP2, or impaired ability facilitating water transport (Kamsteeg et al., 2003; Moeller et al., 2016).

In humans, acquired forms of NDI are much more common. Reduced protein abundance or defective trafficking of AQP2 plays a fundamental role in many acquired forms of NDI, such as sustained lithium intake, ureteral obstruction, hypokalemia, hypercalcemia, acute or chronic kidney injury (Moeller et al., 2016). Each of these conditions is associated with depletion of AQP2 protein from the collecting ducts and a defect in urinary concentration.

7.1.1 Lithium treatment

Lithium-induced NDI is the most common cause of acquired NDI. How lithium causes NDI is still unclear, but studies in animals have helped us understand the potential molecular basis for the disease, it is likely multifactorial. Filtered lithium from glomeruli enters the AQP2 containing collecting duct principal cells via the apical epithelial sodium channel (ENaC) (Christensen et al., 2011; Kortenoeven et al., 2009) resulting in a severe decrease in AQP2 abundance. Several mechanisms appear to contribute to reduced AQP2 protein abundance in lithium-induced NDI. Lithium treatment changes the cellular composition of the collecting duct by decreasing the fraction of principal cells relative to intercalated cells (Christensen et al., 2004), which is likely due to arrest of principal cell proliferation in late G2 phase (de Groot et al., 2014). Lithium is an inhibitor of glycogen synthase kinase type 3β (GSK3β) (Nielsen et al., 2008) which is a critical signaling molecule in urinary concentration (Rao, Patel, Hao, Woodgett, & Harris, 2010). Lithium inactivates adenylyl cyclase and inhibits cAMP production which is likely one of the strongest proposed mechanisms which might be involved in the pathophysiology of lithium-induced NDI (Behl, Kotwani, Kaur, & Goel, 2015). Interestingly, a recent study showed that reduced AQP2 expression was associated with endoplasmic reticulum (ER) stress in the principal cells of the collecting ducts in rats with lithium-induced NDI (Zheng et al., 2016). 4-Phenylbutyric acid is a low molecular weight fatty acid that is approved for clinical treatment of urea cycle disorders. As a molecular chaperon, 4-PBA attenuated ER stress, improved urine concentration capacity, and upregulated AQP2 protein abundance, suggesting the dysfunction of subcellular organelles in lithium-induced NDI (Zheng et al., 2016).

7.1.2 Hypokalemia and hypercalcemia

In these two clinically important causes of NDI, polyuria though is not so marked as that in lithium treatment. Potassium deprivation decreases expression of AQP2 alongside an increase in urine production (Marples, Frokiaer, Dorup, Knepper, & Nielsen, 1996). The precise underlying molecular mechanisms of hypokalemia-induced NDI remain to be resolved. The enhanced autophagic degradation of proteins, e.g., AQP2, is shown to be an early event in hypokalemia-induced NDI, as AQP2 is seen to co-localize with autophagosomes (LC3 as a marker) and autophagolysosomes (LAMP1 as a marker) in the inner medullary collecting duct cells of rats deprived of potassium for only 1 day (Khositseth et al., 2015). In animal models, hypercalcemia is associated with reduced AQP2 expression and increased urine volume (Sands et al., 1997; Wang et al., 2002). The mechanism underlying the decreased AQP2 abundance is proposed to occur via the hypercalciuria which interacts with calcium sensing receptor (CaSR) on the apical membrane of principal cells (Sands et al., 1997). High urinary Ca^{2+} activated CaSR, resulting in decreased sensitivity of the cells to AVP and thus decreased plasma membrane targeting of AQP2 (Bustamante et al., 2008; Procino et al., 2004).

Using a combined proteomics and microscopy approach, autophagy is also shown to contribute to the downregulation of AQP2 in response to hypercalcemia (Khositseth et al., 2017).

7.1.3 Urinary tract obstruction

Animal models of bilateral ureteral obstruction (BUO) or unilateral ureteral obstruction (UUO) have shown reduced AQP2 levels and decreased ability to concentrate urine, which persists for several days after release of the obstruction (Li, Wang, Knepper, Nielsen, & Frokiaer, 2003; Li et al., 2001; Stodkilde et al., 2011). Although the precise mechanisms are unclear, new findings suggest that activation of inflammasome NLRP3 and production of proinflammatory factor IL-1β in the obstructive kidney (by UUO or stone) at least partially induce downregulation of AQP2 protein (Wang et al., 2015, 2018).

7.2 AQP2 in conditions with gain of body water

In many disease conditions, nonosmotic stimulation of AVP release results in alterations of renal AQPs, water retention and hyponatremia. These mainly include congestive heart failure, hepatic cirrhosis, glucocorticoid deficiency, nephrotic syndrome, pregnancy, syndrome of inappropriate ADH secretion (SIADH) (Kortenoeven & Fenton, 2014; Moeller et al., 2016).

7.2.1 Congestive heart failure (CHF) and glucocorticoid deficiency (GD)

In these two conditions (Szatalowicz et al., 1981; Wang, Li, Summer, et al., 2006), AVP is inappropriately released through the activation of a non-osmotic pathway owing to reduced effective circulating blood volume related to arterial underfilling. Increased AVP induces the expression and apical targeting of AQP2 in the collecting duct, thereby facilitating water retention and successive hyponatremia. In rats with CHF or GD, a marked increase in the abundance and apical expression of the AQP2 protein in the collecting ducts is observed. Administration of a V2 receptor antagonist (now it is called tolvaptan) significantly increases urine output and plasma osmolality, in parallel with a significant reduction in AQP2 expression in these animals (Nielsen et al., 1997; Wang, Li, Summer, et al., 2006; Xu et al., 1997).

7.2.2 Hepatic cirrhosis (HC)

HC is a late stage of liver disease, which is usually associated with water retention, hyponatremia and increased AVP levels. The changes in AQP2 protein levels vary considerably among different experimental models. In carbon tetrachloride-induced HC models, increased AQP2 protein and mRNA levels were found, which correlate significantly with the volume of ascites in cirrhotic rats, suggesting a role of AQP2 in abnormal water retention (Asahina et al., 1995; Fernandez-Llama et al., 2000). However, in rats with cirrhosis induced by chronic common bile duct ligation, decreased expression of AQP2 protein was observed, indicating a un-involvement of AQPs in this animal model with excessive water retention and hyponatremia (Fernandez-Llama, Turner, Dibona, & Knepper, 1999). Bile acid, synthesized in the liver, is recently shown to increase AQP2 protein abundance and apical targeting by stimulating its nuclear receptor FXR in collecting ducts (Zhang et al., 2014), whether bile acid play a role in cirrhosis-associated water retention remains unknown.

8. Concluding remarks

To date, at least 13 isoforms of AQPs have been discovered in mammals, which have been thoroughly investigated during the past 30 years after the discovery of the first AQP. We have learned substantial base of knowledge on the structure, cellular localization, biological regulation, physiological function as well as potential pathophysiological significance of these mammalian AQPs, although still some questions need to be solved. Nevertheless, these comprehensive investigations open broad opportunities for the development of aquaporin-based diagnostics and therapeutics, ultimately leading to establish the suitability of AQP-based therapeutics in human diseases.

References

Agre, P., Sasaki, S., & Chrispeels, M. J. (1993). Aquaporins: A family of water channel proteins. *The American Journal of Physiology*, *265*(3 Pt. 2), F461. https://doi.org/10.1152/ajprenal.1993.265.3.F461.

Agutter, P. S., & Wheatley, D. N. (2000). Random walks and cell size. *BioEssays*, *22*(11), 1018–1023. https://doi.org/10.1002/1521-1878(200011)22:11<1018::aid-bies8>3.0.co;2-y.

Alcala, J., Lieska, N., & Maisel, H. (1975). Protein composition of bovine lens cortical fiber cell membranes. *Experimental Eye Research*, *21*(6), 581–595.

Almasalmeh, A., Krenc, D., Wu, B., & Beitz, E. (2014). Structural determinants of the hydrogen peroxide permeability of aquaporins. *The FEBS Journal*, *281*(3), 647–656. https://doi.org/10.1111/febs.12653.

Ando, F., & Uchida, S. (2018). Activation of AQP2 water channels without vasopressin: Therapeutic strategies for congenital nephrogenic diabetes insipidus. *Clinical and Experimental Nephrology*, *22*(3), 501–507. https://doi.org/10.1007/s10157-018-1544-8.

Anthony, T. L., Brooks, H. L., Boassa, D., Leonov, S., Yanochko, G. M., Regan, J. W., et al. (2000). Cloned human aquaporin-1 is a cyclic GMP-gated ion channel. *Molecular Pharmacology*, *57*(3), 576–588.

Aoki, T., Suzuki, T., Hagiwara, H., Kuwahara, M., Sasaki, S., Takata, K., et al. (2012). Close association of aquaporin-2 internalization with caveolin-1. *Acta Histochemica et Cytochemica*, *45*(2), 139–146. https://doi.org/10.1267/ahc.12003.

Arthur, J., Huang, J., Nomura, N., Jin, W. W., Li, W., Cheng, X., et al. (2015). Characterization of the putative phosphorylation sites of the AQP2 C terminus and their role in AQP2 trafficking in LLC-PK1 cells. *American Journal of Physiology. Renal Physiology*, *309*(8), F673–F679. https://doi.org/10.1152/ajprenal.00152.2015.

Asahina, Y., Izumi, N., Enomoto, N., Sasaki, S., Fushimi, K., Marumo, F., et al. (1995). Increased gene expression of water channel in cirrhotic rat kidneys. *Hepatology*, *21*(1), 169–173.

Baumgarten, R., Van De Pol, M. H., Wetzels, J. F., Van Os, C. H., & Deen, P. M. (1998). Glycosylation is not essential for vasopressin-dependent routing of aquaporin-2 in transfected Madin-Darby canine kidney cells. *Journal of the American Society of Nephrology*, *9*(9), 1553–1559.

Behl, T., Kotwani, A., Kaur, I., & Goel, H. (2015). Mechanisms of prolonged lithium therapy-induced nephrogenic diabetes insipidus. *European Journal of Pharmacology*, *755*, 27–33. https://doi.org/10.1016/j.ejphar.2015.02.040.

Benga, G. (2012). On the definition, nomenclature and classification of water channel proteins (aquaporins and relatives). *Molecular Aspects of Medicine*, *33*(5–6), 514–517. https://doi.org/10.1016/j.mam.2012.04.003.

Bienert, G. P., & Chaumont, F. (2014). Aquaporin-facilitated transmembrane diffusion of hydrogen peroxide. *Biochimica et Biophysica Acta*, *1840*(5), 1596–1604. https://doi.org/10.1016/j.bbagen.2013.09.017.

Bijkerk, R., Trimpert, C., van Solingen, C., de Bruin, R. G., Florijn, B. W., Kooijman, S., et al. (2018). MicroRNA-132 controls water homeostasis through regulating MECP2-mediated vasopressin synthesis. *American Journal of Physiology. Renal Physiology*, *315*(4), F1129–F1138. https://doi.org/10.1152/ajprenal.00087.2018.

Bin, K., & Shi-Peng, Z. (2011). Acetazolamide inhibits aquaporin-1 expression and colon cancer xenograft tumor growth. *Hepato-Gastroenterology*, *58*(110 – 111), 1502–1506. https://doi.org/10.5754/hge11154.

Bollag, W. B., Xie, D., Zheng, X., & Zhong, X. (2007). A potential role for the phospholipase D2-aquaporin-3 signaling module in early keratinocyte differentiation: Production of a phosphatidylglycerol signaling lipid. *The Journal of Investigative Dermatology*, *127*(12), 2823–2831. https://doi.org/10.1038/sj.jid.5700921.

Boone, M., Kortenoeven, M., Robben, J. H., & Deen, P. M. (2010). Effect of the cGMP pathway on AQP2 expression and translocation: Potential implications for nephrogenic diabetes insipidus. *Nephrology, Dialysis, Transplantation*, *25*(1), 48–54. https://doi.org/10.1093/ndt/gfp409.

Boone, M., Kortenoeven, M. L., Robben, J. H., Tamma, G., & Deen, P. M. (2011). Counteracting vasopressin-mediated water reabsorption by ATP, dopamine, and phorbol esters: Mechanisms of action. *American Journal of Physiology. Renal Physiology*, *300*(3), F761–F771. https://doi.org/10.1152/ajprenal.00247.2010.

Borgnia, M., Nielsen, S., Engel, A., & Agre, P. (1999). Cellular and molecular biology of the aquaporin water channels. *Annual Review of Biochemistry*, *68*, 425–458. https://doi.org/10.1146/annurev.biochem.68.1.425.

Bouley, R., Breton, S., Sun, T., McLaughlin, M., Nsumu, N. N., Lin, H. Y., et al. (2000). Nitric oxide and atrial natriuretic factor stimulate cGMP-dependent membrane insertion of aquaporin 2 in renal epithelial cells. *The Journal of Clinical Investigation*, *106*(9), 1115–1126. https://doi.org/10.1172/jci9594.

Boury-Jamot, M., Sougrat, R., Tailhardat, M., Le Varlet, B., Bonte, F., Dumas, M., et al. (2006). Expression and function of aquaporins in human skin: Is aquaporin-3 just a glycerol transporter? *Biochimica et Biophysica Acta, 1758*(8), 1034–1042. https://doi.org/10.1016/j.bbamem.2006.06.013.

Brown, D. (2003). The ins and outs of aquaporin-2 trafficking. *American Journal of Physiology. Renal Physiology, 284*(5), F893–F901. https://doi.org/10.1152/ajprenal.00387.2002.

Brown, D., Weyer, P., & Orci, L. (1988). Vasopressin stimulates endocytosis in kidney collecting duct principal cells. *European Journal of Cell Biology, 46*(2), 336–341.

Buck, T. M., Eledge, J., & Skach, W. R. (2004). Evidence for stabilization of aquaporin-2 folding mutants by N-linked glycosylation in endoplasmic reticulum. *American Journal of Physiology. Cell Physiology, 287*(5), C1292–C1299. https://doi.org/10.1152/ajpcell.00561.2003.

Bustamante, M., Hasler, U., Leroy, V., de Seigneux, S., Dimitrov, M., Mordasini, D., et al. (2008). Calcium-sensing receptor attenuates AVP-induced aquaporin-2 expression via a calmodulin-dependent mechanism. *Journal of the American Society of Nephrology, 19*(1), 109–116. https://doi.org/10.1681/asn.2007010092.

Calvanese, L., Pellegrini-Calace, M., & Oliva, R. (2013). In silico study of human aquaporin AQP11 and AQP12 channels. *Protein Science, 22*(4), 455–466. https://doi.org/10.1002/pro.2227.

Cano-Penalver, J. L., Griera, M., Serrano, I., Rodriguez-Puyol, D., Dedhar, S., de Frutos, S., et al. (2014). Integrin-linked kinase regulates tubular aquaporin-2 content and intracellular location: A link between the extracellular matrix and water reabsorption. *The FASEB Journal, 28*(8), 3645–3659. https://doi.org/10.1096/fj.13-249250.

Carbrey, J. M., & Agre, P. (2009). Discovery of the aquaporins and development of the field. *Handbook of Experimental Pharmacology, 190*, 3–28. https://doi.org/10.1007/978-3-540-79885-9_1.

Chandy, G., Zampighi, G. A., Kreman, M., & Hall, J. E. (1997). Comparison of the water transporting properties of MIP and AQP1. *The Journal of Membrane Biology, 159*(1), 29–39.

Cheema, M. U., Irsik, D. L., Wang, Y., Miller-Little, W., Hyndman, K. A., Marks, E. S., et al. (2015). Estradiol regulates AQP2 expression in the collecting duct: A novel inhibitory role for estrogen receptor alpha. *American Journal of Physiology. Renal Physiology, 309*(4), F305–F317. https://doi.org/10.1152/ajprenal.00685.2014.

Chen, Y. C., Cadnapaphornchai, M. A., Summer, S. N., Falk, S., Li, C., Wang, W., et al. (2005). Molecular mechanisms of impaired urinary concentrating ability in glucocorticoid-deficient rats. *Journal of the American Society of Nephrology, 16*(10), 2864–2871. https://doi.org/10.1681/asn.2004110944.

Chou, C. L., Yip, K. P., Michea, L., Kador, K., Ferraris, J. D., Wade, J. B., et al. (2000). Regulation of aquaporin-2 trafficking by vasopressin in the renal collecting duct. Roles of ryanodine-sensitive Ca^{2+} stores and calmodulin. *The Journal of Biological Chemistry, 275*(47), 36839–36846. https://doi.org/10.1074/jbc.M005552200.

Christensen, B. M., Marples, D., Jensen, U. B., Frokiaer, J., Sheikh-Hamad, D., Knepper, M., et al. (1998). Acute effects of vasopressin V2-receptor antagonist on kidney AQP2 expression and subcellular distribution. *The American Journal of Physiology, 275*(2 Pt. 2), F285–F297.

Christensen, B. M., Marples, D., Kim, Y. H., Wang, W., Frokiaer, J., & Nielsen, S. (2004). Changes in cellular composition of kidney collecting duct cells in rats with lithium-induced NDI. *American Journal of Physiology. Cell Physiology, 286*(4), C952–C964. https://doi.org/10.1152/ajpcell.00266.2003.

Christensen, B. M., Zelenina, M., Aperia, A., & Nielsen, S. (2000). Localization and regulation of PKA-phosphorylated AQP2 in response to V(2)-receptor agonist/antagonist treatment. *American Journal of Physiology. Renal Physiology, 278*(1), F29–F42. https://doi.org/10.1152/ajprenal.2000.278.1.F29.

Christensen, B. M., Zuber, A. M., Loffing, J., Stehle, J. C., Deen, P. M., Rossier, B. C., et al. (2011). alphaENaC-mediated lithium absorption promotes nephrogenic diabetes insipidus. *Journal of the American Society of Nephrology*, *22*(2), 253–261. https://doi.org/10.1681/asn.2010070734.

Clemens, D. M., Nemeth-Cahalan, K. L., Trinh, L., Zhang, T., Schilling, T. F., & Hall, J. E. (2013). In vivo analysis of aquaporin 0 function in zebrafish: Permeability regulation is required for lens transparency. *Investigative Ophthalmology & Visual Science*, *54*(7), 5136–5143. https://doi.org/10.1167/iovs.13-12337.

de Groot, T., Alsady, M., Jaklofsky, M., Otte-Holler, I., Baumgarten, R., Giles, R. H., et al. (2014). Lithium causes G2 arrest of renal principal cells. *Journal of the American Society of Nephrology*, *25*(3), 501–510. https://doi.org/10.1681/asn.2013090988.

de Seigneux, S., Nielsen, J., Olesen, E. T., Dimke, H., Kwon, T. H., Frokiaer, J., et al. (2007). Long-term aldosterone treatment induces decreased apical but increased basolateral expression of AQP2 in CCD of rat kidney. *American Journal of Physiology. Renal Physiology*, *293*(1), F87–F99. https://doi.org/10.1152/ajprenal.00431.2006.

Deen, P. M., Verdijk, M. A., Knoers, N. V., Wieringa, B., Monnens, L. A., van Os, C. H., et al. (1994). Requirement of human renal water channel aquaporin-2 for vasopressin-dependent concentration of urine. *Science*, *264*(5155), 92–95.

Denker, B. M., Smith, B. L., Kuhajda, F. P., & Agre, P. (1988). Identification, purification, and partial characterization of a novel Mr 28,000 integral membrane protein from erythrocytes and renal tubules. *The Journal of Biological Chemistry*, *263*(30), 15634–15642.

Domeniconi, R. F., Orsi, A. M., Justulin, L. A., Jr., Leme Beu, C. C., & Felisbino, S. L. (2008). Immunolocalization of aquaporins 1, 2 and 7 in rete testis, efferent ducts, epididymis and vas deferens of adult dog. *Cell and Tissue Research*, *332*(2), 329–335. https://doi.org/10.1007/s00441-008-0592-x.

Ecelbarger, C. A., Terris, J., Frindt, G., Echevarria, M., Marples, D., Nielsen, S., et al. (1995). Aquaporin-3 water channel localization and regulation in rat kidney. *The American Journal of Physiology*, *269*(5 Pt. 2), F663–F672. https://doi.org/10.1152/ajprenal.1995.269.5.F663.

Echevarria, M., Windhager, E. E., Tate, S. S., & Frindt, G. (1994). Cloning and expression of AQP3, a water channel from the medullary collecting duct of rat kidney. *Proceedings of the National Academy of Sciences of the United States of America*, *91*(23), 10997–11001.

Elkjaer, M. L., Nejsum, L. N., Gresz, V., Kwon, T. H., Jensen, U. B., Frokiaer, J., et al. (2001). Immunolocalization of aquaporin-8 in rat kidney, gastrointestinal tract, testis, and airways. *American Journal of Physiology. Renal Physiology*, *281*(6), F1047–F1057. https://doi.org/10.1152/ajprenal.0158.2001.

Eto, K., Noda, Y., Horikawa, S., Uchida, S., & Sasaki, S. (2010). Phosphorylation of aquaporin-2 regulates its water permeability. *The Journal of Biological Chemistry*, *285*(52), 40777–40784. https://doi.org/10.1074/jbc.M110.151928.

Fenton, R. A., Moeller, H. B., Hoffert, J. D., Yu, M. J., Nielsen, S., & Knepper, M. A. (2008). Acute regulation of aquaporin-2 phosphorylation at Ser-264 by vasopressin. *Proceedings of the National Academy of Sciences of the United States of America*, *105*(8), 3134–3139. https://doi.org/10.1073/pnas.0712338105.

Fenton, R. A., Pedersen, C. N., & Moeller, H. B. (2013). New insights into regulated aquaporin-2 function. *Current Opinion in Nephrology and Hypertension*, *22*(5), 551–558. https://doi.org/10.1097/MNH.0b013e328364000d.

Fernandez-Llama, P., Jimenez, W., Bosch-Marce, M., Arroyo, V., Nielsen, S., & Knepper, M. A. (2000). Dysregulation of renal aquaporins and Na-Cl cotransporter in CCl4-induced cirrhosis. *Kidney International*, *58*(1), 216–228. https://doi.org/10.1046/j.1523-1755.2000.00156.x.

Fernandez-Llama, P., Turner, R., Dibona, G., & Knepper, M. A. (1999). Renal expression of aquaporins in liver cirrhosis induced by chronic common bile duct ligation in rats. *Journal of the American Society of Nephrology*, *10*(9), 1950–1957.

Franki, N., Macaluso, F., Gao, Y., & Hays, R. M. (1995). Vesicle fusion proteins in rat inner medullary collecting duct and amphibian bladder. *The American Journal of Physiology*, *268*(3 Pt. 1), C792–C797. https://doi.org/10.1152/ajpcell.1995.268.3.C792.

Fu, D., Libson, A., Miercke, L. J., Weitzman, C., Nollert, P., Krucinski, J., et al. (2000). Structure of a glycerol-conducting channel and the basis for its selectivity. *Science*, *290*(5491), 481–486.

Fuchs, E., & Karakesisoglou, I. (2001). Bridging cytoskeletal intersections. *Genes & Development*, *15*(1), 1–14.

Fushimi, K., Sasaki, S., & Marumo, F. (1997). Phosphorylation of serine 256 is required for cAMP-dependent regulatory exocytosis of the aquaporin-2 water channel. *The Journal of Biological Chemistry*, *272*(23), 14800–14804.

Fushimi, K., Uchida, S., Hara, Y., Hirata, Y., Marumo, F., & Sasaki, S. (1993). Cloning and expression of apical membrane water channel of rat kidney collecting tubule. *Nature*, *361*(6412), 549–552. https://doi.org/10.1038/361549a0.

Gao, M., & Karin, M. (2005). Regulating the regulators: Control of protein ubiquitination and ubiquitin-like modifications by extracellular stimuli. *Molecular Cell*, *19*(5), 581–593. https://doi.org/10.1016/j.molcel.2005.08.017.

Geyer, R. R., Musa-Aziz, R., Qin, X., & Boron, W. F. (2013). Relative CO(2)/NH(3) selectivities of mammalian aquaporins 0-9. *American Journal of Physiology. Cell Physiology*, *304*(10), C985–C994. https://doi.org/10.1152/ajpcell.00033.2013.

Gilbert, M. L., Yang, L., Su, T., & McKnight, G. S. (2015). Expression of a dominant negative PKA mutation in the kidney elicits a diabetes insipidus phenotype. *American Journal of Physiology. Renal Physiology*, *308*(6), F627–F638. https://doi.org/10.1152/ajprenal.00222.2014.

Gonen, T., Sliz, P., Kistler, J., Cheng, Y., & Walz, T. (2004). Aquaporin-0 membrane junctions reveal the structure of a closed water pore. *Nature*, *429*(6988), 193–197. https://doi.org/10.1038/nature02503.

Gorelick, D. A., Praetorius, J., Tsunenari, T., Nielsen, S., & Agre, P. (2006). Aquaporin-11: A channel protein lacking apparent transport function expressed in brain. *BMC Biochemistry*, 7, 14. https://doi.org/10.1186/1471-2091-7-14.

Gouraud, S., Laera, A., Calamita, G., Carmosino, M., Procino, G., Rossetto, O., et al. (2002). Functional involvement of VAMP/synaptobrevin-2 in cAMP-stimulated aquaporin 2 translocation in renal collecting duct cells. *Journal of Cell Science*, *115*(Pt. 18), 3667–3674.

Gunnarson, E., Zelenina, M., & Aperia, A. (2004). Regulation of brain aquaporins. *Neuroscience*, *129*(4), 947–955. https://doi.org/10.1016/j.neuroscience.2004.08.022.

Hara-Chikuma, M., Chikuma, S., Sugiyama, Y., Kabashima, K., Verkman, A. S., Inoue, S., et al. (2012). Chemokine-dependent T cell migration requires aquaporin-3-mediated hydrogen peroxide uptake. *The Journal of Experimental Medicine*, *209*(10), 1743–1752. https://doi.org/10.1084/jem.20112398.

Hara-Chikuma, M., Satooka, H., Watanabe, S., Honda, T., Miyachi, Y., Watanabe, T., et al. (2015). Aquaporin-3-mediated hydrogen peroxide transport is required for NF-kappaB signalling in keratinocytes and development of psoriasis. *Nature Communications*, *6*, 7454. https://doi.org/10.1038/ncomms8454.

Hara-Chikuma, M., Watanabe, S., & Satooka, H. (2016). Involvement of aquaporin-3 in epidermal growth factor receptor signaling via hydrogen peroxide transport in cancer cells. *Biochemical and Biophysical Research Communications*, *471*(4), 603–609. https://doi.org/10.1016/j.bbrc.2016.02.010.

Hasler, U., Jeon, U. S., Kim, J. A., Mordasini, D., Kwon, H. M., Feraille, E., et al. (2006). Tonicity-responsive enhancer binding protein is an essential regulator of aquaporin-2 expression in renal collecting duct principal cells. *Journal of the American Society of Nephrology, 17*(6), 1521–1531. https://doi.org/10.1681/asn.2005121317.

Hasler, U., Leroy, V., Jeon, U. S., Bouley, R., Dimitrov, M., Kim, J. A., et al. (2008). NF-kappaB modulates aquaporin-2 transcription in renal collecting duct principal cells. *The Journal of Biological Chemistry, 283*(42), 28095–28105. https://doi.org/10.1074/jbc.M708350200.

Hendriks, G., Koudijs, M., van Balkom, B. W., Oorschot, V., Klumperman, J., Deen, P. M., et al. (2004). Glycosylation is important for cell surface expression of the water channel aquaporin-2 but is not essential for tetramerization in the endoplasmic reticulum. *The Journal of Biological Chemistry, 279*(4), 2975–2983. https://doi.org/10.1074/jbc.M310767200.

Henn, V., Edemir, B., Stefan, E., Wiesner, B., Lorenz, D., Theilig, F., et al. (2004). Identification of a novel A-kinase anchoring protein 18 isoform and evidence for its role in the vasopressin-induced aquaporin-2 shuttle in renal principal cells. *The Journal of Biological Chemistry, 279*(25), 26654–26665. https://doi.org/10.1074/jbc.M312835200.

Henn, V., Stefan, E., Baillie, G. S., Houslay, M. D., Rosenthal, W., & Klussmann, E. (2005). Compartmentalized cAMP signalling regulates vasopressin-mediated water reabsorption by controlling aquaporin-2. *Biochemical Society Transactions, 33*(Pt. 6), 1316–1318. https://doi.org/10.1042/bst20051316.

Hirano, Y., Okimoto, N., Kadohira, I., Suematsu, M., Yasuoka, K., & Yasui, M. (2010). Molecular mechanisms of how mercury inhibits water permeation through aquaporin-1: Understanding by molecular dynamics simulation. *Biophysical Journal, 98*(8), 1512–1519. https://doi.org/10.1016/j.bpj.2009.12.4310.

Ho, J. D., Yeh, R., Sandstrom, A., Chorny, I., Harries, W. E., Robbins, R. A., et al. (2009). Crystal structure of human aquaporin 4 at 1.8 A and its mechanism of conductance. *Proceedings of the National Academy of Sciences of the United States of America, 106*(18), 7437–7442. https://doi.org/10.1073/pnas.0902725106.

Hoffert, J. D., Fenton, R. A., Moeller, H. B., Simons, B., Tchapyjnikov, D., McDill, B. W., et al. (2008). Vasopressin-stimulated increase in phosphorylation at Ser269 potentiates plasma membrane retention of aquaporin-2. *The Journal of Biological Chemistry, 283*(36), 24617–24627. https://doi.org/10.1074/jbc.M803074200.

Hoffert, J. D., Nielsen, J., Yu, M. J., Pisitkun, T., Schleicher, S. M., Nielsen, S., et al. (2007). Dynamics of aquaporin-2 serine-261 phosphorylation in response to short-term vasopressin treatment in collecting duct. *American Journal of Physiology. Renal Physiology, 292*(2), F691–F700. https://doi.org/10.1152/ajprenal.00284.2006.

Hoffert, J. D., Pisitkun, T., Wang, G., Shen, R. F., & Knepper, M. A. (2006). Quantitative phosphoproteomics of vasopressin-sensitive renal cells: Regulation of aquaporin-2 phosphorylation at two sites. *Proceedings of the National Academy of Sciences of the United States of America, 103*(18), 7159–7164. https://doi.org/10.1073/pnas.0600895103.

Holm, L. M., Klaerke, D. A., & Zeuthen, T. (2004). Aquaporin 6 is permeable to glycerol and urea. *Pflügers Archiv, 448*(2), 181–186. https://doi.org/10.1007/s00424-004-1245-x.

Hua, Y., Ding, S., Zhang, W., Zhou, Q., Ye, W., Chen, M., et al. (2015). Expression of AQP3 protein in hAECs is regulated by cAMP-PKA-CREB signalling pathway. *Frontiers in Bioscience (Landmark edition), 20*, 1047–1055.

Huber, V. J., Tsujita, M., & Nakada, T. (2012). Aquaporins in drug discovery and pharmacotherapy. *Molecular Aspects of Medicine, 33*(5–6), 691–703. https://doi.org/10.1016/j.mam.2012.01.002.

Hunter, T. (2007). The age of crosstalk: Phosphorylation, ubiquitination, and beyond. *Molecular Cell, 28*(5), 730–738. https://doi.org/10.1016/j.molcel.2007.11.019.

Huysseune, S., Kienlen-Campard, P., Hebert, S., Tasiaux, B., Leroy, K., Devuyst, O., et al. (2009). Epigenetic control of aquaporin 1 expression by the amyloid precursor protein. *The FASEB Journal, 23*(12), 4158–4167. https://doi.org/10.1096/fj.09-140012.

Hyndman, K. A., Yang, C. R., Jung, H. J., Umejiego, E. N., Chou, C. L., & Knepper, M. A. (2018). Proteomic determination of the lysine acetylome and phosphoproteome in the rat native inner medullary collecting duct. *Physiological Genomics, 50*(9), 669–679. https://doi.org/10.1152/physiolgenomics.00029.2018.

Ikeda, M., Andoo, A., Shimono, M., Takamatsu, N., Taki, A., Muta, K., et al. (2011). The NPC motif of aquaporin-11, unlike the NPA motif of known aquaporins, is essential for full expression of molecular function. *The Journal of Biological Chemistry, 286*(5), 3342–3350. https://doi.org/10.1074/jbc.M110.180968.

Ikeda, M., Beitz, E., Kozono, D., Guggino, W. B., Agre, P., & Yasui, M. (2002). Characterization of aquaporin-6 as a nitrate channel in mammalian cells. Requirement of pore-lining residue threonine 63. *The Journal of Biological Chemistry, 277*(42), 39873–39879. https://doi.org/10.1074/jbc.M207008200.

Inoue, T., Nielsen, S., Mandon, B., Terris, J., Kishore, B. K., & Knepper, M. A. (1998). SNAP-23 in rat kidney: Colocalization with aquaporin-2 in collecting duct vesicles. *The American Journal of Physiology, 275*(5 Pt 2), F752–F760.

Inoue, Y., Sohara, E., Kobayashi, K., Chiga, M., Rai, T., Ishibashi, K., et al. (2014). Aberrant glycosylation and localization of polycystin-1 cause polycystic kidney in an AQP11 knockout model. *Journal of the American Society of Nephrology, 25*(12), 2789–2799. https://doi.org/10.1681/asn.2013060614.

Ishibashi, K., Hara, S., & Kondo, S. (2009). Aquaporin water channels in mammals. *Clinical and Experimental Nephrology, 13*(2), 107–117. https://doi.org/10.1007/s10157-008-0118-6.

Ishibashi, K., Kuwahara, M., Gu, Y., Kageyama, Y., Tohsaka, A., Suzuki, F., et al. (1997). Cloning and functional expression of a new water channel abundantly expressed in the testis permeable to water, glycerol, and urea. *The Journal of Biological Chemistry, 272*(33), 20782–20786.

Ishibashi, K., Kuwahara, M., Kageyama, Y., Tohsaka, A., Marumo, F., & Sasaki, S. (1997). Cloning and functional expression of a second new aquaporin abundantly expressed in testis. *Biochemical and Biophysical Research Communications, 237*(3), 714–718. https://doi.org/10.1006/bbrc.1997.7219.

Ishibashi, K., Morinaga, T., Kuwahara, M., Sasaki, S., & Imai, M. (2002). Cloning and identification of a new member of water channel (AQP10) as an aquaglyceroporin. *Biochimica et Biophysica Acta, 1576*(3), 335–340.

Ishibashi, K., Sasaki, S., Fushimi, K., Yamamoto, T., Kuwahara, M., & Marumo, F. (1997). Immunolocalization and effect of dehydration on AQP3, a basolateral water channel of kidney collecting ducts. *The American Journal of Physiology, 272*(2 Pt. 2), F235–F241. https://doi.org/10.1152/ajprenal.1997.272.2.F235.

Ishibashi, K., Tanaka, Y., & Morishita, Y. (2014). The role of mammalian superaquaporins inside the cell. *Biochimica et Biophysica Acta, 1840*(5), 1507–1512. https://doi.org/10.1016/j.bbagen.2013.10.039.

Itoh, T., Rai, T., Kuwahara, M., Ko, S. B., Uchida, S., Sasaki, S., et al. (2005). Identification of a novel aquaporin, AQP12, expressed in pancreatic acinar cells. *Biochemical and Biophysical Research Communications, 330*(3), 832–838. https://doi.org/10.1016/j.bbrc.2005.03.046.

Jo, I., Harris, H. W., Amendt-Raduege, A. M., Majewski, R. R., & Hammond, T. G. (1995). Rat kidney papilla contains abundant synaptobrevin protein that participates in the fusion of antidiuretic hormone-regulated water channel-containing endosomes in vitro. *Proceedings of the National Academy of Sciences of the United States of America, 92*(6), 1876–1880.

Jourdain, P., Becq, F., Lengacher, S., Boinot, C., Magistretti, P. J., & Marquet, P. (2014). The human CFTR protein expressed in CHO cells activates aquaporin-3 in a cAMP-dependent pathway: Study by digital holographic microscopy. *Journal of Cell Science, 127*(Pt. 3), 546–556. https://doi.org/10.1242/jcs.133629.

Jung, H. J., & Kwon, T. H. (2016). Molecular mechanisms regulating aquaporin-2 in kidney collecting duct. *American Journal of Physiology. Renal Physiology, 311*(6), F1318–f1328. https://doi.org/10.1152/ajprenal.00485.2016.

Jung, J. S., Preston, G. M., Smith, B. L., Guggino, W. B., & Agre, P. (1994). Molecular structure of the water channel through aquaporin CHIP. The hourglass model. *The Journal of Biological Chemistry, 269*(20), 14648–14654.

Jung, H. J., Raghuram, V., Lee, J. W., & Knepper, M. A. (2018). Genome-wide mapping of DNA accessibility and binding sites for CREB and C/EBPbeta in vasopressin-sensitive collecting duct cells. *Journal of the American Society of Nephrology, 29*(5), 1490–1500. https://doi.org/10.1681/asn.2017050545.

Kamsteeg, E. J., Bichet, D. G., Konings, I. B., Nivet, H., Lonergan, M., Arthus, M. F., et al. (2003). Reversed polarized delivery of an aquaporin-2 mutant causes dominant nephrogenic diabetes insipidus. *The Journal of Cell Biology, 163*(5), 1099–1109. https://doi.org/10.1083/jcb.200309017.

Kamsteeg, E. J., Hendriks, G., Boone, M., Konings, I. B., Oorschot, V., van der Sluijs, P., et al. (2006). Short-chain ubiquitination mediates the regulated endocytosis of the aquaporin-2 water channel. *Proceedings of the National Academy of Sciences of the United States of America, 103*(48), 18344–18349. https://doi.org/10.1073/pnas.0604073103.

Katsura, T., Gustafson, C. E., Ausiello, D. A., & Brown, D. (1997). Protein kinase A phosphorylation is involved in regulated exocytosis of aquaporin-2 in transfected LLC-PK1 cells. *The American Journal of Physiology, 272*(6 Pt. 2), F817–F822.

Khositseth, S., Charngkaew, K., Boonkrai, C., Somparn, P., Uawithya, P., Chomanee, N., et al. (2017). Hypercalcemia induces targeted autophagic degradation of aquaporin-2 at the onset of nephrogenic diabetes insipidus. *Kidney International, 91*(5), 1070–1087. https://doi.org/10.1016/j.kint.2016.12.005.

Khositseth, S., Uawithya, P., Somparn, P., Charngkaew, K., Thippamom, N., Hoffert, J. D., et al. (2015). Autophagic degradation of aquaporin-2 is an early event in hypokalemia-induced nephrogenic diabetes insipidus. *Scientific Reports, 5*, 18311. https://doi.org/10.1038/srep18311.

Kim, H. Y., Choi, H. J., Lim, J. S., Park, E. J., Jung, H. J., Lee, Y. J., et al. (2011). Emerging role of Akt substrate protein AS160 in the regulation of AQP2 translocation. *American Journal of Physiology. Renal Physiology, 301*(1), F151–F161. https://doi.org/10.1152/ajprenal.00519.2010.

Kim, J. E., Jung, H. J., Lee, Y. J., & Kwon, T. H. (2015). Vasopressin-regulated miRNAs and AQP2-targeting miRNAs in kidney collecting duct cells. *American Journal of Physiology. Renal Physiology, 308*(7), F749–F764. https://doi.org/10.1152/ajprenal.00334.2014.

Kittikulsuth, W., Stuart, D., & Kohan, D. E. (2014). Adenylyl cyclase 4 does not regulate collecting duct water and sodium handling. *Physiological Reports, 2*(3), e00277. https://doi.org/10.1002/phy2.277.

Kittikulsuth, W., Stuart, D., Van Hoek, A. N., Stockand, J. D., Bugaj, V., Mironova, E., et al. (2014). Lack of an effect of collecting duct-specific deletion of adenylyl cyclase 3 on renal Na+ and water excretion or arterial pressure. *American Journal of Physiology Renal Physiology, 306*(6), F597–F607. https://doi.org/10.1152/ajprenal.00505.2013.

Klokkers, J., Langehanenberg, P., Kemper, B., Kosmeier, S., von Bally, G., Riethmuller, C., et al. (2009). Atrial natriuretic peptide and nitric oxide signaling antagonizes vasopressin-mediated water permeability in inner medullary collecting duct cells. *American Journal of Physiology. Renal Physiology, 297*(3), F693–F703. https://doi.org/10.1152/ajprenal.00136.2009.

Klussmann, E., Maric, K., Wiesner, B., Beyermann, M., & Rosenthal, W. (1999). Protein kinase A anchoring proteins are required for vasopressin-mediated translocation of aquaporin-2 into cell membranes of renal principal cells. *The Journal of Biological Chemistry, 274*(8), 4934–4938.

Knepper, M. A., Kwon, T. H., & Nielsen, S. (2015). Molecular physiology of water balance. *The New England Journal of Medicine, 372*(14), 1349–1358. https://doi.org/10.1056/NEJMra1404726.

Kortenoeven, M. L., & Fenton, R. A. (2014). Renal aquaporins and water balance disorders. *Biochimica et Biophysica Acta, 1840*(5), 1533–1549. https://doi.org/10.1016/j.bbagen.2013.12.002.

Kortenoeven, M. L., Li, Y., Shaw, S., Gaeggeler, H. P., Rossier, B. C., Wetzels, J. F., et al. (2009). Amiloride blocks lithium entry through the sodium channel thereby attenuating the resultant nephrogenic diabetes insipidus. *Kidney International, 76*(1), 44–53. https://doi.org/10.1038/ki.2009.91.

Kortenoeven, M. L., Trimpert, C., van den Brand, M., Li, Y., Wetzels, J. F., & Deen, P. M. (2012). In mpkCCD cells, long-term regulation of aquaporin-2 by vasopressin occurs independent of protein kinase A and CREB but may involve Epac. *American Journal of Physiology. Renal Physiology, 302*(11), F1395–F1401. https://doi.org/10.1152/ajprenal.00376.2011.

Koyama, N., Ishibashi, K., Kuwahara, M., Inase, N., Ichioka, M., Sasaki, S., et al. (1998). Cloning and functional expression of human aquaporin8 cDNA and analysis of its gene. *Genomics, 54*(1), 169–172. https://doi.org/10.1006/geno.1998.5552.

Lande, M. B., Jo, I., Zeidel, M. L., Somers, M., & Harris, H. W., Jr. (1996). Phosphorylation of aquaporin-2 does not alter the membrane water permeability of rat papillary water channel-containing vesicles. *The Journal of Biological Chemistry, 271*(10), 5552–5557.

Langeberg, L. K., & Scott, J. D. (2015). Signalling scaffolds and local organization of cellular behaviour. *Nature Reviews. Molecular Cell Biology, 16*(4), 232–244. https://doi.org/10.1038/nrm3966.

Lee, Y. J., Lee, J. E., Choi, H. J., Lim, J. S., Jung, H. J., Baek, M. C., et al. (2011). E3 ubiquitin-protein ligases in rat kidney collecting duct: Response to vasopressin stimulation and withdrawal. *American Journal of Physiology. Renal Physiology, 301*(4), F883–F896. https://doi.org/10.1152/ajprenal.00117.2011.

Lei, L., Huang, M., Su, L., Xie, D., Mamuya, F. A., Ham, O., et al. (2018). Manganese promotes intracellular accumulation of AQP2 via modulating F-actin polymerization and reduces urinary concentration in mice. *American Journal of Physiology. Renal Physiology, 314*(2), F306–f316. https://doi.org/10.1152/ajprenal.00391.2017.

Leitch, V., Agre, P., & King, L. S. (2001). Altered ubiquitination and stability of aquaporin-1 in hypertonic stress. *Proceedings of the National Academy of Sciences of the United States of America, 98*(5), 2894–2898. https://doi.org/10.1073/pnas.041616498.

Levine, S. D., Kachadorian, W. A., Levin, D. N., & Schlondorff, D. (1981). Effects of trifluoperazine on function and structure of toad urinary bladder. Role of calmodulin vasopressin-stimulation of water permeability. *The Journal of Clinical Investigation, 67*(3), 662–672. https://doi.org/10.1172/jci110081.

Li, S. Z., McDill, B. W., Kovach, P. A., Ding, L., Go, W. Y., Ho, S. N., et al. (2007). Calcineurin-NFATc signaling pathway regulates AQP2 expression in response to calcium signals and osmotic stress. *American Journal of Physiology. Cell Physiology, 292*(5), C1606–C1616. https://doi.org/10.1152/ajpcell.00588.2005.

Li, J., Patil, R. V., & Verkman, A. S. (2002). Mildly abnormal retinal function in transgenic mice without Muller cell aquaporin-4 water channels. *Investigative Ophthalmology & Visual Science, 43*(2), 573–579.

Li, C., & Wang, W. (2017). Molecular biology of aquaporins. *Advances in Experimental Medicine and Biology, 969*, 1–34. https://doi.org/10.1007/978-94-024-1057-0_1.

Li, C., Wang, W., Knepper, M. A., Nielsen, S., & Frokiaer, J. (2003). Downregulation of renal aquaporins in response to unilateral ureteral obstruction. *American Journal of Physiology. Renal Physiology*, *284*(5), F1066–F1079. https://doi.org/10.1152/ajprenal.00090.2002.

Li, C., Wang, W., Kwon, T. H., Isikay, L., Wen, J. G., Marples, D., et al. (2001). Downregulation of AQP1, -2, and -3 after ureteral obstruction is associated with a long-term urine-concentrating defect. *American Journal of Physiology. Renal Physiology*, *281*(1), F163–F171. https://doi.org/10.1152/ajprenal.2001.281.1.F163.

Li, C., Wang, W., Rivard, C. J., Lanaspa, M. A., Summer, S., & Schrier, R. W. (2011). Molecular mechanisms of angiotensin II stimulation on aquaporin-2 expression and trafficking. *American Journal of Physiology. Renal Physiology*, *300*(5), F1255–F1261. https://doi.org/10.1152/ajprenal.00469.2010.

Li, C., Wang, W., Summer, S. N., Westfall, T. D., Brooks, D. P., Falk, S., et al. (2008). Molecular mechanisms of antidiuretic effect of oxytocin. *Journal of the American Society of Nephrology*, *19*(2), 225–232. https://doi.org/10.1681/asn.2007010029.

Liu, K., Nagase, H., Huang, C. G., Calamita, G., & Agre, P. (2006). Purification and functional characterization of aquaporin-8. *Biology of the Cell*, *98*(3), 153–161. https://doi.org/10.1042/bc20050026.

Lorenz, D., Krylov, A., Hahm, D., Hagen, V., Rosenthal, W., Pohl, P., et al. (2003). Cyclic AMP is sufficient for triggering the exocytic recruitment of aquaporin-2 in renal epithelial cells. *EMBO Reports*, *4*(1), 88–93. https://doi.org/10.1038/sj.embor.embor711.

Lu, H. A., Sun, T. X., Matsuzaki, T., Yi, X. H., Eswara, J., Bouley, R., et al. (2007). Heat shock protein 70 interacts with aquaporin-2 and regulates its trafficking. *The Journal of Biological Chemistry*, *282*(39), 28721–28732. https://doi.org/10.1074/jbc.M611101200.

Lu, X., Wang, F., Xu, C., Soodvilai, S., Peng, K., Su, J., et al. (2016). Soluble (pro)renin receptor via beta-catenin enhances urine concentration capability as a target of liver X receptor. *Proceedings of the National Academy of Sciences of the United States of America*, *113*(13), E1898–E1906. https://doi.org/10.1073/pnas.1602397113.

Lu, D. C., Zhang, H., Zador, Z., & Verkman, A. S. (2008). Impaired olfaction in mice lacking aquaporin-4 water channels. *The FASEB Journal*, *22*(9), 3216–3223. https://doi.org/10.1096/fj.07-104836.

Luo, R., Hu, S., Liu, Q., Han, M., Wang, F., Qiu, M., et al. (2019). Hydrogen sulfide upregulates renal AQP-2 protein expression and promotes urine concentration. *FASEB Journal*, *33*(1), 469–483. https://doi.org/10.1096/fj.201800436R. fj201800436R.

Ma, T., Song, Y., Gillespie, A., Carlson, E. J., Epstein, C. J., & Verkman, A. S. (1999). Defective secretion of saliva in transgenic mice lacking aquaporin-5 water channels. *The Journal of Biological Chemistry*, *274*(29), 20071–20074.

Mamuya, F. A., Cano-Penalver, J. L., Li, W., Rodriguez Puyol, D., Rodriguez Puyol, M., Brown, D., et al. (2016). ILK and cytoskeletal architecture: An important determinant of AQP2 recycling and subsequent entry into the exocytotic pathway. *American Journal of Physiology. Renal Physiology*, *311*(6), F1346–f1357. https://doi.org/10.1152/ajprenal.00336.2016.

Mandon, B., Chou, C. L., Nielsen, S., & Knepper, M. A. (1996). Syntaxin-4 is localized to the apical plasma membrane of rat renal collecting duct cells: Possible role in aquaporin-2 trafficking. *The Journal of Clinical Investigation*, *98*(4), 906–913. https://doi.org/10.1172/jci118873.

Marlar, S., Arnspang, E. C., Koffman, J. S., Locke, E. M., Christensen, B. M., & Nejsum, L. N. (2014). Elevated cAMP increases aquaporin-3 plasma membrane diffusion. *American Journal of Physiology. Cell Physiology*, *306*(6), C598–C606. https://doi.org/10.1152/ajpcell.00132.2013.

Marlar, S., Jensen, H. H., Login, F. H., & Nejsum, L. N. (2017). Aquaporin-3 in Cancer. *International Journal of Molecular Sciences, 18*(10). https://doi.org/10.3390/ijms18102106.

Marples, D., Frokiaer, J., Dorup, J., Knepper, M. A., & Nielsen, S. (1996). Hypokalemia-induced downregulation of aquaporin-2 water channel expression in rat kidney medulla and cortex. *The Journal of Clinical Investigation, 97*(8), 1960–1968. https://doi.org/10.1172/jci118628.

Marples, D., Knepper, M. A., Christensen, E. I., & Nielsen, S. (1995). Redistribution of aquaporin-2 water channels induced by vasopressin in rat kidney inner medullary collecting duct. *The American Journal of Physiology, 269*(3 Pt. 1), C655–C664. https://doi.org/10.1152/ajpcell.1995.269.3.C655.

McDill, B. W., Li, S. Z., Kovach, P. A., Ding, L., & Chen, F. (2006). Congenital progressive hydronephrosis (cph) is caused by an S256L mutation in aquaporin-2 that affects its phosphorylation and apical membrane accumulation. *Proceedings of the National Academy of Sciences of the United States of America, 103*(18), 6952–6957. https://doi.org/10.1073/pnas.0602087103.

McSorley, T., Stefan, E., Henn, V., Wiesner, B., Baillie, G. S., Houslay, M. D., et al. (2006). Spatial organisation of AKAP18 and PDE4 isoforms in renal collecting duct principal cells. *European Journal of Cell Biology, 85*(7), 673–678. https://doi.org/10.1016/j.ejcb.2006.01.005.

Medvar, B., Raghuram, V., Pisitkun, T., Sarkar, A., & Knepper, M. A. (2016). Comprehensive database of human E3 ubiquitin ligases: Application to aquaporin-2 regulation. *Physiological Genomics, 48*(7), 502–512. https://doi.org/10.1152/physiolgenomics.00031.2016.

Michalek, K. (2016). Aquaglyceroporins in the kidney: Present state of knowledge and prospects. *Journal of Physiology and Pharmacology, 67*(2), 185–193.

Milano, S., Carmosino, M., Gerbino, A., Svelto, M., & Procino, G. (2017). Hereditary nephrogenic diabetes insipidus: Pathophysiology and possible treatment. An update. *International Journal of Molecular Sciences, 18*(11). https://doi.org/10.3390/ijms18112385.

Mistry, A. C., Mallick, R., Klein, J. D., Weimbs, T., Sands, J. M., & Frohlich, O. (2009). Syntaxin specificity of aquaporins in the inner medullary collecting duct. *American Journal of Physiology. Renal Physiology, 297*(2), F292–F300. https://doi.org/10.1152/ajprenal.00196.2009.

Moeller, H. B., Aroankins, T. S., Slengerik-Hansen, J., Pisitkun, T., & Fenton, R. A. (2014). Phosphorylation and ubiquitylation are opposing processes that regulate endocytosis of the water channel aquaporin-2. *Journal of Cell Science, 127*(Pt. 14), 3174–3183. https://doi.org/10.1242/jcs.150680.

Moeller, H. B., & Fenton, R. A. (2012). Cell biology of vasopressin-regulated aquaporin-2 trafficking. *Pflügers Archiv, 464*(2), 133–144. https://doi.org/10.1007/s00424-012-1129-4.

Moeller, H. B., Fuglsang, C. H., & Fenton, R. A. (2016). Renal aquaporins and water balance disorders. *Best Practice & Research Clinical Endocrinology & Metabolism, 30*(2), 277–288. https://doi.org/10.1016/j.beem.2016.02.012.

Moeller, H. B., Knepper, M. A., & Fenton, R. A. (2009). Serine 269 phosphorylated aquaporin-2 is targeted to the apical membrane of collecting duct principal cells. *Kidney International, 75*(3), 295–303. https://doi.org/10.1038/ki.2008.505.

Moeller, H. B., MacAulay, N., Knepper, M. A., & Fenton, R. A. (2009). Role of multiple phosphorylation sites in the COOH-terminal tail of aquaporin-2 for water transport: Evidence against channel gating. *American Journal of Physiology. Renal Physiology, 296*(3), F649–F657. https://doi.org/10.1152/ajprenal.90682.2008.

Moeller, H. B., Praetorius, J., Rutzler, M. R., & Fenton, R. A. (2010). Phosphorylation of aquaporin-2 regulates its endocytosis and protein-protein interactions. *Proceedings of the National Academy of Sciences of the United States of America, 107*(1), 424–429. https://doi.org/10.1073/pnas.0910683107.

Moeller, H. B., Rittig, S., & Fenton, R. A. (2013). Nephrogenic diabetes insipidus: Essential insights into the molecular background and potential therapies for treatment. *Endocrine Reviews, 34*(2), 278–301. https://doi.org/10.1210/er.2012-1044.

Mohr, A. M., & Mott, J. L. (2015). Overview of microRNA biology. *Seminars in Liver Disease, 35*(1), 3–11. https://doi.org/10.1055/s-0034-1397344.

Molinas, S. M., Trumper, L., & Marinelli, R. A. (2012). Mitochondrial aquaporin-8 in renal proximal tubule cells: Evidence for a role in the response to metabolic acidosis. *American Journal of Physiology. Renal Physiology, 303*(3), F458–F466. https://doi.org/10.1152/ajprenal.00226.2012.

Moore, M., Ma, T., Yang, B., & Verkman, A. S. (2000). Tear secretion by lacrimal glands in transgenic mice lacking water channels AQP1, AQP3, AQP4 and AQP5. *Experimental Eye Research, 70*(5), 557–562. https://doi.org/10.1006/exer.1999.0814.

Morishita, Y., Matsuzaki, T., Hara-chikuma, M., Andoo, A., Shimono, M., Matsuki, A., et al. (2005). Disruption of aquaporin-11 produces polycystic kidneys following vacuolization of the proximal tubule. *Molecular and Cellular Biology, 25*(17), 7770–7779. https://doi.org/10.1128/mcb.25.17.7770-7779.2005.

Murata, K., Mitsuoka, K., Hirai, T., Walz, T., Agre, P., Heymann, J. B., et al. (2000). Structural determinants of water permeation through aquaporin-1. *Nature, 407*(6804), 599–605. https://doi.org/10.1038/35036519.

Murgatroyd, C., Patchev, A. V., Wu, Y., Micale, V., Bockmuhl, Y., Fischer, D., et al. (2009). Dynamic DNA methylation programs persistent adverse effects of early-life stress. *Nature Neuroscience, 12*(12), 1559–1566. https://doi.org/10.1038/nn.2436.

Nakhoul, N. L., Davis, B. A., Romero, M. F., & Boron, W. F. (1998). Effect of expressing the water channel aquaporin-1 on the CO2 permeability of Xenopus oocytes. *The American Journal of Physiology, 274*(2 Pt. 1), C543–C548.

Nedvetsky, P. I., Tabor, V., Tamma, G., Beulshausen, S., Skroblin, P., Kirschner, A., et al. (2010). Reciprocal regulation of aquaporin-2 abundance and degradation by protein kinase A and p38-MAP kinase. *Journal of the American Society of Nephrology, 21*(10), 1645–1656. https://doi.org/10.1681/asn.2009111190.

Nejsum, L. N., Elkjaer, M., Hager, H., Frokiaer, J., Kwon, T. H., & Nielsen, S. (2000). Localization of aquaporin-7 in rat and mouse kidney using RT-PCR, immunoblotting, and immunocytochemistry. *Biochemical and Biophysical Research Communications, 277*(1), 164–170. https://doi.org/10.1006/bbrc.2000.3638.

Nemeth-Cahalan, K. L., & Hall, J. E. (2000). pH and calcium regulate the water permeability of aquaporin 0. *The Journal of Biological Chemistry, 275*(10), 6777–6782.

Nemeth-Cahalan, K. L., Kalman, K., & Hall, J. E. (2004). Molecular basis of pH and Ca^{2+} regulation of aquaporin water permeability. *The Journal of General Physiology, 123*(5), 573–580. https://doi.org/10.1085/jgp.200308990.

Nico, B., & Ribatti, D. (2010). Aquaporins in tumor growth and angiogenesis. *Cancer Letters, 294*(2), 135–138. https://doi.org/10.1016/j.canlet.2010.02.005.

Nielsen, S., DiGiovanni, S. R., Christensen, E. I., Knepper, M. A., & Harris, H. W. (1993). Cellular and subcellular immunolocalization of vasopressin-regulated water channel in rat kidney. *Proceedings of the National Academy of Sciences of the United States of America, 90*(24), 11663–11667.

Nielsen, S., Frokiaer, J., Marples, D., Kwon, T. H., Agre, P., & Knepper, M. A. (2002). Aquaporins in the kidney: From molecules to medicine. *Physiological Reviews, 82*(1), 205–244. https://doi.org/10.1152/physrev.00024.2001.

Nielsen, J., Hoffert, J. D., Knepper, M. A., Agre, P., Nielsen, S., & Fenton, R. A. (2008). Proteomic analysis of lithium-induced nephrogenic diabetes insipidus: Mechanisms for aquaporin 2 down-regulation and cellular proliferation. *Proceedings of the National Academy of Sciences of the United States of America*, *105*(9), 3634–3639. https://doi.org/10.1073/pnas.0800001105.

Nielsen, S., King, L. S., Christensen, B. M., & Agre, P. (1997). Aquaporins in complex tissues. II. Subcellular distribution in respiratory and glandular tissues of rat. *The American Journal of Physiology*, *273*(5 Pt 1), C1549–C1561.

Nielsen, J., Kwon, T. H., Praetorius, J., Frokiaer, J., Knepper, M. A., & Nielsen, S. (2006). Aldosterone increases urine production and decreases apical AQP2 expression in rats with diabetes insipidus. *American Journal of Physiology. Renal Physiology*, *290*(2), F438–F449. https://doi.org/10.1152/ajprenal.00158.2005.

Nielsen, S., Marples, D., Birn, H., Mohtashami, M., Dalby, N. O., Trimble, M., et al. (1995). Expression of VAMP-2-like protein in kidney collecting duct intracellular vesicles. Colocalization with Aquaporin-2 water channels. *The Journal of Clinical Investigation*, *96*(4), 1834–1844. https://doi.org/10.1172/jci118229.

Nielsen, S., Terris, J., Andersen, D., Ecelbarger, C., Frokiaer, J., Jonassen, T., et al. (1997). Congestive heart failure in rats is associated with increased expression and targeting of aquaporin-2 water channel in collecting duct. *Proceedings of the National Academy of Sciences of the United States of America*, *94*(10), 5450–5455.

Noda, Y., Horikawa, S., Furukawa, T., Hirai, K., Katayama, Y., Asai, T., et al. (2004). Aquaporin-2 trafficking is regulated by PDZ-domain containing protein SPA-1. *FEBS Letters*, *568*(1–3), 139–145. https://doi.org/10.1016/j.febslet.2004.05.021.

Noda, Y., Horikawa, S., Kanda, E., Yamashita, M., Meng, H., Eto, K., et al. (2008). Reciprocal interaction with G-actin and tropomyosin is essential for aquaporin-2 trafficking. *The Journal of Cell Biology*, *182*(3), 587–601. https://doi.org/10.1083/jcb.200709177.

Noda, Y., Horikawa, S., Katayama, Y., & Sasaki, S. (2005). Identification of a multiprotein "motor" complex binding to water channel aquaporin-2. *Biochemical and Biophysical Research Communications*, *330*(4), 1041–1047. https://doi.org/10.1016/j.bbrc.2005.03.079.

Ohshiro, K., Yaoita, E., Yoshida, Y., Fujinaka, H., Matsuki, A., Kamiie, J., et al. (2001). Expression and immunolocalization of AQP6 in intercalated cells of the rat kidney collecting duct. *Archives of Histology and Cytology*, *64*(3), 329–338.

Okutsu, R., Rai, T., Kikuchi, A., Ohno, M., Uchida, K., Sasaki, S., et al. (2008). AKAP220 colocalizes with AQP2 in the inner medullary collecting ducts. *Kidney International*, *74*(11), 1429–1433. https://doi.org/10.1038/ki.2008.402.

Olesen, E. T., & Fenton, R. A. (2017). Aquaporin-2 membrane targeting: Still a conundrum. *American Journal of Physiology. Renal Physiology*, *312*(4), F744–f747. https://doi.org/10.1152/ajprenal.00010.2017.

Olesen, E. T., Rutzler, M. R., Moeller, H. B., Praetorius, H. A., & Fenton, R. A. (2011). Vasopressin-independent targeting of aquaporin-2 by selective E-prostanoid receptor agonists alleviates nephrogenic diabetes insipidus. *Proceedings of the National Academy of Sciences of the United States of America*, *108*(31), 12949–12954. https://doi.org/10.1073/pnas.1104691108.

Oshikawa, S., Sonoda, H., & Ikeda, M. (2016). Aquaporins in urinary extracellular vesicles (exosomes). *International Journal of Molecular Sciences*, *17*(6). https://doi.org/10.3390/ijms17060957.

Prasad, G. V., Coury, L. A., Finn, F., & Zeidel, M. L. (1998). Reconstituted aquaporin 1 water channels transport CO_2 across membranes. *The Journal of Biological Chemistry*, *273*(50), 33123–33126.

Preston, G. M., & Agre, P. (1991). Isolation of the cDNA for erythrocyte integral membrane protein of 28 kilodaltons: Member of an ancient channel family. *Proceedings of the National Academy of Sciences of the United States of America, 88*(24), 11110–11114.

Preston, G. M., Carroll, T. P., Guggino, W. B., & Agre, P. (1992). Appearance of water channels in Xenopus oocytes expressing red cell CHIP28 protein. *Science, 256*(5055), 385–387.

Preston, G. M., Jung, J. S., Guggino, W. B., & Agre, P. (1993). The mercury-sensitive residue at cysteine 189 in the CHIP28 water channel. *The Journal of Biological Chemistry, 268*(1), 17–20.

Procino, G., Barbieri, C., Tamma, G., De Benedictis, L., Pessin, J. E., Svelto, M., et al. (2008). AQP2 exocytosis in the renal collecting duct—Involvement of SNARE isoforms and the regulatory role of Munc18b. *Journal of Cell Science, 121*(Pt. 12), 2097–2106. https://doi.org/10.1242/jcs.022210.

Procino, G., Carmosino, M., Marin, O., Brunati, A. M., Contri, A., Pinna, L. A., et al. (2003). Ser-256 phosphorylation dynamics of aquaporin 2 during maturation from the ER to the vesicular compartment in renal cells. *The FASEB Journal, 17*(13), 1886–1888. https://doi.org/10.1096/fj.02-0870fje.

Procino, G., Carmosino, M., Tamma, G., Gouraud, S., Laera, A., Riccardi, D., et al. (2004). Extracellular calcium antagonizes forskolin-induced aquaporin 2 trafficking in collecting duct cells. *Kidney International, 66*(6), 2245–2255. https://doi.org/10.1111/j.1523-1755.2004.66036.x.

Procino, G., Gerbino, A., Milano, S., Nicoletti, M. C., Mastrofrancesco, L., Carmosino, M., et al. (2015). Rosiglitazone promotes AQP2 plasma membrane expression in renal cells via a Ca-dependent/cAMP-independent mechanism. *Cellular Physiology and Biochemistry, 35*(3), 1070–1085. https://doi.org/10.1159/000373933.

Procino, G., Mastrofrancesco, L., Sallustio, F., Costantino, V., Barbieri, C., Pisani, F., et al. (2011). AQP5 is expressed in type-B intercalated cells in the collecting duct system of the rat, mouse and human kidney. *Cellular Physiology and Biochemistry, 28*(4), 683–692. https://doi.org/10.1159/000335762.

Rao, R., Patel, S., Hao, C., Woodgett, J., & Harris, R. (2010). GSK3beta mediates renal response to vasopressin by modulating adenylate cyclase activity. *Journal of the American Society of Nephrology, 21*(3), 428–437. https://doi.org/10.1681/asn.2009060672.

Rieg, T., Tang, T., Murray, F., Schroth, J., Insel, P. A., Fenton, R. A., et al. (2010). Adenylate cyclase 6 determines cAMP formation and aquaporin-2 phosphorylation and trafficking in inner medulla. *Journal of the American Society of Nephrology, 21*(12), 2059–2068. https://doi.org/10.1681/asn.2010040409.

Roos, K. P., Strait, K. A., Raphael, K. L., Blount, M. A., & Kohan, D. E. (2012). Collecting duct-specific knockout of adenylyl cyclase type VI causes a urinary concentration defect in mice. *American Journal of Physiology. Renal Physiology, 302*(1), F78–F84. https://doi.org/10.1152/ajprenal.00397.2011.

Sachdeva, R., & Singh, B. (2014). Phosphorylation of Ser-180 of rat aquaporin-4 shows marginal affect on regulation of water permeability: Molecular dynamics study. *Journal of Biomolecular Structure & Dynamics, 32*(4), 555–566. https://doi.org/10.1080/07391102.2013.780981.

Saito, T., Ishikawa, S. E., Ando, F., Higashiyama, M., Nagasaka, S., & Sasaki, S. (2000). Vasopressin-dependent upregulation of aquaporin-2 gene expression in glucocorticoid-deficient rats. *American Journal of Physiology. Renal Physiology, 279*(3), F502–F508. https://doi.org/10.1152/ajprenal.2000.279.3.F502.

Sandoval, P. C., Slentz, D. H., Pisitkun, T., Saeed, F., Hoffert, J. D., & Knepper, M. A. (2013). Proteome-wide measurement of protein half-lives and translation rates in vasopressin-sensitive collecting duct cells. *Journal of the American Society of Nephrology, 24*(11), 1793–1805. https://doi.org/10.1681/asn.2013030279.

Sands, J. M., Naruse, M., Baum, M., Jo, I., Hebert, S. C., Brown, E. M., et al. (1997). Apical extracellular calcium/polyvalent cation-sensing receptor regulates vasopressin-elicited water permeability in rat kidney inner medullary collecting duct. *The Journal of Clinical Investigation*, *99*(6), 1399–1405. https://doi.org/10.1172/jci119299.

Saparov, S. M., Liu, K., Agre, P., & Pohl, P. (2007). Fast and selective ammonia transport by aquaporin-8. *The Journal of Biological Chemistry*, *282*(8), 5296–5301. https://doi.org/10.1074/jbc.M609343200.

Satooka, H., & Hara-Chikuma, M. (2016). Aquaporin-3 controls breast Cancer cell migration by regulating hydrogen peroxide transport and its downstream cell signaling. *Molecular and Cellular Biology*, *36*(7), 1206–1218. https://doi.org/10.1128/mcb.00971-15.

Schenk, L. K., Bolger, S. J., Luginbuhl, K., Gonzales, P. A., Rinschen, M. M., Yu, M. J., et al. (2012). Quantitative proteomics identifies vasopressin-responsive nuclear proteins in collecting duct cells. *Journal of the American Society of Nephrology*, *23*(6), 1008–1018. https://doi.org/10.1681/asn.2011070738.

Schrade, K., Troger, J., Eldahshan, A., Zuhlke, K., Abdul Azeez, K. R., Elkins, J. M., et al. (2018). An AKAP-Lbc-RhoA interaction inhibitor promotes the translocation of aquaporin-2 to the plasma membrane of renal collecting duct principal cells. *PLoS One*, *13*(1). e0191423https://doi.org/10.1371/journal.pone.0191423.

Simon, H., Gao, Y., Franki, N., & Hays, R. M. (1993). Vasopressin depolymerizes apical F-actin in rat inner medullary collecting duct. *The American Journal of Physiology*, *265*(3 Pt. 1), C757–C762. https://doi.org/10.1152/ajpcell.1993.265.3.C757.

Smith, B. L., & Agre, P. (1991). Erythrocyte Mr 28,000 transmembrane protein exists as a multisubunit oligomer similar to channel proteins. *The Journal of Biological Chemistry*, *266*(10), 6407–6415.

Sohara, E., Rai, T., Miyazaki, J., Verkman, A. S., Sasaki, S., & Uchida, S. (2005). Defective water and glycerol transport in the proximal tubules of AQP7 knockout mice. *American Journal of Physiology. Renal Physiology*, *289*(6), F1195–F1200. https://doi.org/10.1152/ajprenal.00133.2005.

Song, Y., Sonawane, N., & Verkman, A. S. (2002). Localization of aquaporin-5 in sweat glands and functional analysis using knockout mice. *The Journal of Physiology*, *541*(Pt. 2), 561–568.

Song, Y., & Verkman, A. S. (2001). Aquaporin-5 dependent fluid secretion in airway submucosal glands. *The Journal of Biological Chemistry*, *276*(44), 41288–41292. https://doi.org/10.1074/jbc.M107257200.

Soria, L. R., Fanelli, E., Altamura, N., Svelto, M., Marinelli, R. A., & Calamita, G. (2010). Aquaporin-8-facilitated mitochondrial ammonia transport. *Biochemical and Biophysical Research Communications*, *393*(2), 217–221. https://doi.org/10.1016/j.bbrc.2010.01.104.

Stodkilde, L., Norregaard, R., Fenton, R. A., Wang, G., Knepper, M. A., & Frokiaer, J. (2011). Bilateral ureteral obstruction induces early downregulation and redistribution of AQP2 and phosphorylated AQP2. *American Journal of Physiology. Renal Physiology*, *301*(1), F226–F235. https://doi.org/10.1152/ajprenal.00664.2010.

Su, W., Huang, S. Z., Gao, M., Kong, X. M., Gustafsson, J. A., Xu, S. J., et al. (2017). Liver X receptor beta increases aquaporin 2 protein level via a posttranscriptional mechanism in renal collecting ducts. *American Journal of Physiology. Renal Physiology*, *312*(4), F619–f628. https://doi.org/10.1152/ajprenal.00564.2016.

Sui, H., Han, B. G., Lee, J. K., Walian, P., & Jap, B. K. (2001). Structural basis of water-specific transport through the AQP1 water channel. *Nature*, *414*(6866), 872–878. https://doi.org/10.1038/414872a.

Sun, P., Enslen, H., Myung, P. S., & Maurer, R. A. (1994). Differential activation of CREB by Ca2+/calmodulin-dependent protein kinases type II and type IV involves phosphorylation of a site that negatively regulates activity. *Genes & Development*, *8*(21), 2527–2539.

Sun, T. X., Van Hoek, A., Huang, Y., Bouley, R., McLaughlin, M., & Brown, D. (2002). Aquaporin-2 localization in clathrin-coated pits: Inhibition of endocytosis by dominant-negative dynamin. *American Journal of Physiology. Renal Physiology*, *282*(6), F998–1011. https://doi.org/10.1152/ajprenal.00257.2001.

Szatalowicz, V. L., Arnold, P. E., Chaimovitz, C., Bichet, D., Berl, T., & Schrier, R. W. (1981). Radioimmunoassay of plasma arginine vasopressin in hyponatremic patients with congestive heart failure. *The New England Journal of Medicine*, *305*(5), 263–266. https://doi.org/10.1056/nejm198107303050506.

Taguchi, D., Takeda, T., Kakigi, A., Takumida, M., Nishioka, R., & Kitano, H. (2007). Expressions of aquaporin-2, vasopressin type 2 receptor, transient receptor potential channel vanilloid (TRPV)1, and TRPV4 in the human endolymphatic sac. *Laryngoscope*, *117*(4), 695–698. https://doi.org/10.1097/mlg.0b013e318031c802.

Tamma, G., Carmosino, M., Svelto, M., & Valenti, G. (2005). Bradykinin signaling counteracts cAMP-elicited aquaporin 2 translocation in renal cells. *Journal of the American Society of Nephrology*, *16*(10), 2881–2889. https://doi.org/10.1681/asn.2005020190.

Tamma, G., Klussmann, E., Procino, G., Svelto, M., Rosenthal, W., & Valenti, G. (2003). cAMP-induced AQP2 translocation is associated with RhoA inhibition through RhoA phosphorylation and interaction with RhoGDI. *Journal of Cell Science*, *116*(Pt. 8), 1519–1525.

Tamma, G., Ranieri, M., Di Mise, A., Centrone, M., Svelto, M., & Valenti, G. (2014). Glutathionylation of the aquaporin-2 water channel: A novel post-translational modification modulated by the oxidative stress. *The Journal of Biological Chemistry*, *289*(40), 27807–27813. https://doi.org/10.1074/jbc.M114.586024.

Tamma, G., Robben, J. H., Trimpert, C., Boone, M., & Deen, P. M. (2011). Regulation of AQP2 localization by S256 and S261 phosphorylation and ubiquitination. *American Journal of Physiology. Cell Physiology*, *300*(3), C636–C646. https://doi.org/10.1152/ajpcell.00433.2009.

Tamma, G., Wiesner, B., Furkert, J., Hahm, D., Oksche, A., Schaefer, M., et al. (2003). The prostaglandin E2 analogue sulprostone antagonizes vasopressin-induced antidiuresis through activation of Rho. *Journal of Cell Science*, *116*(Pt. 16), 3285–3294. https://doi.org/10.1242/jcs.00640.

Tiwari, S., Blasi, E. R., Heyen, J. R., McHarg, A. D., & Ecelbarger, C. M. (2008). Time course of AQP-2 and ENaC regulation in the kidney in response to PPAR agonists associated with marked edema in rats. *Pharmacological Research*, *57*(5), 383–392. https://doi.org/10.1016/j.phrs.2008.03.013.

Tornroth-Horsefield, S., Hedfalk, K., Fischer, G., Lindkvist-Petersson, K., & Neutze, R. (2010). Structural insights into eukaryotic aquaporin regulation. *FEBS Letters*, *584*(12), 2580–2588. https://doi.org/10.1016/j.febslet.2010.04.037.

Tornroth-Horsefield, S., Wang, Y., Hedfalk, K., Johanson, U., Karlsson, M., Tajkhorshid, E., et al. (2006). Structural mechanism of plant aquaporin gating. *Nature*, *439*(7077), 688–694. https://doi.org/10.1038/nature04316.

Tournaire-Roux, C., Sutka, M., Javot, H., Gout, E., Gerbeau, P., Luu, D. T., et al. (2003). Cytosolic pH regulates root water transport during anoxic stress through gating of aquaporins. *Nature*, *425*(6956), 393–397. https://doi.org/10.1038/nature01853.

Trimpert, C., Wesche, D., de Groot, T., Pimentel Rodriguez, M. M., Wong, V., van den Berg, D. T. M., et al. (2017). NDFIP allows NEDD4/NEDD4L-induced AQP2 ubiquitination and degradation. *PLoS One*, *12*(9), e0183774. https://doi.org/10.1371/journal.pone.0183774.

Tsukaguchi, H., Weremowicz, S., Morton, C. C., & Hediger, M. A. (1999). Functional and molecular characterization of the human neutral solute channel aquaporin-9. *The American Journal of Physiology*, *277*(5 Pt. 2), F685–F696.

Valenti, G., Laera, A., Gouraud, S., Pace, G., Aceto, G., Penza, R., et al. (2002). Low-calcium diet in hypercalciuric enuretic children restores AQP2 excretion and improves clinical symptoms. *American Journal of Physiology. Renal Physiology*, *283*(5), F895–F903. https://doi.org/10.1152/ajprenal.00354.2001.

Valenti, G., Laera, A., Pace, G., Aceto, G., Lospalluti, M. L., Penza, R., et al. (2000). Urinary aquaporin 2 and calciuria correlate with the severity of enuresis in children. *Journal of the American Society of Nephrology*, *11*(10), 1873–1881.

van Hoek, A. N., Ma, T., Yang, B., Verkman, A. S., & Brown, D. (2000). Aquaporin-4 is expressed in basolateral membranes of proximal tubule S3 segments in mouse kidney. *American Journal of Physiology. Renal Physiology*, *278*(2), F310–F316. https://doi.org/10.1152/ajprenal.2000.278.2.F310.

Vukicevic, T., Schulz, M., Faust, D., & Klussmann, E. (2016). The trafficking of the water channel aquaporin-2 in renal principal cells—a potential target for pharmacological intervention in cardiovascular diseases. *Frontiers in Pharmacology*, 7, 23. https://doi.org/10.3389/fphar.2016.00023.

Wang, W., Li, C., Kwon, T. H., Knepper, M. A., Frokiaer, J., & Nielsen, S. (2002). AQP3, p-AQP2, and AQP2 expression is reduced in polyuric rats with hypercalcemia: Prevention by cAMP-PDE inhibitors. *American Journal of Physiology. Renal Physiology*, *283*(6), F1313–F1325. https://doi.org/10.1152/ajprenal.00040.2002.

Wang, W., Li, C., Nejsum, L. N., Li, H., Kim, S. W., Kwon, T. H., et al. (2006). Biphasic effects of ANP infusion in conscious, euvolumic rats: Roles of AQP2 and ENaC trafficking. *American Journal of Physiology. Renal Physiology*, *290*(2), F530–F541. https://doi.org/10.1152/ajprenal.00070.2005.

Wang, W., Li, C., Summer, S. N., Falk, S., Cadnapaphornchai, M. A., Chen, Y. C., et al. (2006). Molecular analysis of impaired urinary diluting capacity in glucocorticoid deficiency. *American Journal of Physiology. Renal Physiology*, *290*(5), F1135–F1142. https://doi.org/10.1152/ajprenal.00356.2005.

Wang, F., Liu, Q., Jin, L., Hu, S., Luo, R., Han, M., et al. (2018). Combination exposure of melamine and cyanuric acid is associated with polyuria and activation of NLRP3 inflammasome in rats. *American Journal of Physiology. Renal Physiology*, *315*(2), F199–f210. https://doi.org/10.1152/ajprenal.00609.2017.

Wang, W., Luo, R., Lin, Y., Wang, F., Zheng, P., Levi, M., et al. (2015). Aliskiren restores renal AQP2 expression during unilateral ureteral obstruction by inhibiting the inflammasome. *American Journal of Physiology. Renal Physiology*, *308*(8), F910–F922. https://doi.org/10.1152/ajprenal.00649.2014.

Wang, C. C., Ng, C. P., Shi, H., Liew, H. C., Guo, K., Zeng, Q., et al. (2010). A role for VAMP8/endobrevin in surface deployment of the water channel aquaporin 2. *Molecular and Cellular Biology*, *30*(1), 333–343. https://doi.org/10.1128/mcb.00814-09.

Whiting, J. L., Ogier, L., Forbush, K. A., Bucko, P., Gopalan, J., Seternes, O. M., et al. (2016). AKAP220 manages apical actin networks that coordinate aquaporin-2 location and renal water reabsorption. *Proceedings of the National Academy of Sciences of the United States of America*, *113*(30), E4328–E4337. https://doi.org/10.1073/pnas.1607745113.

Xie, X., Lu, J., Kulbokas, E. J., Golub, T. R., Mootha, V., Lindblad-Toh, K., et al. (2005). Systematic discovery of regulatory motifs in human promoters and 3' UTRs by comparison of several mammals. *Nature*, *434*(7031), 338–345. https://doi.org/10.1038/nature03441.

Xu, D. L., Martin, P. Y., Ohara, M., St John, J., Pattison, T., Meng, X., et al. (1997). Upregulation of aquaporin-2 water channel expression in chronic heart failure rat. *The Journal of Clinical Investigation, 99*(7), 1500–1505. https://doi.org/10.1172/jci119312.

Yakata, K., Hiroaki, Y., Ishibashi, K., Sohara, E., Sasaki, S., Mitsuoka, K., et al. (2007). Aquaporin-11 containing a divergent NPA motif has normal water channel activity. *Biochimica et Biophysica Acta, 1768*(3), 688–693. https://doi.org/10.1016/j.bbamem.2006.11.005.

Yamamura, Y., Motegi, K., Kani, K., Takano, H., Momota, Y., Aota, K., et al. (2012). TNF-alpha inhibits aquaporin 5 expression in human salivary gland acinar cells via suppression of histone H4 acetylation. *Journal of Cellular and Molecular Medicine, 16*(8), 1766–1775. https://doi.org/10.1111/j.1582-4934.2011.01456.x.

Yang, B., & Verkman, A. S. (1997). Water and glycerol permeabilities of aquaporins 1-5 and MIP determined quantitatively by expression of epitope-tagged constructs in Xenopus oocytes. *The Journal of Biological Chemistry, 272*(26), 16140–16146.

Yasui, M., Hazama, A., Kwon, T. H., Nielsen, S., Guggino, W. B., & Agre, P. (1999). Rapid gating and anion permeability of an intracellular aquaporin. *Nature, 402*(6758), 184–187. https://doi.org/10.1038/46045.

Yasui, M., Kwon, T. H., Knepper, M. A., Nielsen, S., & Agre, P. (1999). Aquaporin-6: An intracellular vesicle water channel protein in renal epithelia. *Proceedings of the National Academy of Sciences of the United States of America, 96*(10), 5808–5813.

Yasui, M., Zelenin, S. M., Celsi, G., & Aperia, A. (1997). Adenylate cyclase-coupled vasopressin receptor activates AQP2 promoter via a dual effect on CRE and AP1 elements. *The American Journal of Physiology, 272*(4 Pt. 2), F443–F450. https://doi.org/10.1152/ajprenal.1997.272.4.F443.

Yu, M. J., Miller, R. L., Uawithya, P., Rinschen, M. M., Khositseth, S., Braucht, D. W., et al. (2009). Systems-level analysis of cell-specific AQP2 gene expression in renal collecting duct. *Proceedings of the National Academy of Sciences of the United States of America, 106*(7), 2441–2446. https://doi.org/10.1073/pnas.0813002106.

Yu, L., Moriguchi, T., Souma, T., Takai, J., Satoh, H., Morito, N., et al. (2014). GATA2 regulates body water homeostasis through maintaining aquaporin 2 expression in renal collecting ducts. *Molecular and Cellular Biology, 34*(11), 1929–1941. https://doi.org/10.1128/mcb.01659-13.

Yui, N., Lu, H. J., Bouley, R., & Brown, D. (2012). AQP2 is necessary for vasopressin- and forskolin-mediated filamentous actin depolymerization in renal epithelial cells. *Biology Open, 1*(2), 101–108. https://doi.org/10.1242/bio.2011042.

Zeidel, M. L., Ambudkar, S. V., Smith, B. L., & Agre, P. (1992). Reconstitution of functional water channels in liposomes containing purified red cell CHIP28 protein. *Biochemistry, 31*(33), 7436–7440.

Zelenina, M., Christensen, B. M., Palmer, J., Nairn, A. C., Nielsen, S., & Aperia, A. (2000). Prostaglandin E(2) interaction with AVP: Effects on AQP2 phosphorylation and distribution. *American Journal of Physiology. Renal Physiology, 278*(3), F388–F394. https://doi.org/10.1152/ajprenal.2000.278.3.F388.

Zelenina, M., Zelenin, S., Bondar, A. A., Brismar, H., & Aperia, A. (2002). Water permeability of aquaporin-4 is decreased by protein kinase C and dopamine. *American Journal of Physiology. Renal Physiology, 283*(2), F309–F318. https://doi.org/10.1152/ajprenal.00260.2001.

Zhang, X., Huang, S., Gao, M., Liu, J., Jia, X., Han, Q., et al. (2014). Farnesoid X receptor (FXR) gene deficiency impairs urine concentration in mice. *Proceedings of the National Academy of Sciences of the United States of America, 111*(6), 2277–2282. https://doi.org/10.1073/pnas.1323977111.

Zhang, X. Y., Wang, B., & Guan, Y. F. (2016). Nuclear receptor regulation of aquaporin-2 in the kidney. *International Journal of Molecular Sciences*, *17*(7). https://doi.org/10.3390/ijms17071105.

Zheng, P., Lin, Y., Wang, F., Luo, R., Zhang, T., Hu, S., et al. (2016). 4-PBA improves lithium-induced nephrogenic diabetes insipidus by attenuating ER stress. *American Journal of Physiology. Renal Physiology*, *311*(4), F763–f776. https://doi.org/10.1152/ajprenal.00225.2016.

Zhou, L., Panasiuk, A., Downton, M., Zhao, D., Yang, B., Jia, Z., et al. (2015). Systemic PPARgamma deletion causes severe disturbance in fluid homeostasis in mice. *Physiological Genomics*, *47*(11), 541–547. https://doi.org/10.1152/physiolgenomics.00066.2015.

CHAPTER SEVEN

The vasotocinergic system and its role in the regulation of stress in birds

Wayne J. Kuenzel[a,*], Seong W. Kang[a], Alexander Jurkevich[b]
[a]Poultry Science Center, University of Arkansas, Fayetteville, AR, United States
[b]Molecular Cytology Research Core Facility, University of Missouri, Columbia, MO, United States
*Corresponding author: e-mail address: wkuenzel@uark.edu

Contents

Vitamins and Hormones, Volume 113
ISSN 0083-6729
https://doi.org/10.1016/bs.vh.2019.10.004

Abstract

The regulation of stress in birds includes a complex interaction of neural systems affecting the hypothalamic-pituitary-adrenal (HPA) axis. In addition to the hypothalamic paraventricular nucleus, a structure called the nucleus of the hippocampal commissure likewise affects the output of pituitary stress hormones and appears to be unique to avian species. Within the anterior pituitary, the avian V1a and V1b receptors were found in corticotropes. Based on our studies with central administration of hormones in the chicken, corticotropic releasing hormone (CRH) is a more potent ACTH secretagogue than arginine vasotocin (AVT). In contrast, when applied peripherally, AVT is more efficacious. Co-administration of AVT and CRH peripherally, resulted in a synergistic stimulation of corticosterone release. Data suggest receptor oligomerization as one possible mechanism. In birds, vasotocin receptors associated with stress responses include the V1a and V1b receptors. Three-dimensional, homology-based structural models of the avian V1aR were built to test agonists and antagonists for each receptor that were screened by molecular docking to map their binding sites on each receptor. Additionally, binding affinity values for each available peptide antagonist to the V1aR and V1bR were determined. An anterior pituitary primary culture system was developed to determine how effective each antagonist blocked the function of each receptor in culture when stimulated by a combination of AVT/CRH administration. Use of an antagonist in subsequent *in vivo* studies identified the V1aR in regulating food intake in birds. The V1aR was likewise found in circumventricular organs of the brain, suggesting a possible function in stress.

1. Introduction

Stress is an important organismal response to environmental changes. It plays a critical role in adaptation to variable and often unpredictable environmental challenges. Among all vertebrates, physiological mechanisms of stress responses are remarkably conserved. At the heart of these mechanisms lies the HPA axis, a set of structures interconnected by signaling molecules into a neuroendocrine network. Two hypothalamic hormones, corticotropic releasing hormone (CRH) and arginine vasotocin (AVT), substituted by vasopressin (AVP) in mammals, play a central role in the regulation of the HPA axis. AVT/AVP and CRH belong to two ancient families of peptides: the AVT/AVP/mesotocin/oxytocin and CRH/urocortin families. Both peptide families developed as antagonistic regulators of an essential homeostatic mechanism associated with osmoregulation, the former controlled retention of water while the latter controlled excretion of body fluids (Acher, 1993; Fekete & Zorrilla, 2007). Their coupling with the regulation of the HPA axis may be a mechanism that evolved later and is unique to chordates (Chang & Hsu, 2004; Lovejoy, Chang, Lovejoy, & del Castillo, 2014). Vasotocin-dependent mechanisms of stress responses will be addressed in this review with an emphasis on avian species.

Investigations utilizing birds have provided exciting opportunities to explore a variety of regulatory processes underlying stress responses in vertebrates. In their natural habitats birds are exposed to stressors that occur unexpectedly due to sudden storms, extreme weather effects, environmental pollution, global warming and human disturbance, among others (Silverin, 1998). In contrast, a few domesticated avian species, raised commercially for food, face different challenges associated with their nutritional requirements, metabolic responses, diseases, crowding and extensive farming management practices. Integration of data obtained studying stress responses of birds living in natural settings and in poultry farms can reveal general principles of the avian stress response as well as specific adaptation mechanisms associated with life in natural habitats and agricultural settings. These findings can be applied to benefit welfare of commercially raised birds that encounter a variety of stressors in their life cycle, and also improve conservation efforts to maintain rich species diversity of free-living birds in natural areas.

This review will focus on recent advances in defining the role of AVT and its receptors in the control of stress responses in birds. The avian stress response, particularly its neuroendocrine component, has recently been shown to be quite complex. Although the avian HPA axis is well conserved when compared to mammalian species, there have been some important recent findings addressing its uniqueness as well as the advantage of using avian species to understand the complexity of the stress response in vertebrates.

2. Anatomy of the avian HPA axis, with emphasis on arginine vasotocin (AVT)

2.1 Paraventricular nucleus

The paraventricular nucleus (PVN) is the primary hypothalamic nucleus of the stress pathway in birds. Its wing-shaped boundaries are delineated by neurons producing AVT (Fig. 1A). Throughout the PVN, AVT can be found in magnocellular and parvocellular neurons as well as in neurons of variable sizes that are difficult to attribute definitively as magno- or parvocellular. They do not appear in distinct sub-nuclei as have been described in the rat or mouse PVN (Biag et al., 2012; Simmons & Swanson, 2009; Swanson & Sawchenko, 1980), rather AVT cells of various sizes are intermingled (Fig. 1A). In the chicken, neurons toward the medial part of the PVN, near midline, tend to be parvocellular; however, mixed-sized AVT neurons are seen there as well.

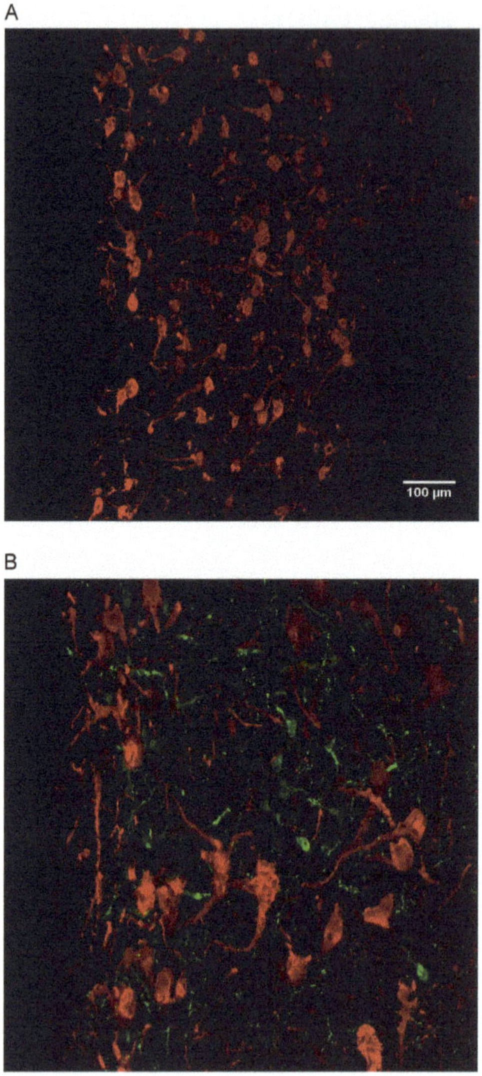

Fig. 1 The paraventricular nucleus (PVN) in the avian hypothalamus. (A) Cross section of arginine vasotocin (AVT) neurons in the core region of the PVN. (B) AVT neurons (red) and CRH neurons (green) within the PVN. A sub-nucleus of CRH neurons is observed. Scale bars: (A) 100 μm; (B) 50 μm.

With dual-labeling, both AVT and CRH neurons are seen within the PVN (Fig. 1B). CRH neurons are parvocellular (green) and appear to occupy a sub-nuclear area within the PVN.

The magnocellular (mc) AVT neurons located in the PVN project to the mediobasal region of the hypothalamus and head toward the median

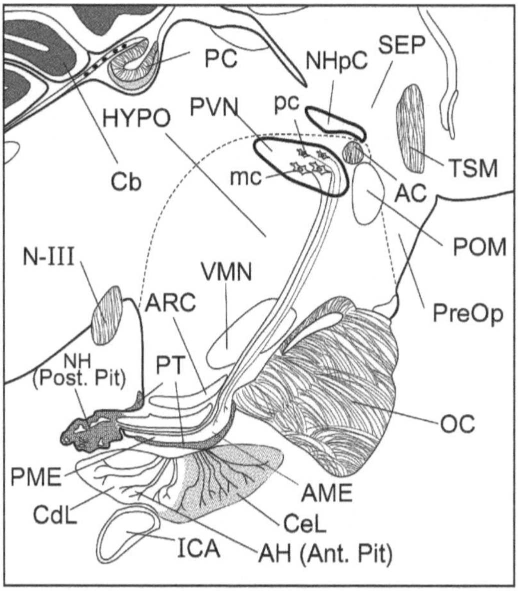

Fig. 2 Schematic, sagittal section of chick ventral septum and diencephalon. Magnocellular (mc) and parvocellular (pc) AVT neurons are seen within the PVN projecting to the median eminence (ME). The ME comprises two divisions, an anterior (AME) and posterior (PME) division. Axons of mc neurons pass through the internal zone of ME and terminate in the neurohypophysis (NH) or posterior pituitary. Axons of pc AVT and CRH neurons terminate in the AME. Surrounding the ME and ventral portion of the mediobasal hypothalamus is the pars turberalis (PT). The adenohypothesis (AH) or anterior pituitary of birds has two lobes, the cephalic (CeL) and caudal (CdL) lobes. Two markers separate the two lobes: (1) corticotropes and (2) the internal carotid artery (ICA). Both the avian V1aR and V1bR are found in corticotropes and are shown by fine stippling and represent the CeL. Past data (see text) show that there appears to be separate capillary networks serving the CeL and CdL. Abbreviations: AC, anterior commissure; ARC, arcuate nucleus; Cb, cerebellum; HYPO, hypothalamus; N III, third cranial nerve; NHpC, nucleus of the hippocampal commissure; OC, optic chiasma; PC, posterior commissure; POM, medial preoptic nucleus; PreOp, preoptic area; PVN, paraventricular nucleus; SEP, septum; TSM, septopallial mesencephalic tract; VMN, ventromedial nucleus of hypothalamus.

eminence (ME; Fig. 2). In sagittal sections, their thick axons are seen passing through the internal zone of the ME (Fig. 2) and terminate in the posterior pituitary. In contrast, parvocellular (pc) AVT and CRH neurons project delicate, thin axons from the PVN to the external zone of the ME (Fig. 2), primarily in the anterior region of the ME (AME, Fig. 2). Of interest, in avian species, the ME has distinct anterior and posterior (AME, PME) divisions that are linked to almost completely segregated anterior and posterior capillary networks (Fig. 2) supplying blood to the cephalic and caudal lobes of the anterior pituitary, respectively (Dominic & Singh, 1969; Oksche & Farner, 1974; Vitums, Mikami, Oksche, & Farner, 1964). Importantly, corticotropes occupy the cephalic lobe of the anterior pituitary (Jurkevich, Berghman, Cornett, & Kuenzel, 2005, 2008; Mikami, 1986); therefore, the terminal field of AVT and CRH parvocellular neurons in the AME insures that their secreted peptides are transported via portal capillaries from the ME to the cephalic lobe of the anterior pituitary region where corticotropes reside. In the PVN of rats and humans, colocalization of AVP and CRH was observed exclusively in parvocellular neurons (Raadsheer, Sluiter, Ravid, Tilders, & Swaab, 1993; Sawchenko, Swanson, & Vale, 1984). Almost half of CRH-immunoreactive (ir) parvocellular neurons also contain vasopressin (Whitnall, 1988, 1989). We also have evidence that some of the magnocellular AVT neurons contain both CRH and AVT immunoreactivity (Fig. 3) suggesting for the first time that in the chicken magnocellular neurons in the PVN have the capacity to produce both AVT and CRH neuropeptides. Interestingly, AVT and CRH in these cells appear to be co-packaged in the same secretory vesicles suggesting that both neuropeptides can be released simultaneously at the same target site.

In the internal zone (iz) of the chicken ME, some AVT-ir axonal terminals of magnocellular neurons form Herring bodies (Fig. 4A). Herring bodies were previously reported in the ME of the zebra finch (Oksche, Farner, Serventy, Wolff, & Nicholls, 1963). Although the function of these axonal varicosities in birds is not known, studies in mammals have shown that at the Herring bodies, AVP is released in the network of portal capillaries supplying the anterior pituitary (Holmes, Antoni, Aguilera, & Catt, 1986;

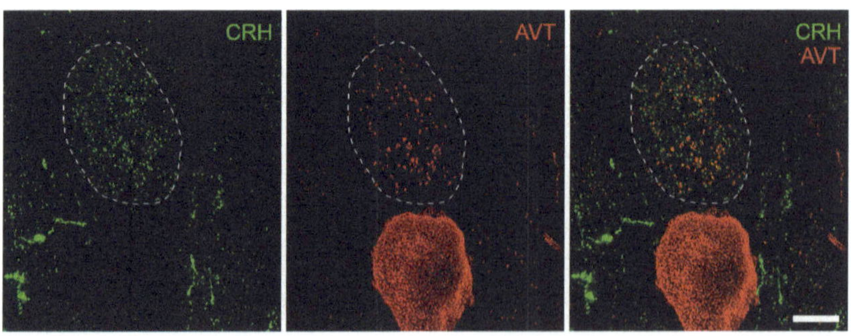

Fig. 3 (Panel 1, left) AVT mc neuron in PVN containing CRH (green) in vesicles. (Panel 2, middle and Panel 3, right) Two types of mc AVT neurons, ventral mc neuron contains solely AVT (red), dorsal mc AVT contains AVT (red), CRH (green) and co-localized AVT/CRH (yellow) vesicles. Scale bar = 10 μm.

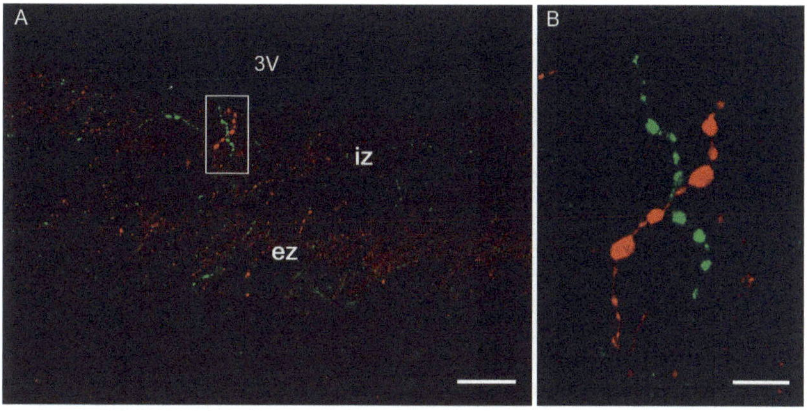

Fig. 4 (A) Herring bodies observed in internal zone (iz) of ME; both AVT (red) and CRH (green) enlarged swellings (see text) from magnocellular neurons. (B) Higher magnification showing Herring bodies. Scale bars: (A) 50 μm; (B) 10 μm.

Recht, Hoffman, Haldar, Silverman, & Zimmerman, 1981). These data suggest that AVT/AVP originating from at least some magnocellular neurons can contribute to the regulation of anterior pituitary functions. For the first time, we also report that Herring bodies of CRH axonal terminals (Fig. 4B, green) are also observed within the iz of the ME.

2.2 Nucleus of the hippocampal commissure

It has recently been demonstrated that the nucleus of the hippocampal commissure (NHpC) of the chicken brain contains a large group of CRH neurons (Nagarajan, Jurkevich, Kang, & Kuenzel, 2017). The anterior portion of the NHpC is located directly dorsal to the anterior commissure (AC) and continues posteriorly for nearly 1.0 mm (Fig. 2). Injection of colchicine was necessary to visualize CRH immunoreactivity in these cells probably due to the rapid transport of CRH from the nucleus down into the axon and into its terminal field. Curiously, no similar nucleus containing CRH neurons has been found in a corresponding location of either the rat or mouse brain. In addition to showing CRH neurons in the dorsal region of the NHpC (red cells, Fig. 5A), the avian vasotocin 1a receptor (V1aR/AVPR1a) (green cells, Fig. 5B) resides in glia and a terminal field of AVT fibers (purple, Fig. 5B) occurs predominantly in the dorsal region (Nagarajan, Jurkevich, et al., 2017).

A feed restriction (FR), *in vivo* study showed that following initiation of the stressor, the first responder to FR was CRH neurons in the NHpC

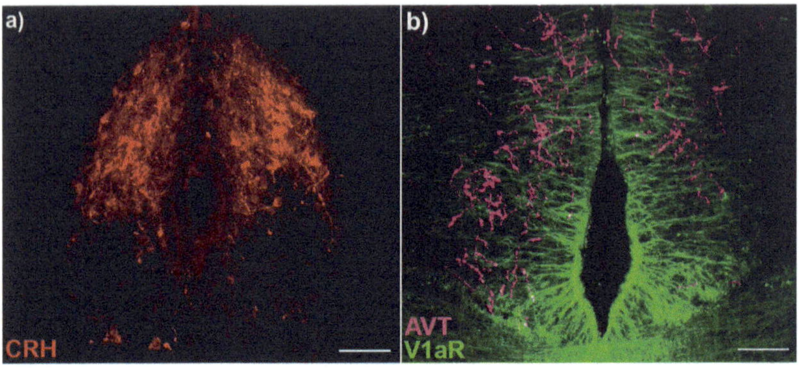

Fig. 5 Presence of a vasotocinergic system and CRH neurons in the nucleus of the hippocampal commissure (NHpC). (A) CRH neurons (red) in dorsal region of NHpC. (B) Vasotocin immunoreactive terminals (magenta, pseudo color) and V1aR immunoreactivity (green) in glial cells. Scale Bars = 100 μm. *Modified from Nagarajan, G., Jurkevich, A., Kang, S. W., & Kuenzel, W. J. (2017). Anatomical and functional implications of corticotrophin-releasing hormone neurones in a septal nucleus of the avian brain: An emphasis on glial-neuronal interaction via V1a receptors in vitro.* Journal of Neuroendocrinology, *29(7), 1–11. doi:10.1111/jne.12494, reproduced with permission from Wiley.*

displaying significantly elevated CRH mRNA at 2h. Thereafter CRH mRNA in the PVN was significantly elevated at 4h while AVT mRNA within the PVN was increased at 12h (Nagarajan, Kang, & Kuenzel, 2017). A subsequent *in vitro* study involving slices of the avian septum containing the NHpC showed that CRH neuronal activity is modulated by AVT via V1aRs involving brain-derived neurotrophic factor (BDNF) and TrkB in glia within the NHpC (Nagarajan, Jurkevich, et al., 2017).

2.3 The avian V1a receptor, location in circumventricular organs, possible role in stress

Avian AVT receptors are encoded by four different genes—*AVPR1A (VT4/V1aR)*, *AVPR1B (VT2/V1bR)*, *AVPR2 (VT1/V2BR)* and *OXTR (VT3/MTR)* (Table 1). Specific location of avian AVT receptors in the brain is not well known. To date, the avian V1bR has not been found in the brain of the chicken (Jurkevich et al., 2005, 2008) nor in two songbird species (Leung et al., 2011). In contrast, the distribution of the avian V1aR mRNA was successfully mapped by *in situ* hybridization histochemistry in the white-throated sparrow (*Zonotrichia albicollis*) and zebra finch (*Taeniopygia guttata*) (Leung et al., 2011). An immunohistochemical study was also completed and mapped the distribution of the avian V1aR protein

Table 1 Avian vasotocin receptors and their known functions.

Gene name[a]	Receptor subtype	Alternative name	Tissue distribution	Physiological function
AVPR1A	V1aR	VT4	Brain, vascular smooth muscle, liver, anterior pituitary	Blood pressure control, glucose homeostasis, social behavior
AVPR1B	V1bR	VT2	Anterior pituitary	ACTH release
AVPR2	V2BR[b]	VT1	Shell gland, endometrium, brain	Water balance (?)
OXTR	OTR	VT3, MTR[c]	Oviduct (shell gland) myometrium	Egg laying

[a]Gene names according to Chicken Gene Nomenclature Consortium (CGNC).
[b]Ocampo Daza, Lewicka, & Larhammar, 2012.
[c]Mesotocin receptor.

in the chicken brain (Selvam, Jurkevich, & Kuenzel, 2015). A striking finding was detection of V1aR immunoreactivity in structures associated with 10 circumventricular organs (CVOs). The predominant structures immunostained for V1aR in CVOs were glia. CVOs throughout the central nervous system have been studied in detail in a diversity of vertebrate species, including mammals (Weindl & Joynt, 1973; Weindl & Sofroniew, 1981) and birds (Vigh, 1971; Vigh & Vigh-Teichmann, 1973) nearly 50 years ago. They are characterized as specialized organs commonly lacking a blood-brain barrier (BBB) and most often located in periventricular midline sites. They usually have cerebrospinal fluid-contacting neurons and specialized glia and/or neurons that appear to make contact with cerebrospinal fluid and nearby blood vessels. Could it be that V1aR located in avian CVOs assist in monitoring AVT in cerebrospinal fluid as well as in plasma and thereby function in stress and/or osmoregulation? The V1aR immunoreactivity was particularly striking in the median eminence, organum vasculosum of the lamina terminalis (OVLT), subseptal organ (SSO) homologous to the mammalian subfornical organ (SFO), pineal, choroid plexus, paraventricular organ, subcommissural organ and subtrochlear organ. Currently it is unknown whether avian CVOs are utilized to monitor AVT within the cerebrospinal fluid. CVOs have been proposed to possess specialized glial cells that can serve as chemosensors (Bolborea & Dale, 2013). The OVLT, SFO and area postrema have been termed sensory CVOs and proposed to function collectively to sustain body fluid homeostasis (Johnson & Gross, 1993) and respond to osmotic stress.

The OVLT and SFO in mammals have been shown to possess osmoreceptors and the products of the renin-angiotensin system, respectively. Specifically the *trpv1* gene has been shown to be involved in body fluid osmoregulation, particularly hypertonicity in mammals (Ciura & Bourque, 2006; Liedtke et al., 2000; Sharif Naeini, Witty, Seguela, & Bourque, 2006) and encodes a mechanically regulated channel that confers intrinsic osmosensitivity to OVLT neurons (Ciura, Liedtke, & Bourque, 2011). The renin-angiotensin system has been shown to play a key role in regulating water intake (Fitzsimons & Simons, 1969). A product of that physiological system, angiotensin II, acts centrally in the SFO to stimulate water intake (Simpson & Routtenberg, 1973). The receptor subtype specific for angiotensin II was shown to be the AT_1 and found to be located in the SFO (Bunnemann et al., 1992; Giles et al., 1999; Lenkei, Palkovits, Corvol, & Llorens-Cortes, 1997).

A recent study (Aman, Nagarajan, Kang, Hancock, & Kuenzel, 2016) revealed that chickens administered with hypertonic saline, a physical stressor, showed significantly increased levels of *V1aR* mRNA in the SSO/SFO and OVLT. Concurrently, both the AT_1R mRNA and *TRPV1* mRNA were significantly increased in the SSO/SFO. In contrast, both genes were significantly downregulated in the OVLT. Birds subjected to the psychogenic stressor, immobilization, showed a significant increase of the V1aR in the SSO/SFO compared to controls, but no significant change in AT_1R mRNA nor *TRPV1* mRNA. In contrast, the OVLT of immobilized birds showed significantly increased *V1aR* mRNA, AT_1R mRNA and *TRPV1* mRNA compared to the non-treated controls. In summary, results suggest that the avian SSO/SFO and OVLT are involved in osmoregulation and probably in osmotic stress, while the OVLT serves an additional function related to psychological stress (Aman et al., 2016). Usually, CVOs are not discussed in reviews of the HPA axis; however, when one considers that the OVLT, SSO/SFO and ME all border the diencephalon of birds, mammals and other vertebrates, they appear relevant structures that contribute to the normal function of the HPA axis.

2.4 Anterior pituitary

AVT and CRH released in the hypophyseal portal system reach corticotropes in the anterior pituitary where they bind specific receptors triggering complex mechanisms of adrenocorticotropic hormone (ACTH) secretion and re-synthesis (Jurkevich, Mikhailova, Madison, Cornett, & Kuenzel, 2010).

Immunohistochemical studies demonstrated ample presence of AVPR1b/V1bR within the cephalic lobe of the anterior pituitary. Utilizing dual-labeled immunocytochemistry, it was shown that virtually all cells producing pro-opiomelanocortin (POMC), a precursor of ACTH and α-melanocyte-stimulating hormone (αMSH), contained the VT2R/V1bR and were located in the cephalic lobe. Approximately 15% of cells in the cephalic lobe were shown to be lactotropes that likewise contained the VT2R/V1bR (Jurkevich et al., 2005, 2008).

Immunolabeling the chicken VT4R/V1aR showed that corticotropes in the cephalic lobe likewise contained the VT4R/V1aR. A majority of the VT4/V1aR were found within the cytoplasm suggesting possible receptor internalization. A smaller percentage was present in the peripheral region of the cell body where the cell membrane would reside (Selvam et al., 2013). In contrast, results from immunocytochemistry for the VT2R/V1bR showed most receptors were located in the peripheral region of each cell body in the location of the cell membrane (Jurkevich et al., 2005, 2008). The VT4R/V1aR was found only in POMC containing cells (corticotropes), but not gonadotropes (luteinizing hormone containing cells), somatotropes, or lactotropes (Selvam et al., 2013).

The internal carotid artery (ICA) ventral to the anterior pituitary (Fig. 2) and an antibody to the avian V1bR (Jurkevich et al., 2005) and V1aR (Selvam et al., 2013) are useful markers identifying the cephalic from the caudal lobes of the avian adenohypophysis. The neuroendocrine anatomy nicely shows that in birds, stress-related peptides, AVT and CRH, from the hypothalamus are transported into the AME. Here, a separate network of portal capillaries delivers AVT and CRH to the cephalic lobe of the anterior pituitary where corticotropes and stress-related receptors are located (Fig. 2), hence efficiently activating the specialized cells to synthesize and release the pituitary stress hormone, adrenocorticotropic hormone, into the vasculature where it will be transported to the adrenal gland.

2.5 Adrenal gland

In contrast to mammals that have a distinct adrenal cortex responsible for the synthesis and release of the stress hormone cortisol (humans, many mammalian species and fish) or corticosterone (rodents, birds, reptiles and amphibians) and an adrenal medullary region that contains cells that produce epinephrine and norepinephrine, avian adrenal glands comprise cell types that are interspersed. The anatomical result is an adrenal gland without an

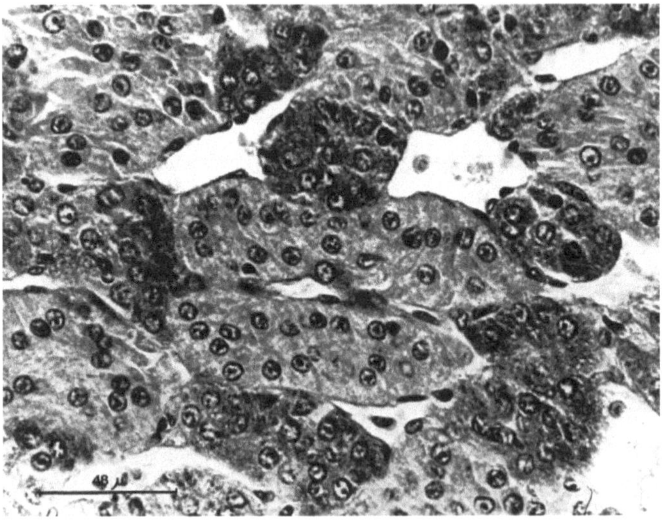

Fig. 6 The avian adrenal gland has interrenal cells interspersed with chromaffin cells. Interrenal (cortical-like) cells are shown as elongated strands of columnar cells that produce glucocorticoids. Chromaffin (medullary-type) cells are polygonal in shape (Hartman, Knouff, McNutt, & Carver, 1947) and produce catecholamines. *Reproduced from Wells, J.W., Wight, P.A.L. (1971). The adrenal glands. In D.J. Bell & B.M. Freeman, (Eds.),* Physiology and biochemistry of the domestic fowl *(pp. 489–520). London: Academic Press, with permission from Academic Press.*

external cortical and central, medullary region. In birds, small groups of columnar cortical cells with their nuclei near the center of each cell form a double row of elongated strands of cells (Fig. 6) that were proposed to be called interrenal tissue (Hartman et al., 1947). These cells produce corticosterone. They can be distinguished from smaller groups of polygonal cells called chromaffin cells (Hartman et al., 1947) that produce the two catecholamines, epinephrine and norepinephrine (Fig. 6).

3. Roles of AVT and CRH in regulating blood concentration of corticosterone in birds

Among mammalian species, the relative abilities of AVP and CRH to stimulate ACTH release are variable. In rat and man, CRH is the most potent ACTH secretagogue (Brubaker, Baird, Bennett, Browne, & Solomon, 1982; Gillies, Linton, & Lowry, 1982; Vale et al., 1983). In bovine species, AVP and CRH are equipotent in releasing ACTH (Schwartz & Vale, 1988) while AVP is more potent in sheep (Familari, Smith, Smith, & Funder, 1989;

Kemppainen, Clark, Sartin, & Zerbe, 1993; Liu, Robinson, Funder, & Engler, 1990). Overall, in mammals, there appears to be species differences in the effectiveness of AVP compared to CRH to release the stress hormone corticosterone (CORT). Are AVT and CRH equipotent stimulators of the HPA axis in birds? Are there species differences similar to mammalian studies? Results to date appear to depend upon whether these peptides were administered centrally or peripherally.

3.1 Central administration

The effects of central administration of AVT and CRH were examined by injecting each compound intracerebroventricularly (*icv*) in 4-week-old chicks and taking blood samples every 15 min for 2 h. To minimize stress of animal handling, blood was sampled remotely through an implanted catheter from a room compartment separated from birds by a one-way glass partition. In males, administering 100 pmol of AVT or CRH in each bird showed a significant elevation of CORT at each time point compared to saline injected controls. CRH, however, was clearly more effective than AVT in releasing CORT (Madison, Jurkevich, & Kuenzel, 2008). The data confirmed previous results showing that CRH injected *icv* induces greater levels of CORT in circulation compared to AVT in young chicks (Tachibana et al., 2004).

Importantly, relative abilities of AVT and CRH to elevate plasma CORT are sex-dependent. In response to the 100 pmol dose of AVT or CRH administered *icv*, males showed significantly higher levels of CORT throughout the 2 h period of post-injection compared to females (Madison et al., 2008). In general, male chickens have been shown to be more fearful, and therefore may be more stress-susceptible than females (Banks, Wood-Gush, Hughes, & Mankovich, 1979; Jones, 1987). In seasonal wild, free-living avian species, differential responsiveness of the HPA axis in males and females have been associated with a specific part of the annual cycle, particularly during the breeding season (Romero, 2002). Abert's towhees in the Sonoran desert (Wingfield, Vleck, & Moore, 1992) and an arctic-breeding bird, the white-crowned sparrow (*Zonotrichia leucophrys gambelii*) showed that males had significantly higher levels of glucocorticoids than females during the breeding season (Astheimer, Buttemer, & Wingfield, 1995). Free-living house sparrow males responded to capture and restraint stress with higher total CORT levels than females during breeding (Romero, Cyr, & Romero, 2006); however, during other stages of the annual cycle no sex differences in response to stress were detected. The mechanisms of sex-dependent responsiveness of the avian HPA axis to stressors are poorly understood. Possible explanations could be the presence of sex differences in AVT- and CRH-expressing neurons (Iwasaki-Sekino, Mano-Otagiri, Ohata, Yamauchi, & Shibasaki, 2009; Jurkevich, Barth, & Grossmann, 1997; Rivalland, Iqbal, Clarke, Turner, & Tilbrook, 2005) and/or distribution of AVT, CRH and glucocorticoid receptors (Dumais & Veenema, 2016; Koch, Wingfield, & Buntin, 2002; Lim, Nair, & Young, 2005).

Overall, at the level of the brain, CRH is a more potent stimulator of the HPA axis than AVT in chickens and induces higher levels of CORT in males compared to females. Once released within the brain, CRH and AVT exert enduring modulatory effects on the pattern of CORT secretion (Madison et al., 2008). These modulatory effects lasted well beyond the half-life of peptide clearance from cerebrospinal fluid [11–12.4 min (Banks, Kastin, Horvath, & Michals, 1987; Martins, Kastin, & Banks, 1996)] and suggested a likely involvement of secondary mechanisms. One such mechanism could be the presence of ultra-short positive feedback loops found in the PVN of rats (Imaki et al., 1996; Liposits, Paull, Setalo, & Vigh, 1985). Acting within such a loop, CRH may upregulate its own synthesis and secretion by increasing the expression of CRH and CRH1 type receptor genes (Imaki et al., 1996; Parkes, Rivest, Lee, Rivier, & Vale, 1993). Of interest, the presence of similar positive feedback loops involving AVP release was confirmed in a different brain region, the lateral septum, of rats (Landgraf, Ramirez, & Ramirez, 1991).

3.2 Peripheral administration

In contrast to *icv* administration, AVT injected *iv* induced a significant elevation of plasma CORT in 5-week-old male chickens while CRH had no effect on CORT levels (Fig. 7) (Mikhailova et al., 2007). Similarly, intravenously administered AVT was more efficacious than CRH in releasing ACTH and CORT in pigeons *Columba livia* (Romero & Wingfield, 2001; Westerhof, Lumeij, Mol, van den Brom, & Rijnberk, 1992). In European starlings, a single subcutaneous injection of AVT significantly increased CORT release whereas subcutaneously or *iv* injected CRH had no effect on CORT release (Rich & Romero, 2005). In cultured pituitary cells of duck, *Anas platyrhynchos,* AVT was significantly more effective than ovine CRH in releasing ACTH; (Castro, Estivariz, & Iturriza, 1986). In summary, peripherally administered AVT appears to be a more potent CORT

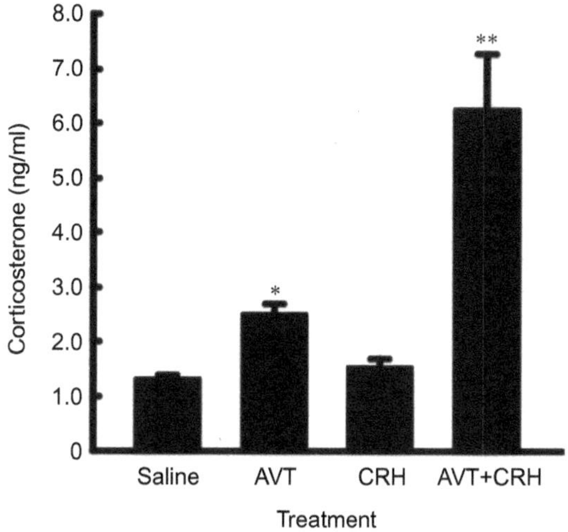

Fig. 7 Plasma CORT levels in chicks, 5 weeks of age, determined in blood samples taken 30 min. After *iv* administration of either physiological saline, AVT, CRH, or AVT+CRH. Data presented as means ± SEM, $*P \leq 0.05$ from saline controls, $**P \leq 0.01$ from other treatments. *Modified from Mikhailova, M. V., Mayeux, P. R., Jurkevich, A., Kuenzel, W. J., Madison, F., Periasamy, A., … Cornett, L. E. (2007). Heterooligomerization between vasotocin and corticotropin-releasing hormone (CRH) receptors augments CRH-stimulated 3′,5′-cyclic adenosine monophosphate production.* Molecular Endocrinology, *21(9), 2178–2188. https://doi.org/10.1210/me.2007-0160 (web archive link), reproduced with permission from the Endocrine Society.*

secretagogue in most investigated bird species than CRH. However, intramuscular administration of AVT to the captive chukar partridge, a bird species from arid regions, did not change plasma CORT levels while CRH significantly increased plasma CORT (Dickens, Earle, & Romero, 2009).

3.3 Peripheral administration of both AVT and CRH

When a combination of nearly equimolar doses of AVT and CRH was given by *iv* injections to male chickens, the plasma concentration of CORT increased remarkably compared to levels induced by each injected peptide alone (Fig. 7). Similar greater than additive effects of combined administration of AVT/AVP and CRH on ACTH and glucocorticoid release were reported *in vivo* in humans and rats (DeBold et al., 1984; Rivier & Vale, 1983) and *in vitro* (on ACTH release) in a number of various species from fish to mammals (Baker, Bird, & Buckingham, 1996; Castro et al., 1986; Evans et al., 1988; Familari et al., 1989; Gillies et al., 1982; Liu et al., 1990).

Data indicate that cooperative interactions between the two hypothalamic peptides have evolved early during phylogeny. The primary synergistic interaction between AVT/AVP and CRH occurs most likely at the level of pituitary corticotropes stimulating sub-additive release of ACTH that in turn results in augmented secretion of glucocorticoids from the adrenals. A possible molecular mechanism of this super-additive effect may be due to the interaction of G protein-coupled receptors (GPCR) that function together for a particular molecular response, a process called GPCR oligomerization (reviewed in Park, Filipek, Wells, & Palczewski, 2004; Milligan, 2009) and briefly presented in the following Section 3.4.

In summary, central vs peripheral administration resulted in CRH being more efficacious in the former and AVT in the latter for the release of CORT in chickens. The result in chickens may likewise occur in other avian species. Concurrent targeting of the pituitary corticotropes by both peptides may result in pronounced super-additive stimulation of the glucocorticoid release.

3.4 Receptor dimerization, as a possible mechanism for the synergistic effect of AVT and CRH in the HPA axis

The use of immunocytochemistry showed that both the avian V1bR and CRHR1 proteins are co-expressed in corticotropes in the rostral lobe of the anterior pituitary gland (Jurkevich et al., 2010; Mikhailova et al., 2007). Perhaps protein-protein interactions of the two receptor types might be responsible for the more than additive effect of plasma CORT levels after peripheral co-administration of AVT and CRH. A synergistic effect on the release of ACTH was reported following both AVP and CRH administration in mammalian systems (Arsenijevic et al., 1994; Brooks & Challis, 1989; Leong, 1988). FRET analysis revealed dimerization between the mammalian V1b receptor and CRH type-one receptors (Young, Griffante, & Aguilera, 2007). Utilizing HeLa cells expressing both the avian V1aR and CRHR1, it was found that combined administration of AVT and CRH significantly augmented cAMP production compared to CRH given alone (Mikhailova et al., 2007). FRET analysis showed that the avian V1aR and CRHR1 may form heterodimers, particularly in the presence of AVT + CRH (Cornett, Kang, & Kuenzel, 2013; Jurkevich et al., 2010; Mikhailova et al., 2007). Attention was therefore directed to receptors within the avian anterior pituitary involved in the stress response.

4. Receptors associated with stress within the anterior pituitary and vasotocin antagonists specific for the avian V1a and V1bRs

4.1 Stress-related receptors in the anterior pituitary and their response to immobilization

In the anterior pituitary of the chicken, the effects of AVT and CRH are mediated by V1bR and V1aR, and CRHR1 and CRHR2, respectively (De Groef et al., 2004; Jurkevich et al., 2005, 2008; Kuenzel, Kang, & Jurkevich, 2013; Selvam et al., 2013). Changes in expression of the avian *V1bR*, *V1aR*, *CRHR1* and *CRHR2* in the anterior pituitary have been documented in birds subjected to acute and chronic immobilization (Fig. 8).

Both acute and chronic stress decreased *V1aR* expression, however, the reduction in *V1aR* mRNA levels was greater following chronic stress (Fig. 8A). *CRHR1* expression was significantly decreased in both acute and chronic stress conditions compared to controls (Fig. 8C). The levels of *V1bR* and *CRHR2* mRNAs increased in acutely stressed birds but returned to the level of controls in birds exposed to 1 h immobilization during 10 consecutive days (Fig. 8B and D). Hence, transcriptional activation of *V1b* and *CRHR2* genes in the anterior pituitary induced by acute stress was shown to return to their control levels following chronic stress. Results suggest that the two receptors become non-responsive to continued immobilization stress (Kang & Kuenzel, 2014; Kuenzel et al., 2013; Selvam et al., 2013). Lack of responsiveness of *V1bR* and *CRHR2* gene expression during chronic stress may play a critical role for the homeostatic control of the avian HPA axis as shown by the reduced expression of the *V1aR* and *CRHR1* following chronic stress. The continued downregulation of V1aR expression following chronic stress compared to that of acute stressed birds and the consistent reduction of the *CRHR1* expression during both acute and chronic stress conditions suggest a key role of these two receptors in the negative feedback inhibition occurring in the avian anterior pituitary.

Regulatory mechanisms controlling functional expression of AVT and CRH receptors in the pituitary gland of chickens remain unclear. One possible explanation is the negative/positive feedback interaction between glucocorticoid receptors and VTRs/CRHRs. Negative feedback regulation of the POMC gene by the glucocorticoid receptor is a well-accepted mechanism (Autelitano, Lundblad, Blum, & Roberts, 1989; Biebermann, Kuhnen,

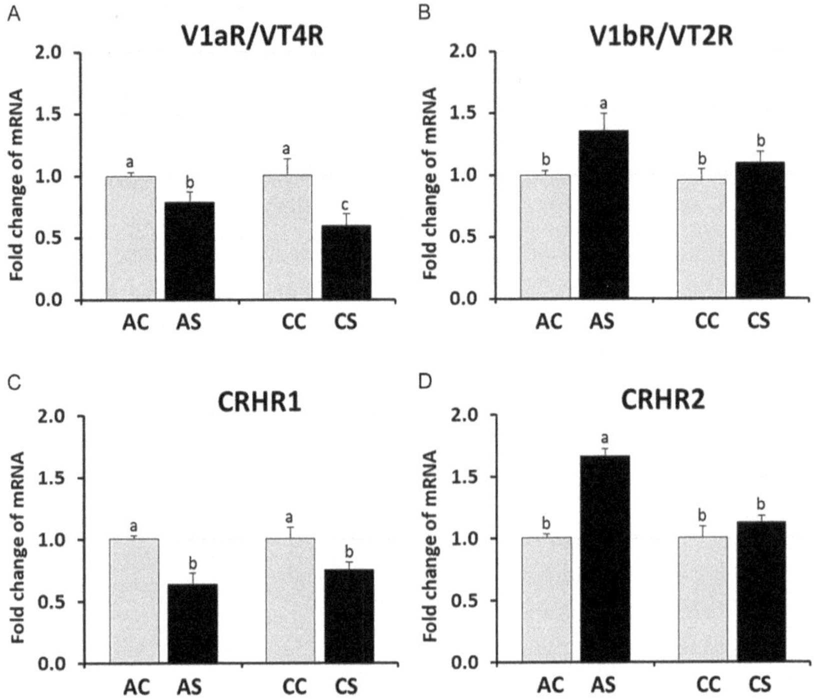

Fig. 8 Differential changes in expression of (A) *V1a*, (B) *V1b*, (C) *CRH1* and (D) *CRH2* receptors following acute and chronic immobilization stress. Total RNA from the anterior pituitary gland was extracted after 1 h acute and after 10 consecutive days of 1 h chronic immobilization. Data (mean ± SEM) were expressed from a value set for 1.0 for the acute control (AC) for each receptor. Significant differences among means are denoted by different letters ($P \leq 0.05$). Abbreviations: AC, acute control; AS, acute stress; CC, chronic control; CS, chronic stress. *Modified from Kang, S. W., Madkour, M., & Kuenzel, W. J. (2017). Tissue-specific expression of DNA methyltransferases involved in early-life nutritional stress of chicken,* Gallus gallus. Frontiers in Genetics, *8, 204. https://doi.org/10.3389/fgene.2017.00204 (web archive link), with permission from Elsevier.*

Kleinau, & Krude, 2012; Bilodeau et al., 2006). The human CRH-R2 gene contains the putative glucocorticoid receptor binding site at −2755 and −2243 of its promoter (Catalano, Kyriakou, Chen, Easton, & Hillhouse, 2003) suggesting a role of direct binding of the glucocorticoid receptor to the promoter of this receptor. Our choice, however, was to focus on the sequence of the two vasotocin receptors, V1aR and V1bR and to develop a realistic 3D model of each of their structures in order to determine the key binding sites for the agonist, AVT, as well as potential antagonists. The first receptor studied was the avian V1aR.

4.2 Relcovaptan (SR-49059), avian V1aR antagonist

The relationship between structure and functional activity of AVP receptors in mammals was examined in several past studies using three-dimensional (3D) homology modeling techniques, chimeric receptor approaches and site-directed mutagenesis studies (Acharjee et al., 2004; Cho et al., 2007; Cotte et al., 2000; Czaplewski, Kazmierkiewicz, & Ciarkowski, 1998a, 1998b; Hausmann et al., 1996; Mahlmann et al., 1994; Mouillac, Chini, Balestre, Elands, et al., 1995). Our recent study, using *in silico* 3D modeling/docking analyses of the chicken V1aR (cV1aR) identified several compounds with potential antagonistic properties. These compounds were then assayed using an *in vitro* chicken primary anterior pituitary (AP) cell culture assay (Jayanthi et al., 2014). SR-49059, a known antagonist of the mammalian V1aR, was found to be a most potent antagonist of the avian V1aR. SR-49059 appeared to be a more specific and effective antagonist than the Manning compound used in several studies involving birds (Baran, Sklar, & Adkins-Regan, 2016; Goodson & Evans, 2004). Our studies used 3D homology models of the chicken V1aR by building common modeling templates and performing docking analyses with a series of known, effective mammalian antagonists. An *in vitro* primary anterior pituitary cell culture procedure was developed to examine the effect(s) of antagonists on the chicken anterior pituitary stress response by monitoring the specificity of selected antagonists to attenuate expression of the cV1aR.

Based on the literature and our previous study (Jayanthi et al., 2014; Kang, Jayanthi, Nagarajan, Suresh Kumar, & Kuenzel, 2019), a list of potential mammalian V1aR antagonists was screened against the cV1aR and relcovaptan (SR-49059) was identified as an effective avian V1aR antagonist. Three-dimensional modeling and binding affinity analyses revealed 10 unique amino acid resides on the cV1aR model that showed molecular interaction with specific sites on the antagonist, SR-49059 (Fig. 9A).

The molecular basis underlying ligand-receptor interactions has been extensively investigated in the mammalian V1aR (Conner et al., 2007; Hawtin et al., 2006; Mouillac, Chini, Balestre, Elands, et al., 1995; Mouillac, Chini, Balestre, Jard, et al., 1995). In avian species, Gln (Q) residues in TM II, III, IV, VI (Q115, Q138, Q192, Q313) and a Lys (K) in TM III (K135) of the cV1aR interacted with the agonist AVT (Kang et al., 2019). Some of the Q residues (Q138 and Q313 of cV1aR) also interacted with antagonists (Fig. 10A).

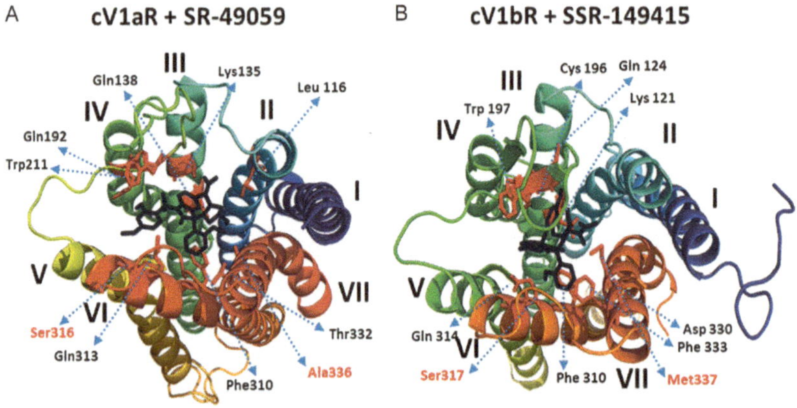

Fig. 9 Three-dimensional (3D) homology structures and docking of antagonists onto the modeled structures of the (A) chicken V1aR and (B) chicken V1bR. Homology models of cV1aR and cV1bR built using opioid receptor (PDB ID: 4EA3) as a template. Seven transmembrane helices (TM-I-VII), each shown with a different spectral color are labeled with Roman numbers. Top views of the binding amino acid residues (shown as sticks: magenta) of the modeled cV1aR (A) and cV1bR (B) structures with SR-49059 and SSR-149415 (black), respectively. The residues are numbered according to their position in the primary sequence of the cV1aR and cV1bR (Kang et al., 2019). All of the residues potentially interacting with the different parts of antagonists are shown.

Likewise, a mutagenesis study on the human V1aR (Mouillac, Chini, Balestre, Jard, et al., 1995) showed that important amino acids involved in agonist binding were Gln in transmembrane (TM) II, IV (Q108, Q185), and a Lys that is located in the spatial vicinity of TM III (K128). Substitution of these amino acids in positions Q108A, Q185A, and K128A with Ala, a neutral amino acid, significantly decreased binding affinity of agonists. Interestingly, these substitutions only caused minor differences in the binding affinities of an antagonist, suggesting a similar mechanism of ligand-receptor binding interactions between the avian V1aR and human V1aR (Kang et al., 2019).

4.2.1 Application of the V1aR blocker, SR-49059 and its effect on stress and food intake

In vitro studies using anterior pituitary cell cultures and *in silico* 3D modeling and docking analysis of the avian V1aR suggested that SR-49059 was a specific antagonist of that receptor (Jayanthi et al., 2014). An attempt was made to apply SR-49059 *in vivo* to determine whether it would be effective in inhibiting the stress response when birds were subjected to the stressor immobilization. Related to the stress response, neuropeptide Y (NPY), a potent compound for stimulating food intake, was likewise administered

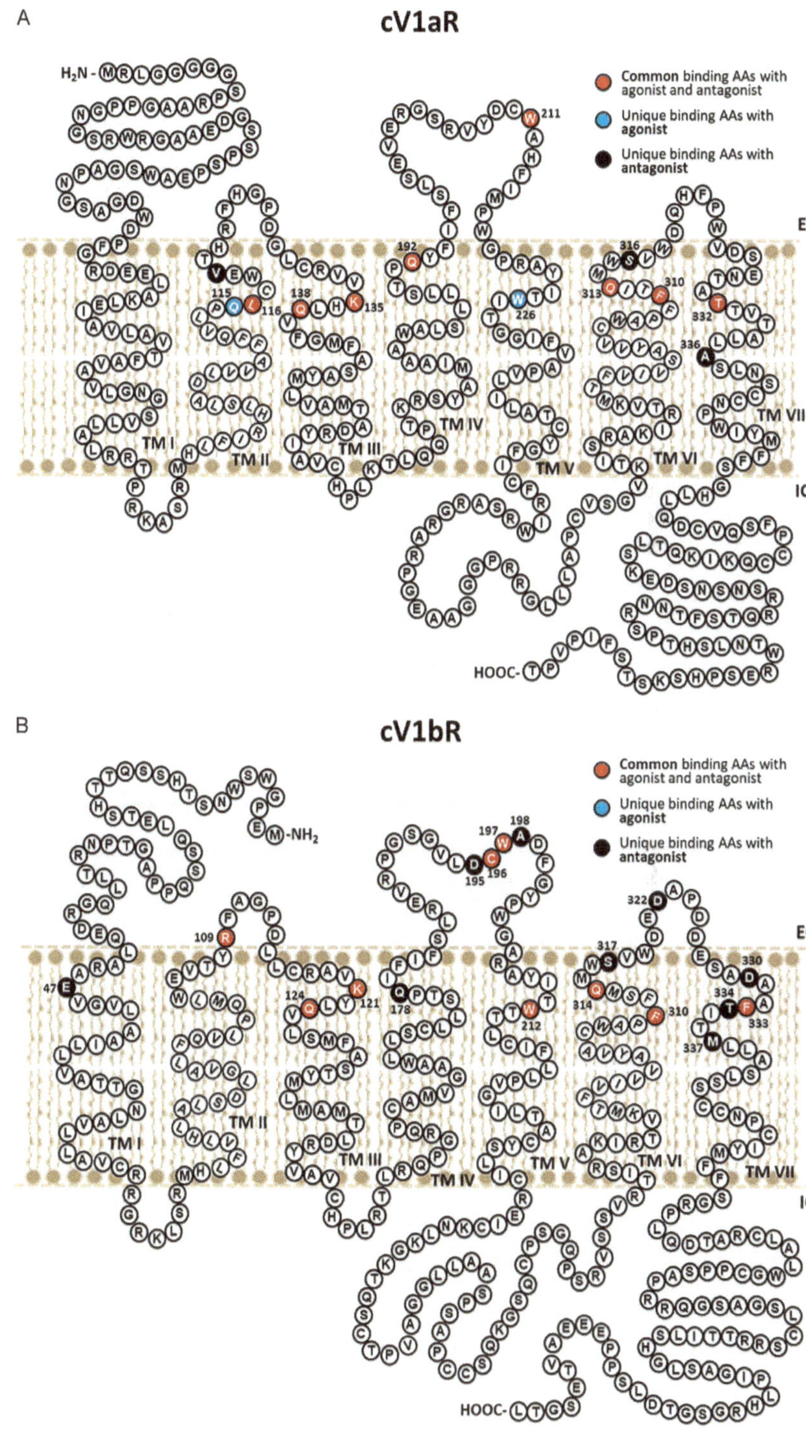

Fig. 10 See legend on next page.

to chicks as it is known to release high levels of vasopressin and CORT in rats when given *icv* (Leibowitz, Sladek, Spencer, & Tempel, 1988) and causes secretion of vasopressin in unanesthetized rats when injected into the hypothalamic supraoptic nucleus (Willoughby & Blessing, 1987). Since vasopressin in mammals (Ikemura, Matsuwaki, Yamanouchi, & Nishihara, 2004; Meyer, Langhans, & Scharrer, 1989) and AVT administered to chicks (Tachibana et al., 2004) significantly reduces food intake. Administering SR-49059 prior to subjecting chicks to immobilization stress for 30 min resulted in a significant reduction in plasma CORT compared to controls given physiological saline and subjected to the same stressor (Kuenzel, Hancock, Nagarajan, Aman, & Kang, 2016). When chicks were concurrently administered either saline and NPY or SR-49059 and NPY compared to controls given saline alone, both NPY administered positive controls consumed more food than saline controls; however, the combination of SR-49050 + NPY resulted in significantly more food intake than the saline + NPY group (Kuenzel et al., 2016). A follow-up study utilizing the V1aR blocker SR-49059 and FOS-immunoreactive cells showed that four structures in the diencephalon containing populations of AVT magnocellular neurons including the supraoptic nucleus, preoptic periventricular n., ventral periventricular hypothalamic n., and paraventricular core n. all showed significantly reduced FOS activation of that type neuron due to the antagonist blocking activation of the magnocellular neurons in those four nuclei. Results suggest that similar to mammals, NPY administered centrally to birds releases AVT and CORT and that SR-49059 is effective in blocking V1aR sites such that the AVT is unable to bind to neural sites that have been suggested to inhibit food intake when activated by AVT (Nagarajan, Jurkevich, Kang, & Kuenzel, 2016). Additionally, the AVT transported by blood vessels to the anterior pituitary would likewise be blocked from binding to V1aR sites on corticotropes thereby inhibiting pituitary ACTH and adrenal CORT release that likewise would result in decreased food intake.

Fig. 10 Two-dimensional schematic diagrams of (A) chicken V1aR and (B) chicken V1bR showing the amino acid residues involved in interacting with agonist and antagonists. Common binding sites with agonist and antagonists in structures are shown in red and unique binding AAs with agonist and antagonists are shown in blue and dark purple, respectively. *Reproduced from Kang, S. W., Jayanthi, S., Nagarajan, G., Suresh Kumar, T. K., & Kuenzel, W. J. (2019). Identification of avian vasotocin receptor subtype-specific antagonists involved in the stress response of the chicken,* Gallus gallus. Journal of Biomolecular Structure & Dynamics, *37, 1685–1699. https://doi.org/10.1080/07391102.2018.1464957 (web archive link), with permission from Taylor & Francis.*

4.3 Nelivaptan (SSR-149415), avian V1bR antagonist

Similar to our previous study using a 3D homology model of the cV1aR, a 3D homology model of the chicken V1bR (cV1bR), was built using a common modeling template and docking analyses were performed with a series of known effective mammalian antagonists. An *in vitro* primary anterior pituitary cell culture was used to examine the effect(s) of antagonists on the stress response of those cells. The specificity of selected antagonists to attenuate expression of the cV1bR gene (Kang et al., 2019) was determined. In the past, SSR-149415 was the first selective, orally active vasopressin V1bR antagonist in mammals (Serradeil-Le Gal et al., 2005). A list of potential mammalian V1bR antagonists was screened against the cV1bR structure and nine ligands with the highest binding affinity were identified (Fig. 9B). Gln (Q) residues in TM III, VI (Q124, Q314) and a Lys (K) residue in TM III (K121) of the cV1bR interacted with the agonist AVT (Kang et al., 2019). Some of the Q residues (Q124 and Q314 of cV1bR) interacted with SSR-149415 (Fig. 10B).

Extracellular loop (ECL)2 of GPCRs has attracted increased interest, because the X-ray structure of bovine rhodopsin revealed that ECL2 projects into the binding crevice within the TM bundle (Barwell, Woolley, Wheatley, Conner, & Poyner, 2012; Hawtin et al., 2006). In mammals, analysis of residues in or close to ECL2 by mutagenesis revealed that four aromatic residues Phe189 (F189), Trp206 (W206), Phe209 (F209), and Tyr218 (Y218) appear to be important for agonist binding and receptor activation (Conner et al., 2007). These aromatic resides are highly conserved throughout the neurohypophysial hormone sub-family of peptide-GPCRs (Conner et al., 2007). Specifically, the highly conserved aromatic residues W206 and F209 were suggested to project into the binding crevice, and F189 and Y218, located at the extreme ends of ECL2 were suggested to have a critical role for determining the position of the ECL2 cap over the binding crevice in the human V1aR (Conner et al., 2007; Hawtin et al., 2006). In rhodopsin, class A GPCRs, ECL2 is usually the longest ECL and comprises two β-sheets which form a twisted β-hairpin that projects down into the TM bundle (Wheatley et al., 2012). In the ECL2 of cV1bR, there are two β-sheets and a highly conserved Cys-Trp (CW) motif (Cys 196 and Trp 197 in Fig. 10B) that resides in the middle of the second β-sheet. The CW motif is a common binding site of the avian agonist AVT and antagonists (SR-49059, SSR-149415, and the Manning compound) (Kang et al., 2019), suggesting the significance of the CW motif of the ECL2 in avian species.

In mammals, including humans, the CW motif in the ECL2 of the V1aR and V1bR is well conserved (Conner et al., 2007; Hawtin et al., 2006). In this conserved CW motif, cV1bR showed the interaction of C196 and W197 with AVT and three tested antagonists (one peptide and two non-peptides), likewise indicating a critical role of the CW motif in ECL2 in avian species. In the interaction of cV1bR with the Manning compound, two adjacent conserved Asp195 (D195) and Ala198 (A198) were involved in the interaction, perhaps contributing to the highest binding affinity shown by the Manning compound with cV1bR obtained in the docking study (Kang et al., 2019). These results may indicate that the differences in the interaction(s) between the agonist/antagonists and the conserved CW motif in ECL2 of cV1aR and cV1bR probably accounts for the observed differences in specific ligand recognition. The conserved Cys residue in the middle of ECL2 (C210 in cV1aR and C196 in cV1bR) was suggested as the reference point for tertiary folding of ECL2 (Unal, Jagannathan, Bhat, & Karnik, 2010; Wheatley et al., 2007). Importantly, it has been shown that Cys (C205) in the mammalian ECL2 can form a disulfide bond with C124 near the extracellular region of TM III, and consequently affecting the ECL2 cap over the binding pocket crevice involving TM III, TM IV and other TMs (Hawtin et al., 2006). Similarly, both cV1aR and cV1bR have a Cys near their extracellular border of their TM III suggesting strongly that a disulfide bond can form with their Cys residue in the ECL2 forming an ECL2 cap over the binding pocket crevice involving TM III (Fig. 10).

Despite the similarity of binding residues in cV1aR and cV1bR with SR-49059 and SSR-149059, respectively, uniqueness was observed in the interacting residues on receptors. The residues K135, Q138, W211, F310, and Q313 in the cV1aR interacted with SR-49059 and similar residues K121, Q124, W197, F310, and Q314 in the cV1bR were found to interact with SSR-149415 (Fig. 10). Residues S316 and A336 in the cV1aR are unique interacting residues with SR-49059 compared to the agonist AVT and other tested antagonists. Additionally, residues S317 and M337 in the cV1bR were unique compared to the agonist AVT, but common with other tested antagonists.

In summary, examining interactions between cV1aR and SR-49059, and between cV1bR and SSR-149415, may give an insight for possible structural demands for the cV1aR and cV1bR to use as a concurrent pair of effective receptor specific antagonists to attenuate chronic stress responses in a given avian species to allow birds to cope better with and recover from stressors that would otherwise overwhelm the bird's natural responses.

Another alternative approach, used in past mammalian and avian studies to enhance the understanding of stress responses is to examine negative/positive feedback interactions between glucocorticoid receptors and vasopressin/vasotocin/CRH receptors (Bale & Vale, 2004; Burford, Webster, & Cruz-Topete, 2017; Herman, McKlveen, Solomon, Carvalho-Netto, & Myers, 2012; Iredale & Duman, 1997; Kang, Madkour, & Kuenzel, 2017; Keller-Wood, 2015; Yoshida, 2008).

5. Conclusions and future directions

Here, we summarized anatomical, cellular and molecular evidence elucidating the roles of AVT at different levels of the HPA axis of birds. Almost at all levels of the HPA axis the vasotocin system interacts profoundly with the CRH system. In the PVN, a subpopulation of magnocellular neurons synthesizes both AVT and CRH that are co-deposited in the same secretory vesicles of their cytoplasm. In the internal zone of ME, Herring bodies of AVT axons of magnocellular neurons co-exist with Herring bodies containing CRH. On the cellular level, the NHpC contains a dense population of CRH neurons intermingled with glial cells expressing the vasotocin V1a receptor. At the molecular level, this nucleus shows early gene expression of *CRH* mRNA after exposure to a stressor. The avian V1aR is located in glia of circumventricular organs and may function in osmotic and psychogenic stress. Effective antagonist of the avian V1aR and V1bR were identified by a combination of 3D modeling and pituitary primary cell culture techniques. Future studies are needed to determine the function and connectivity of the NHpC in order to establish whether it is a functional part of the avian HPA axis. Additionally pursing the functional role of AVT magnocellular neurons producing both CRH and AVT peptides needs to be explored. The role of V1aR in glial cells, and in tanycytes and ependymal cells in particular, is almost completely unknown. Tanycytes can internalize and transport molecules between the cerebrospinal fluid and neurons. Studies of the roles of tanycytes in neuroendocrine control of stress responses could close some gaps in our knowledge about feedback mechanism within the HPA axis.

Acknowledgments

The authors acknowledge past support from the National Science Foundation IOS-0842937, the Arkansas Biosciences Institute (ABI) and the Poultry Science Department.

References

Acharjee, S., Do-Rego, J. L., Oh, D. Y., Ahn, R. S., Choe, H., Vaudry, H., et al. (2004). Identification of amino acid residues that direct differential ligand selectivity of mammalian and nonmammalian V1a type receptors for arginine vasopressin and vasotocin. Insights into molecular coevolution of V1a type receptors and their ligands. *The Journal of Biological Chemistry*, *279*(52), 54445–54453.

Acher, R. (1993). Neurohypophysial peptide systems: Processing machinery, hydroosmotic regulation, adaptation and evolution. *Regulatory Peptides*, *45*(1–2), 1–13.

Aman, N. A., Nagarajan, G., Kang, S. W., Hancock, M., & Kuenzel, W. J. (2016). Differential responses of the vasotocin 1a receptor (V1aR) and osmoreceptors to immobilization and osmotic stress in sensory circumventricular organs of the chicken (*Gallus gallus*) brain. *Brain Research*, *1649*(Pt. A), 67–78. https://doi.org/10.1016/j.brainres.2016.08.028.

Arsenijevic, Y., Dubois-Dauphin, M., Tribollet, E., Manning, M., Sawyer, W. H., & Dreifuss, J. J. (1994). Vasopressin-binding sites in the pig pituitary gland: Competition by novel vasopressin antagonists suggests the existence of an unusual receptor subtype in the anterior lobe. *The Journal of Endocrinology*, *141*(3), 383–391.

Astheimer, L. B., Buttemer, W. A., & Wingfield, J. C. (1995). Seasonal and acute changes in adrenocortical responsiveness in an arctic-breeding bird. *Hormones and Behavior*, *29*, 442–457.

Autelitano, D. J., Lundblad, J. R., Blum, M., & Roberts, J. L. (1989). Hormonal regulation of POMC gene expression. *Annual Review of Physiology*, *51*, 715–726.

Baker, B. I., Bird, D. J., & Buckingham, J. C. (1996). In the trout, CRH and AVT synergize to stimulate ACTH release. *Regulatory Peptides*, *67*(3), 207–210.

Bale, T. L., & Vale, W. W. (2004). CRF and CRF receptors: Role in stress responsivity and other behaviors. *Annual Review of Pharmacology and Toxicology*, *44*, 525–557. https://doi.org/10.1146/annurev.pharmtox.44.101802.121410.

Banks, W. A., Kastin, A. J., Horvath, A., & Michals, E. A. (1987). Carrier-mediated transport of vasopressin across the blood-brain barrier of the mouse. *Journal of Neuroscience Research*, *18*(2), 326–332. https://doi.org/10.1002/jnr.490180209.

Banks, E. M., Wood-Gush, D. G., Hughes, B. O., & Mankovich, N. J. (1979). Social rank and priority of access to resources in domestic fowl. *Behavioural Processes*, *4*(3), 197–209. https://doi.org/10.1016/0376-6357(79)90001-9.

Baran, N. M., Sklar, N. C., & Adkins-Regan, E. (2016). Developmental effects of vasotocin and nonapeptide receptors on early social attachment and affiliative behavior in the zebra finch. *Hormones and Behavior*, *78*, 20–31. https://doi.org/10.1016/j.yhbeh.2015.10.005.

Barwell, J., Woolley, M. J., Wheatley, M., Conner, A. C., & Poyner, D. R. (2012). The role of the extracellular loops of the CGRP receptor, a family B GPCR. *Biochemical Society Transactions*, *40*(2), 433–437. https://doi.org/10.1042/BST20110726.

Biag, J., Huang, Y., Gou, L., Hintiryan, H., Askarinam, A., Hahn, J. D., et al. (2012). Cyto- and chemoarchitecture of the hypothalamic paraventricular nucleus in the C57BL/6J male mouse: A study of immunostaining and multiple fluorescent tract tracing. *The Journal of Comparative Neurology*, *520*(1), 6–33. https://doi.org/10.1002/cne.22698.

Biebermann, H., Kuhnen, P., Kleinau, G., & Krude, H. (2012). The neuroendocrine circuitry controlled by POMC, MSH, and AGRP. *Handbook of Experimental Pharmacology*, *209*, 47–75. https://doi.org/10.1007/978-3-642-24716-3_3.

Bilodeau, S., Vallette-Kasic, S., Gauthier, Y., Figarella-Branger, D., Brue, T., Berthelet, F., et al. (2006). Role of Brg1 and HDAC2 in GR trans-repression of the pituitary POMC gene and misexpression in Cushing disease. *Genes & Development*, *20*(20), 2871–2886. https://doi.org/10.1101/gad.1444606.

Bolborea, M., & Dale, N. (2013). Hypothalamic tanycytes: Potential roles in the control of feeding and energy balance. *Trends in Neurosciences*, *36*(2), 91–100. https://doi.org/10.1016/j.tins.2012.12.008.
Brooks, A. N., & Challis, J. R. (1989). Effects of CRF, AVP and opioid peptides on pituitary-adrenal responses in sheep. *Peptides*, *10*(6), 1291–1293.
Brubaker, P. L., Baird, A. C., Bennett, H. P., Browne, C. A., & Solomon, S. (1982). Corticotropic peptides in the human fetal pituitary. *Endocrinology*, *111*(4), 1150–1155. https://doi.org/10.1210/endo-111-4-1150.
Bunnemann, B., Iwai, N., Metzger, R., Fuxe, K., Inagami, T., & Ganten, D. (1992). The distribution of angiotensin II AT1 receptor subtype mRNA in the rat brain. *Neuroscience Letters*, *142*(2), 155–158.
Burford, N. G., Webster, N. A., & Cruz-Topete, D. (2017). Hypothalamic-pituitary-adrenal axis modulation of glucocorticoids in the cardiovascular system. *International Journal of Molecular Sciences*, *18*(10). https://doi.org/10.3390/ijms18102150.
Castro, M. G., Estivariz, F. E., & Iturriza, F. C. (1986). The regulation of the corticomelanotropic cell activity in Aves—II. Effect of various peptides on the release of ACTH from dispersed, perfused duck pituitary cells. *Comparative Biochemistry and Physiology A, Comparative Physiology*, *83*(1), 71–75.
Catalano, R. D., Kyriakou, T., Chen, J., Easton, A., & Hillhouse, E. W. (2003). Regulation of corticotropin-releasing hormone type 2 receptors by multiple promoters and alternative splicing: Identification of multiple splice variants. *Molecular Endocrinology*, *17*(3), 395–410. https://doi.org/10.1210/me.2002-0302.
Chang, C. L., & Hsu, S. Y. (2004). Ancient evolution of stress-regulating peptides in vertebrates. *Peptides*, *25*(10), 1681–1688. https://doi.org/10.1016/j.peptides.2004.05.022.
Cho, H. J., Acharjee, S., Moon, M. J., Oh, D. Y., Vaudry, H., Kwon, H. B., et al. (2007). Molecular evolution of neuropeptide receptors with regard to maintaining high affinity to their authentic ligands. *General and Comparative Endocrinology*, *153*(1–3), 98–107.
Ciura, S., & Bourque, C. W. (2006). Transient receptor potential vanilloid 1 is required for intrinsic osmoreception in organum vasculosum lamina terminalis neurons and for normal thirst responses to systemic hyperosmolality. *The Journal of Neuroscience*, *26*(35), 9069–9075. https://doi.org/10.1523/JNEUROSCI.0877-06.2006.
Ciura, S., Liedtke, W., & Bourque, C. W. (2011). Hypertonicity sensing in organum vasculosum lamina terminalis neurons: A mechanical process involving TRPV1 but not TRPV4. *The Journal of Neuroscience*, *31*(41), 14669–14676. https://doi.org/10.1523/JNEUROSCI.1420-11.2011.
Conner, M., Hawtin, S. R., Simms, J., Wootten, D., Lawson, Z., Conner, A. C., et al. (2007). Systematic analysis of the entire second extracellular loop of the V(1a) vasopressin receptor: Key residues, conserved throughout a G-protein-coupled receptor family, identified. *The Journal of Biological Chemistry*, *282*(24), 17405–17412. https://doi.org/10.1074/jbc.M702151200.
Cornett, L. E., Kang, S. W., & Kuenzel, W. J. (2013). A possible mechanism contributing to the synergistic action of vasotocin (VT) and corticotropin-releasing hormone (CRH) receptors on corticosterone release in birds. *General and Comparative Endocrinology*, *188*, 46–53. https://doi.org/10.1016/j.ygcen.2013.02.032.
Cotte, N., Balestre, M. N., Aumelas, A., Mahe, E., Phalipou, S., Morin, D., et al. (2000). Conserved aromatic residues in the transmembrane region VI of the V1a vasopressin receptor differentiate agonist vs. antagonist ligand binding. *European Journal of Biochemistry*, *267*(13), 4253–4263.
Czaplewski, C., Kazmierkiewicz, R., & Ciarkowski, J. (1998a). Molecular modeling of the human vasopressin V2 receptor/agonist complex. *Journal of Computer-Aided Molecular Design*, *12*(3), 275–287.

Czaplewski, C., Kazmierkiewicz, R., & Ciarkowski, J. (1998b). Molecular modelling of the vasopressin V2 receptor/antagonist interactions. *Acta Biochimica Polonica*, *45*(1), 19–26.

De Groef, B., Grommen, S. V., Mertens, I., Schoofs, L., Kuhn, E. R., & Darras, V. M. (2004). Cloning and tissue distribution of the chicken type 2 corticotropin-releasing hormone receptor. *General and Comparative Endocrinology*, *138*(1), 89–95.

DeBold, C. R., Sheldon, W. R., DeCherney, G. S., Jackson, R. V., Alexander, A. N., Vale, W., et al. (1984). Arginine vasopressin potentiates adrenocorticotropin release induced by ovine corticotropin-releasing factor. *The Journal of Clinical Investigation*, *73*(2), 533–538. https://doi.org/10.1172/JCI111240.

Dickens, M. J., Earle, K. A., & Romero, L. M. (2009). Initial transference of wild birds to captivity alters stress physiology. *General and Comparative Endocrinology*, *160*(1), 76–83. https://doi.org/10.1016/j.ygcen.2008.10.023.

Dominic, C. J., & Singh, R. M. (1969). Anterior and posterior groups of portal vessels in the avian pituitary. *General and Comparative Endocrinology*, *13*(1), 22–26.

Dumais, K. M., & Veenema, A. H. (2016). Vasopressin and oxytocin receptor systems in the brain: Sex differences and sex-specific regulation of social behavior. *Frontiers in Neuroendocrinology*, *40*, 1–23. https://doi.org/10.1016/j.yfrne.2015.04.003.

Evans, M. J., Brett, J. T., McIntosh, R. P., McIntosh, J. E., McLay, J. L., Livesey, J. H., et al. (1988). Characteristics of the ACTH response to repeated pulses of corticotrophin-releasing factor and arginine vasopressin in vitro. *The Journal of Endocrinology*, *117*(3), 387–395.

Familari, M., Smith, A. I., Smith, R., & Funder, J. W. (1989). Arginine vasopressin is a much more potent stimulus to ACTH release from ovine anterior pituitary cells than ovine corticotropin-releasing factor. 1. In vitro studies. *Neuroendocrinology*, *50*(2), 152–157. https://doi.org/10.1159/000125214.

Fekete, E. M., & Zorrilla, E. P. (2007). Physiology, pharmacology, and therapeutic relevance of urocortins in mammals: Ancient CRF paralogs. *Frontiers in Neuroendocrinology*, *28*(1), 1–27. https://doi.org/10.1016/j.yfrne.2006.09.002.

Fitzsimons, J. T., & Simons, B. J. (1969). The effect on drinking in the rat of intravenous infusion of angiotensin, given alone or in combination with other stimuli of thirst. *The Journal of Physiology*, *203*(1), 45–57.

Giles, M. E., Fernley, R. T., Nakamura, Y., Moeller, I., Aldred, G. P., Ferraro, T., et al. (1999). Characterization of a specific antibody to the rat angiotensin II AT1 receptor. *The Journal of Histochemistry and Cytochemistry*, *47*(4), 507–516. https://doi.org/10.1177/002215549904700409.

Gillies, G. E., Linton, E. A., & Lowry, P. J. (1982). Corticotropin releasing activity of the new CRF is potentiated several times by vasopressin. *Nature*, *299*(5881), 355–357.

Goodson, J. L., & Evans, A. K. (2004). Neural responses to territorial challenge and nonsocial stress in male song sparrows: Segregation, integration, and modulation by a vasopressin V1 antagonist. *Hormones and Behavior*, *46*(4), 371–381. https://doi.org/10.1016/j.yhbeh.2004.02.008.

Hartman, F. A., Knouff, R. A., McNutt, A. W., & Carver, J. E. (1947). Chromaffin patterns in bird adrenals. *Anatomical Record*, *97*(2), 211–221.

Hausmann, H., Richters, A., Kreienkamp, H. J., Meyerhof, W., Mattes, H., Lederis, K., et al. (1996). Mutational analysis and molecular modeling of the nonapeptide hormone binding domains of the. *Proceedings of the National Academy of Sciences of the United States of America*, *93*(14), 6907–6912.

Hawtin, S. R., Simms, J., Conner, M., Lawson, Z., Parslow, R. A., Trim, J., et al. (2006). Charged extracellular residues, conserved throughout a G-protein-coupled receptor family, are required for ligand binding, receptor activation, and cell-surface expression. *The Journal of Biological Chemistry*, *281*(50), 38478–38488. https://doi.org/10.1074/jbc.M607639200.

Herman, J. P., McKlveen, J. M., Solomon, M. B., Carvalho-Netto, E., & Myers, B. (2012). Neural regulation of the stress response: Glucocorticoid feedback mechanisms. *Brazilian Journal of Medical and Biological Research*, *45*(4), 292–298.

Holmes, M. C., Antoni, F. A., Aguilera, G., & Catt, K. J. (1986). Magnocellular axons in passage through the median eminence release vasopressin. *Nature*, *319*(6051), 326–329. https://doi.org/10.1038/319326a0.

Ikemura, R., Matsuwaki, T., Yamanouchi, K., & Nishihara, M. (2004). Involvement of endogenous vasopressin in high plasma osmolality-induced anorexia via V1 receptor-mediated mechanism. *The Journal of Veterinary Medical Science*, *66*(8), 951–955.

Imaki, T., Naruse, M., Harada, S., Chikada, N., Imaki, J., Onodera, H., et al. (1996). Corticotropin-releasing factor up-regulates its own receptor mRNA in the paraventricular nucleus of the hypothalamus. *Brain Research. Molecular Brain Research*, *38*(1), 166–170.

Iredale, P. A., & Duman, R. S. (1997). Glucocorticoid regulation of corticotropin-releasing factor1 receptor expression in pituitary-derived AtT-20 cells. *Molecular Pharmacology*, *51*(5), 794–799.

Iwasaki-Sekino, A., Mano-Otagiri, A., Ohata, H., Yamauchi, N., & Shibasaki, T. (2009). Gender differences in corticotropin and corticosterone secretion and corticotropin-releasing factor mRNA expression in the paraventricular nucleus of the hypothalamus and the central nucleus of the amygdala in response to footshock stress or psychological stress in rats. *Psychoneuroendocrinology*, *34*(2), 226–237. https://doi.org/10.1016/j.psyneuen.2008.09.003.

Jayanthi, S., Kang, S. W., Bingham, D., Tessaro, B. A., Suresh Kumar, T. K., & Kuenzel, W. J. (2014). Identification of antagonists to the vasotocin receptor sub-type 4 (VT4R) involved in stress by molecular modelling and verification using anterior pituitary cells. *Journal of Biomolecular Structure & Dynamics*, *32*(4), 648–660. https://doi.org/10.1080/07391102.2013.787025.

Johnson, A. K., & Gross, P. M. (1993). Sensory circumventricular organs and brain homeostatic pathways. *The FASEB Journal*, 7(8), 678–686.

Jones, R. B. (1987). Assessment of fear in adult laying hens: Correlational analysis of methods and measures. *British Poultry Science*, *28*(2), 319–326. https://doi.org/10.1080/00071668708416964.

Jurkevich, A., Barth, S. W., & Grossmann, R. (1997). Sexual dimorphism of arg-vasotocin gene expressing neurons in the telencephalon and dorsal diencephalon of the domestic fowl. An immunocytochemical and in situ hybridization study. *Cell and Tissue Research*, *287*(1), 69–77.

Jurkevich, A., Berghman, L. R., Cornett, L. E., & Kuenzel, W. J. (2005). Characterization and immunohistochemical visualization of the vasotocin VT2 receptor in the pituitary gland of the chicken, Gallus gallus. *General and Comparative Endocrinology*, *143*(1), 82–91. https://doi.org/10.1016/j.ygcen.2005.02.017.

Jurkevich, A., Berghman, L. R., Cornett, L. E., & Kuenzel, W. J. (2008). Immunohistochemical characterization of chicken pituitary cells containing the vasotocin VT2 receptor. *Cell and Tissue Research*, *333*(2), 253–262. https://doi.org/10.1007/s00441-008-0636.

Jurkevich, A., Mikhailova, M. V., Madison, F. N., Cornett, L. E., & Kuenzel, W. J. (2010). Molecular processes underlying neuroendocrine regulation of stress responses in an avian species. In N. S. Penkava & L. R. Haight (Eds.), *Neuroendocrinology research developments* (pp. 1–33). New York: Nova Science Publishers, Inc.

Kang, S. W., Jayanthi, S., Nagarajan, G., Suresh Kumar, T. K., & Kuenzel, W. J. (2019). Identification of avian vasotocin receptor subtype-specific antagonists involved in the stress response of the chicken, *Gallus gallus*. *Journal of Biomolecular Structure & Dynamics*, *37*, 1685–1699. https://doi.org/10.1080/07391102.2018.1464957.

Kang, S. W., & Kuenzel, W. J. (2014). Regulation of gene expression of vasotocin and corticotropin-releasing hormone receptors in the avian anterior pituitary by corticosterone. *General and Comparative Endocrinology*, *204*, 25–32. https://doi.org/10.1016/j.ygcen.2014.04.018.

Kang, S. W., Madkour, M., & Kuenzel, W. J. (2017). Tissue-specific expression of DNA methyltransferases involved in early-life nutritional stress of chicken, *Gallus gallus*. *Frontiers in Genetics*, *8*, 204. https://doi.org/10.3389/fgene.2017.00204.

Keller-Wood, M. (2015). Hypothalamic-pituitary—adrenal axis-feedback control. *Comprehensive Physiology*, *5*(3), 1161–1182. https://doi.org/10.1002/cphy.c140065.

Kemppainen, R. J., Clark, T. P., Sartin, J. L., & Zerbe, C. A. (1993). Hypothalamic peptide regulation of ACTH secretion from sheep pituitary. *The American Journal of Physiology*, *265*(4 Pt. 2), R840–R845. https://doi.org/10.1152/ajpregu.1993.265.4.R840.

Koch, K. A., Wingfield, J. C., & Buntin, J. D. (2002). Glucocorticoids and parental hyperphagia in ring doves (Streptopelia risoria). *Hormones and Behavior*, *41*(1), 9–21. https://doi.org/10.1006/hbeh.2001.1726.

Kuenzel, W. J., Hancock, M., Nagarajan, G., Aman, N. A., & Kang, S. W. (2016). Central effect of vasotocin 4 receptor (VT4R/V1aR) antagonists on the stress response and food intake in chicks given neuropeptide Y (NPY). *Neuroscience Letters*, *620*, 57–61. https://doi.org/10.1016/j.neulet.2016.03.036.

Kuenzel, W. J., Kang, S. W., & Jurkevich, A. (2013). Neuroendocrine regulation of stress in birds with an emphasis on vasotocin receptors (VTRs). *General and Comparative Endocrinology*, *190*, 18–23. https://doi.org/10.1016/j.ygcen.2013.02.029.

Landgraf, R., Ramirez, A. D., & Ramirez, V. D. (1991). The positive feedback action of vasopressin on its own release from rat septal tissue in vitro is receptor-mediated. *Brain Research*, *545*(1–2), 137–141.

Leibowitz, S. F., Sladek, C., Spencer, L., & Tempel, D. (1988). Neuropeptide Y, epinephrine and norepinephrine in the paraventricular nucleus: Stimulation of feeding and the release of corticosterone, vasopressin and glucose. *Brain Research Bulletin*, *21*(6), 905–912.

Lenkei, Z., Palkovits, M., Corvol, P., & Llorens-Cortes, C. (1997). Expression of angiotensin type-1 (AT1) and type-2 (AT2) receptor mRNAs in the adult rat brain: A functional neuroanatomical review. *Frontiers in Neuroendocrinology*, *18*(4), 383–439. https://doi.org/10.1006/frne.1997.0155.

Leong, D. A. (1988). A complex mechanism of facilitation in pituitary ACTH cells: Recent single-cell studies. *The Journal of Experimental Biology*, *139*, 151–168.

Leung, C. H., Abebe, D. F., Earp, S. E., Goode, C. T., Grozhik, A. V., Mididoddi, P., et al. (2011). Neural distribution of vasotocin receptor mRNA in two species of songbird. *Endocrinology*, *152*(12), 4865–4881. https://doi.org/10.1210/en.2011-1394.

Liedtke, W., Choe, Y., Marti-Renom, M. A., Bell, A. M., Denis, C. S., Sali, A., et al. (2000). Vanilloid receptor-related osmotically activated channel (VR-OAC), a candidate vertebrate osmoreceptor. *Cell*, *103*(3), 525–535.

Lim, M. M., Nair, H. P., & Young, L. J. (2005). Species and sex differences in brain distribution of corticotropin-releasing factor receptor subtypes 1 and 2 in monogamous and promiscuous vole species. *The Journal of Comparative Neurology*, *487*(1), 75–92. https://doi.org/10.1002/cne.20532.

Liposits, Z., Paull, W. K., Setalo, G., & Vigh, S. (1985). Evidence for local corticotropin releasing factor (CRF)-immunoreactive neuronal circuits in the paraventricular nucleus of the rat hypothalamus. An electron microscopic immunohistochemical analysis. *Histochemistry*, *83*(1), 5–16.

Liu, J. P., Robinson, P. J., Funder, J. W., & Engler, D. (1990). The biosynthesis and secretion of adrenocorticotropin by the ovine anterior pituitary is predominantly regulated by arginine vasopressin (AVP). Evidence that protein kinase C mediates the action of AVP. *The Journal of Biological Chemistry*, *265*(24), 14136–14142.

Lovejoy, D. A., Chang, B. S., Lovejoy, N. R., & del Castillo, J. (2014). Molecular evolution of GPCRs: CRH/CRH receptors. *Journal of Molecular Endocrinology*, *52*(3), T43–T60. https://doi.org/10.1530/JME-13-0238.

Madison, F. N., Jurkevich, A., & Kuenzel, W. J. (2008). Sex differences in plasma corticosterone release in undisturbed chickens (*Gallus gallus*) in response to arginine vasotocin and corticotropin releasing hormone. *General and Comparative Endocrinology*, *155*(3), 566–573. https://doi.org/10.1016/j.ygcen.2007.08.014.

Mahlmann, S., Meyerhof, W., Hausmann, H., Heierhorst, J., Schonrock, C., Zwiers, H., et al. (1994). Structure, function, and phylogeny of [Arg8]vasotocin receptors from teleost fish and toad. *Proceedings of the National Academy of Sciences of the United States of America*, *91*(4), 1342–1345.

Martins, J. M., Kastin, A. J., & Banks, W. A. (1996). Unidirectional specific and modulated brain to blood transport of corticotropin-releasing hormone. *Neuroendocrinology*, *63*(4), 338–348. https://doi.org/10.1159/000126974.

Meyer, A. H., Langhans, W., & Scharrer, E. (1989). Vasopressin reduces food intake in goats. *Quarterly Journal of Experimental Physiology*, *74*(4), 465–473.

Mikami, S. (1986). Immunocytochemistry of the avian hypothalamus and adenohypophysis. *International Review of Cytology*, *103*, 189–248.

Mikhailova, M. V., Mayeux, P. R., Jurkevich, A., Kuenzel, W. J., Madison, F., Periasamy, A., et al. (2007). Heterooligomerization between vasotocin and corticotropin-releasing hormone (CRH) receptors augments CRH-stimulated 3′,5′-cyclic adenosine monophosphate production. *Molecular Endocrinology*, *21*(9), 2178–2188. https://doi.org/10.1210/me.2007-0160.

Milligan, G. (2009). G protein-coupled receptor hetero-dimerization: Contribution to pharmacology and function. *British Journal of Pharmacology*, *158*(1), 5–14. https://doi.org/10.1111/j.1476-5381.2009.00169.x.

Mouillac, B., Chini, B., Balestre, M. N., Elands, J., Trumpp-Kallmeyer, S., Hoflack, J., et al. (1995). The binding site of neuropeptide vasopressin V1a receptor. Evidence for a major localization within transmembrane regions. *The Journal of Biological Chemistry*, *270*(43), 25771–25777.

Mouillac, B., Chini, B., Balestre, M. N., Jard, S., Barberis, C., Manning, M., et al. (1995). Identification of agonist binding sites of vasopressin and oxytocin receptors. *Advances in Experimental Medicine and Biology*, *395*, 301–310.

Nagarajan, G., Jurkevich, A., Kang, S. W., & Kuenzel, W. J. (2016). Diencephalic and septal structures containing the avian vasotocin receptor (V1aR) involved in the regulation of food intake in chickens, *Gallus gallus*. *Physiology & Behavior*, *164*(Pt. A), 268–276. https://doi.org/10.1016/j.physbeh.2016.06.018.

Nagarajan, G., Jurkevich, A., Kang, S. W., & Kuenzel, W. J. (2017). Anatomical and functional implications of corticotrophin-releasing hormone neurones in a septal nucleus of the avian brain: An emphasis on glial-neuronal interaction via V1a receptors in vitro. *Journal of Neuroendocrinology*, *29*(7), 1–11. https://doi.org/10.1111/jne.12494.

Nagarajan, G., Kang, S. W., & Kuenzel, W. J. (2017). Functional evidence that the nucleus of the hippocampal commissure shows an earlier activation from a stressor than the paraventricular nucleus: Implication of an additional structural component of the avian hypothalamo-pituitary-adrenal axis. *Neuroscience Letters*, *642*, 14–19. https://doi.org/10.1016/j.neulet.2017.01.064.

Ocampo Daza, D., Lewicka, M., & Larhammar, D. (2012). The oxytocin/vasopressin receptor family has at least five members in the gnathostome lineage, inclucing two distinct V2 subtypes. *General and Comparative Endocrinology*, *175*(1), 135–143. https://doi.org/10.1016/j.ygcen.2011.10.011.

Oksche, A., & Farner, D. S. (1974). Neurohistological studies of the hypothalamo-hypophysial system of *Zonotrichia leucophrys gambelii* (Aves, Passeriformes). *Advances in Anatomy, Embryology, and Cell Biology*, *48*(4), 1–136.

Oksche, A., Farner, D. S., Serventy, D. L., Wolff, F., & Nicholls, C. A. (1963). The hypothalamo-hypophysial neurosecretory system of the zebra finch. *Taeniopygia castanotis*. *Zeitschrift für Zellforschung und Mikroskopische Anatomie*, *58*, 846–914.

Park, P. S., Filipek, S., Wells, J. W., & Palczewski, K. (2004). Oligomerization of G protein-coupled receptors: Past, present, and future. *Biochemistry*, *43*(50), 15643–15656. https://doi.org/10.1021/bi047907k.

Parkes, D., Rivest, S., Lee, S., Rivier, C., & Vale, W. (1993). Corticotropin-releasing factor activates c-fos, NGFI-B, and corticotropin-releasing factor gene expression within the paraventricular nucleus of the rat hypothalamus. *Molecular Endocrinology*, 7(10), 1357–1367. https://doi.org/10.1210/mend.7.10.8264665.

Raadsheer, F. C., Sluiter, A. A., Ravid, R., Tilders, F. J., & Swaab, D. F. (1993). Localization of corticotropin-releasing hormone (CRH) neurons in the paraventricular nucleus of the human hypothalamus; age-dependent colocalization with vasopressin. *Brain Research*, *615*(1), 50–62.

Recht, L. D., Hoffman, D. L., Haldar, J., Silverman, A. J., & Zimmerman, E. A. (1981). Vasopressin concentrations in hypophysial portal plasma: Insignificant reduction following removal of the posterior pituitary gland. *Neuroendocrinology*, *33*(2), 88–90. https://doi.org/10.1159/000123208.

Rich, E. L., & Romero, L. M. (2005). Exposure to chronic stress downregulates corticosterone responses to acute stressors. *American Journal of Physiology. Regulatory, Integrative and Comparative Physiology*, *288*(6), R1628–R1636. https://doi.org/10.1152/ajpregu.00484.2004.

Rivalland, E. T., Iqbal, J., Clarke, I. J., Turner, A. I., & Tilbrook, A. J. (2005). Co-localization and distribution of corticotrophin-releasing hormone, arginine vasopressin and enkephalin in the paraventricular nucleus of sheep: A sex comparison. *Neuroscience*, *132*(3), 755–766. https://doi.org/10.1016/j.neuroscience.2005.01.045.

Rivier, C., & Vale, W. (1983). Interaction of corticotropin-releasing factor and arginine vasopressin on adrenocorticotropin secretion in vivo. *Endocrinology*, *113*(3), 939–942. https://doi.org/10.1210/endo-113-3-939.

Romero, L. M. (2002). Seasonal changes in plasma glucocorticoid concentrations in free-living vertebrates. *General and Comparative Endocrinology*, *128*(1), 1–24.

Romero, L. M., Cyr, N. E., & Romero, R. C. (2006). Corticosterone responses change seasonally in free-living house sparrows (*Passer domesticus*). *General and Comparative Endocrinology*, *149*(1), 58–65. https://doi.org/10.1016/j.ygcen.2006.05.004.

Romero, L. M., & Wingfield, J. C. (2001). Regulation of the hypothalamic-pituitary-adrenal axis in free-living pigeons. *Journal of Comparative Physiology B*, *171*(3), 231–235.

Sawchenko, P. E., Swanson, L. W., & Vale, W. W. (1984). Corticotropin-releasing factor: Co-expression within distinct subsets of oxytocin-, vasopressin-, and neurotensin-immunoreactive neurons in the hypothalamus of the male rat. *The Journal of Neuroscience*, *4*(4), 1118–1129.

Schwartz, J., & Vale, W. (1988). Dissociation of the adrenocorticotropin secretory responses to corticotropin-releasing factor (CRF) and vasopressin or oxytocin by using a specific cytotoxic analog of CRF. *Endocrinology*, *122*(4), 1695–1700. https://doi.org/10.1210/endo-122-4-1695.

Selvam, R., Jurkevich, A., Kang, S. W., Mikhailova, M. V., Cornett, L. E., & Kuenzel, W. J. (2013). Distribution of the vasotocin subtype four receptor (VT4R) in the anterior pituitary gland of the chicken, *Gallus gallus*, and its possible role in the avian stress response. *Journal of Neuroendocrinology*, *25*(1), 56–66.

Selvam, R., Jurkevich, A., & Kuenzel, W. J. (2015). Distribution of the vasotocin type 4 receptor throughout the brain of the chicken, *Gallus gallus*. *The Journal of Comparative Neurology*, *523*(2), 335–358. https://doi.org/10.1002/cne.23684.

Serradeil-Le Gal, C., Wagnon, J., 3rd, Tonnerre, B., Roux, R., Garcia, G., Griebel, G., et al. (2005). An overview of SSR149415, a selective nonpeptide vasopressin V(1b) receptor antagonist for the treatment of stress-related disorders. *CNS Drug Reviews*, *11*(1), 53–68.

Sharif Naeini, R., Witty, M. F., Seguela, P., & Bourque, C. W. (2006). An N-terminal variant of Trpv1 channel is required for osmosensory transduction. *Nature Neuroscience*, *9*(1), 93–98. https://doi.org/10.1038/nn1614.

Silverin, B. (1998). Behavioural and hormonal responses of the pied flycatcher to environmental stressors. *Animal Behaviour*, *55*(6), 1411–1420. https://doi.org/10.1006/anbe.1997.0717.

Simmons, D. M., & Swanson, L. W. (2009). Comparison of the spatial distribution of seven types of neuroendocrine neurons in the rat paraventricular nucleus: Toward a global 3D model. *The Journal of Comparative Neurology*, *516*(5), 423–441. https://doi.org/10.1002/cne.22126.

Simpson, J. B., & Routtenberg, A. (1973). Subfornical organ: Site of drinking elicitation by angiotensin II. *Science*, *181*(4105), 1172–1175.

Swanson, L. W., & Sawchenko, P. E. (1980). Paraventricular nucleus: A site for the integration of neuroendocrine and autonomic mechanisms. *Neuroendocrinology*, *31*(6), 410–417. https://doi.org/10.1159/000123111.

Tachibana, T., Saito, E. S., Saito, S., Tomonaga, S., Denbow, D. M., & Furuse, M. (2004). Comparison of brain arginine-vasotocin and corticotrophin-releasing factor for physiological responses in chicks. *Neuroscience Letters*, *360*(3), 165–169. https://doi.org/10.1016/j.neulet.2004.02.055.

Unal, H., Jagannathan, R., Bhat, M. B., & Karnik, S. S. (2010). Ligand-specific conformation of extracellular loop-2 in the angiotensin II type 1 receptor. *The Journal of Biological Chemistry*, *285*(21), 16341–16350. https://doi.org/10.1074/jbc.M109.094870.

Vale, W., Vaughan, J., Smith, M., Yamamoto, G., Rivier, J., & Rivier, C. (1983). Effects of synthetic ovine corticotropin-releasing factor, glucocorticoids, catecholamines, neurohypophysial peptides, and other substances on cultured corticotropic cells. *Endocrinology*, *113*(3), 1121–1131. https://doi.org/10.1210/endo-113-3-1121.

Vigh, B. (1971). Das Paraventrikularorgan und das zirkumventrikuläre System des Gehirns. In *Studia biologica hungarica*: *Vol. 10*. Budapest: Akad Kiadó.

Vigh, B., & Vigh-Teichmann, I. (1973). Comparative ultrastructure of the cerebrospinal fluid-contacting neurons. *International Review of Cytology*, *35*, 189–251.

Vitums, A., Mikami, S. I., Oksche, A., & Farner, D. S. (1964). Vascularization of the hypothalamo-hypophysial-complex in the white-crowned sparrow, *Zonotrichia leucophrys gambelii*. *Zeitschrift für Zellforschung und Mikroskopische Anatomie*, *64*, 541–569.

Weindl, A., & Joynt, R. J. (1973). Barrier properties of the subcommissural organ. *Archives of Neurology*, *29*(1), 16–22.

Weindl, A., & Sofroniew, M. V. (1981). Relation of neuropeptides to mammalian circumventricular organs. *Advances in Biochemical Psychopharmacology*, *28*, 303–320.

Westerhof, I., Lumeij, J. T., Mol, J. A., van den Brom, W. E., & Rijnberk, A. (1992). In vivo studies on the effects of ovine corticotrophin-releasing hormone, arginine vasotocin, arginine vasopressin, and haloperidol on adrenocortical function in the racing pigeon (*Columba livia domestica*). *General and Comparative Endocrinology*, *88*(1), 76–82.

Wheatley, M., Simms, J., Hawtin, S. R., Wesley, V. J., Wootten, D., Conner, M., et al. (2007). Extracellular loops and ligand binding to a subfamily of family A G-protein-coupled receptors. *Biochemical Society Transactions*, *35*(Pt. 4), 717–720. https://doi.org/10.1042/BST0350717.

Wheatley, M., Wootten, D., Conner, M. T., Simms, J., Kendrick, R., Logan, R. T., et al. (2012). Lifting the lid on GPCRs: The role of extracellular loops. *British Journal of Pharmacology*, *165*(6), 1688–1703. https://doi.org/10.1111/j.1476-5381.2011.01629.x.

Whitnall, M. H. (1988). Distributions of pro-vasopressin expressing and pro-vasopressin deficient CRH neurons in the paraventricular hypothalamic nucleus of colchicine-treated normal and adrenalectomized rats. *The Journal of Comparative Neurology*, *275*(1), 13–28. https://doi.org/10.1002/cne.902750103.
Whitnall, M. H. (1989). Stress selectively activates the vasopressin-containing subset of corticotropin-releasing hormone neurons. *Neuroendocrinology*, *50*(6), 702–707. https://doi.org/10.1159/000125302.
Willoughby, J. O., & Blessing, W. W. (1987). Neuropeptide Y injected into the supraoptic nucleus causes secretion of vasopressin in the unanesthetized rat. *Neuroscience Letters*, *75*(1), 17–22.
Wingfield, J. C., Vleck, C. M., & Moore, M. C. (1992). Seasonal changes of the adrenocortical response to stress in birds of the Sonoran Desert. *The Journal of Experimental Zoology*, *264*(4), 419–428. https://doi.org/10.1002/jez.1402640407.
Yoshida, M. (2008). Gene regulation system of vasopressin and corticotropin-releasing hormone. *Gene Regulation and Systems Biology*, *2*, 71–88.
Young, S. F., Griffante, C., & Aguilera, G. (2007). Dimerization between vasopressin V1b and corticotropin releasing hormone type 1 receptors. *Cellular and Molecular Neurobiology*, *27*(4), 439–461. https://doi.org/10.1007/s10571-006-9135-8.

Further reading

Aguilera, G., Rabadan-Diehl, C., & Nikodemova, M. (2001). Regulation of pituitary corticotropin releasing hormone receptors. *Peptides*, *22*(5), 769–774.
Carsia, R. V., Weber, H., & Perez, F. M. J. (1986). Corticotropin-releasing factor stimulates the release of adrenocorticotropin from domestic fowl pituitary cells. *Endocrinology*, *118*(1), 143–148.
Cornett, L. E., Kirby, J. D., Vizcarra, J. A., Ellison, J. C., Thrash, J., Mayeux, P. R., et al. (2003). Molecular cloning and functional characterization of a vasotocin receptor subtype expressed in the pituitary gland of the domestic chicken (*Gallus domesticus*): Avian homolog of the mammalian V1b-vasopressin receptor. *Regulatory Peptides*, *110*(3), 231–239.
Seasholtz, A. (2000). Regulation of adrenocorticotropic hormone secretion: Lessons from mice deficient in corticotropin-releasing hormone. *The Journal of Clinical Investigation*, *105*(9), 1187–1188. https://doi.org/10.1172/JCI10002.
Smith, S. M., & Vale, W. W. (2006). The role of the hypothalamic-pituitary-adrenal axis in neuroendocrine responses to stress. *Dialogues in Clinical Neuroscience*, *8*(4), 383–395.

CHAPTER EIGHT

Vasopressin actions in the kidney renin angiotensin system and its role in hypertension and renal disease

Alexis A. Gonzalez*, Nicolas Salinas-Parra, Flavia Cifuentes-Araneda, Cristian Reyes-Martinez
Instituto de Química Pontificia Universidad Católica de Valparaíso, Valparaíso, Chile
*Corresponding author: e-mail address: alexis.gonzalez@pucv.cl

Contents

Abstract

Vasopressin, also named antidiuretic hormone (ADH), arginine vasopressin (AVP) is the main hormone responsible for water maintenance in the body through the antidiuretic actions in the kidney. The posterior pituitary into the blood releases vasopressin formed in the hypothalamus. Hypothalamic osmotic neurons are responsible to initiate the cascade for AVP actions. The effects of AVP peptide includes activation of V2 receptors which stimulate the formation of cyclic AMP (cAMP) and phosphorylation of water channels aquaporin 2 (AQP2) in the collecting duct. AVP also has vasoconstrictor effects through V1a receptors in the vasculature, while V1b is found in the nervous system. V1a and b receptors increases intracellular Ca^{2+} while activation of V2 receptors of signaling pathways are related to cAMP-dependent phosphorylation in kidney collecting ducts acting in coordination to stimulate water and electrolyte homeostasis. AVP potentiate

Vitamins and Hormones, Volume 113
ISSN 0083-6729
https://doi.org/10.1016/bs.vh.2019.09.003

formation of intratubular angiotensin II (Ang II) through V2 receptors-dependent distal tubular renin formation, contributing to Na+ reabsorption. On the same way, Ang II receptors are able to potentiate the effects of V2-dependent stimulation of AQP2 abundance in the plasma membrane. The role of AVP in hypertension and renal disease has been demonstrated in pathological states with the involvement of V2 receptors in the progression of kidney damage in diabetes and also on the stimulation of intracellular pathways linked to the development of polycystic kidney.

1. Introduction

Despite variations in water intake and excretion, the black box illustrated by our body remains in equilibrium. This exact control of water and electrolyte depends on actions of an integrated system between sensory elements in the nervous system with final effects on the renal mechanisms able to regulate water excretion and reabsorption. These mechanisms depend on the antidiuretic hormone (ADH) also known commonly as arginine vasopressin (AVP), which is the key player in all this complex process. AVP is secreted from the posterior pituitary in response to osmolality on hypothalamic neurons (Prager-Khoutorsky & Bourque, 2010). Among physiologic effects of vasopressin, its antidiuretic effect in the kidney collecting duct is the most described well known pathway. In conditions of high plasma angiotensin II (Ang II), a vasoconstrictor hormone, AVP is released to the circulation exerting vasoconstrictor effects on vasculature indicating that AVP is an active contributor to high blood pressure in conditions of renin angiotensin system (RAS) activation (Gonzalez et al., 2016) and models of salt-sensitive hypertension (Prager-Khoutorsky, Choe, Levi, & Bourque, 2017).

AVP exerts its effects on G protein coupled receptors (GPCR) V1a, V1b and V2 receptors (Lozic, Sarenac, Murphy, & Japundzic-Zigon, 2018). Among them, V1a are widely present in the vasculature, exerting vasoconstrictor effects through increases in Intracellular Ca^{2+} and the phosphatidylinositol-bisphosphonate cascade (Koshimizu et al., 2012). AVP increases distal sodium reabsorption (Nicco et al., 2001) demonstrating that AVP might be as important as RAS in the control of arterial blood pressure (Burrell et al., 1994).

V1b are present in the brain (Phillips, Abrahams, et al., 1988) and also activates Ca^{2+} related pathways. AVP effects are classically known by its effects on water retention through the activation of V2 receptors that are abundantly expressed in the kidney collecting ducts of rat, mouse and humans (Phillips, Kelly, et al., 1988). V2 receptors are linked to cAMP accumulation and activation of protein kinase A, which in turn phosphorylates AQP2 proteins and also the cyclic AMP response binding protein (CREB) leading to CREB translocation to the nucleus increasing gene expression (Promeneur, Kwon, Frokiaer, Knepper, & Nielsen, 2000). Phosphorylation of AQP2 causes AQP2 translocation to the apical cell plasma membrane in the principal cells of the collecting ducts increasing solute-free water transport through this tubular cells into blood (Sands, Nonoguchi, & Knepper, 1987). The final effect is the reduced plasma osmolality and increased urine osmolality. AVP stimulation of the V2/PKA/CREB pathway in the principal cells of the collecting duct is also responsible for the stimulation of gene expression of renin, the rate limiting step in Ang II formation, which finally will stimulate sodium reabsorption (Bankir, Bichet, & Bouby, 2010) by activation of intratubular renin angiotensin system (RAS) (Gonzalez et al., 2016).

2. Vasopressin and its receptors in the kidney

AVP is a nine-amino acid polypeptide exerts its effects by three distinct subtypes of receptors, V1a, V1b and V2 receptors, which belong to the large family of G protein coupled receptors (GPCR) (Bankir, Bichet, & Morgenthaler, 2017). The V1a receptor is found on vascular smooth muscle. Additionally, it is expressed in testis, superior cervical ganglion, liver, blood vessels, renal medulla and in the brain (Phillips et al., 1990), however, is the most abundant receptor subtype in the brain (Lozic et al., 2018). The V1b receptor is found in the brain (pituitary-specific subtype), adrenal glands, kidneys and pancreas (Lozic et al., 2018). Both receptors (V1a and V1b) increases intracellular calcium via the phosphatidyl-inositol-bisphosphonate cascade (Koshimizu et al., 2012). V2 receptor is principally found in the kidney (Knepper, Nielsen, & Chou, 2001; Knepper, Nielsen, Chou, & Digiovanni, 1994). Mutig et al. (2007) demonstrate that gene expression of V2 is more abundant in medullary and cortical thick ascending limb (MTAL), macula densa (MD) and collecting duct (CD) of rat, mouse and human kidney (Fig. 1; Mutig et al., 2007).

In other study realized for Fenton, Brond, Nielsen, and Praetorius (2007) using a polyclonal antibody against the N-terminal of V2 receptor demonstrate that this protein is abundantly in CD of rat and mouse with not labeling of V2 in other vascular structure or renal tubules (Fig. 2; Fenton et al., 2007).

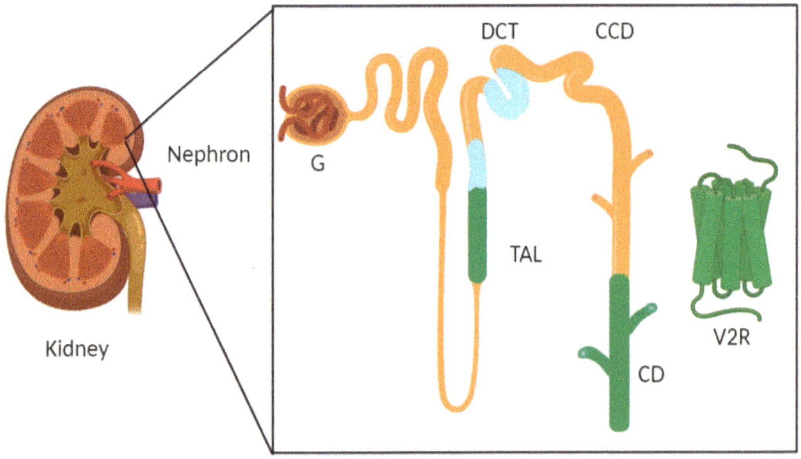

Fig. 1 Vasopressin receptor 2 (V2R) mRNA expression in rat, mouse, and human epithelia. The green regions indicates high expression and light blue low expression of V2R mRNA. G: Glomerulus; TAL: Thick ascending limb; DCT: Distal convoluted tubule; CCD: Cortical collecting duct; CD: Collecting Duct.

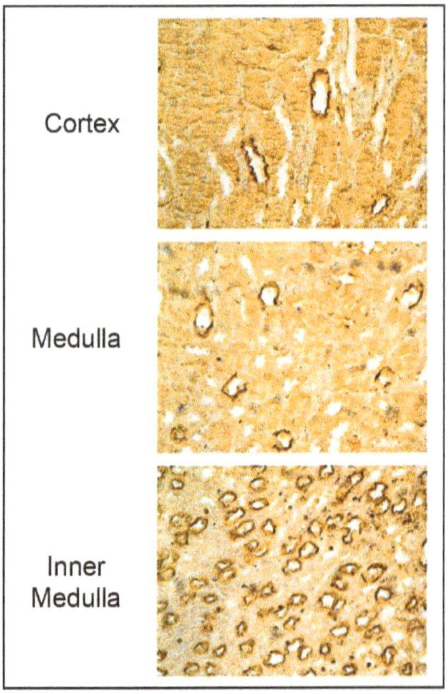

Fig. 2 Immunolocalization of V2R in rat kidney in the cortex, medulla and inner medulla.

The gene is in q28 from X chromosome of *Homo sapiens* (Seibold, Rosenthal, Barberis, & Birnbaumer, 1993; Seibold, Rosenthal, Bichet, & Birnbaumer, 1993). V2R exerts its effects trough Gs proteins, leading to increase cAMP levels and activation of protein kinase A (Chou, Knepper, & Layton, 1993; Hoffert, Chou, & Knepper, 2009).

3. Structural characterization of vasopressin receptors

Bioinformatic analysis associated to the measurement of the hydrophobicity index of a protein allows to establish the transmembrane segments of this, allowing inferring that the V2 receptor would belong to the long family of GPCRs composed of seven-transmembrane helices, as well as it would belong to type 1 of membrane proteins (with an extracellular N-terminal end and an intracellular C-terminal end) (Fig. 3A; High, Flint, & Dobberstein, 1991). An analysis of the protein sequence of V2 from different species with evidence at the protein level allowed determining that this protein shares a high percentage of identity (>80%), which means that it is a highly conserved protein, as well as that this is mainly composed of alpha helix (Fig. 3B).

In 1997 Schulein and collaborators determined through the Pho/LacZ gene fusion system, that a fragment of 71 amino acids near the amino terminus of the protein was enough for its insertion and correct orientation in prokaryotic cells such as eukaryotes.

Although studies establish the topology of the protein and bioinformatics tools allow predicting the behavior of the sequence at membrane level, currently there are no crystallographic structures of the complete V2R protein. A search in the RCSB protein structure database allows to identify the existence of a 3D structure corresponding to i3 loop from human V2 receptor (PDB: 2JX4), as well as the structures of various proteins belonging to the GPCRs family, but due to the importance of this receptor in the renal water reabsorption process a three-dimensional structure of the V2 receptor has been developed by comparative modeling (Fig. 4A) that has allowed to clarify the interaction that takes place between ligand and its receptor (Slusarz, Slusarz, & Ciarkowski, 2006). The study carried out by Slusarz and collaborators determined by molecular dynamics, situating the receptor in a membrane of 1-palmitoyl-2-oleoyl-sn-glycero-3-phosphatidylcholine (POPC), that the residues Q92, Q96, Q119, Q291 involved in the interaction between the ligand and receptor were especially important for ligand binding (Fig. 4B), as well as the residue S315, which in a "non-active" state

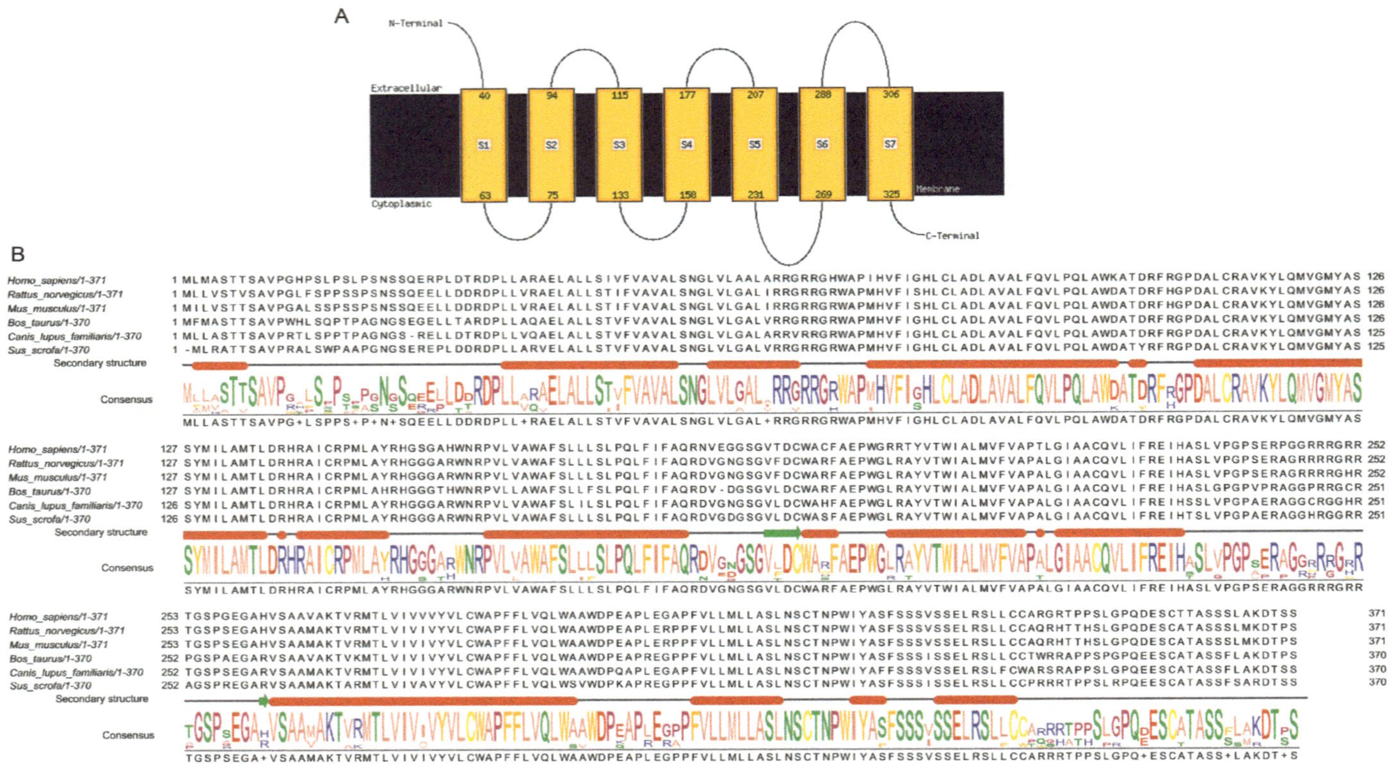

Fig. 3 Topology and sequence analysis of vasopressin V2 receptor. Composition and topographic organization in membrane of V2R from *Homo sapiens* (A) the topology analysis was realized with MEMSAT, a PSIPRED algorithm (http://bioinf.cs.ucl.ac.uk/psipred/). Multiple alignment constructed with Clustal Omega and visualized with Jalview (B). A logo is shown indicating conservation of residues and the prediction of secondary structure indicate that this protein is composed principally by α-helix.

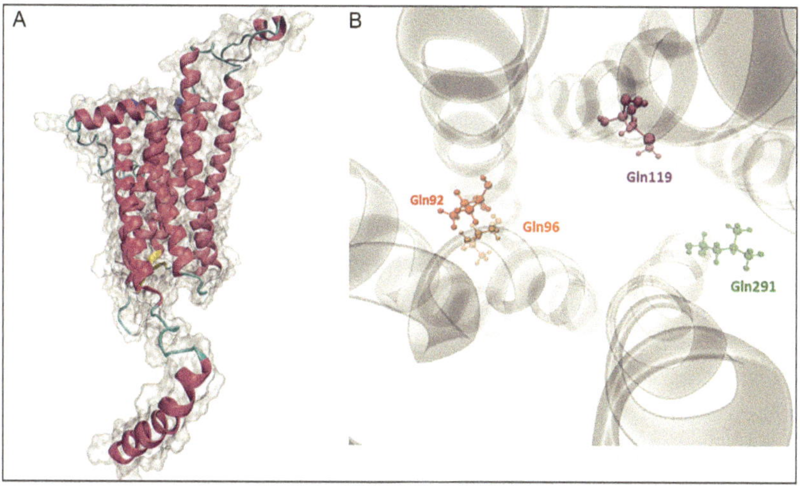

Fig. 4 3D model of vasopressin V2 receptor. A 3D model was constructed using the I-TASSER server (https://zhanglab.ccmb.med.umich.edu/I-TASSER/). For the evaluation of the model, we use PROSA (https://prosa.services.came.sbg.ac.at/prosa.php) and Ramachandran plot (http://mordred.bioc.cam.ac.uk/~rapper/rampage.php). For visualization we used VMD. The structure present predominantly α-helix structure. α-Helix are purple, β-sheet are yellow and random coil are cyan (A). The residues of glutamine involucrate in the ligand binding represented by color red (Q96), orange (Q96), purple (Q119), green (291) are present in the ligand pocket and their contribution would be associated with the formation of hydrogen bonds with AVP (B).

(absence of ligand) would be interacting with the glutamine residues by hydrogen bonds (Slusarz et al., 2006). This serine residue would be involved in the nephrogenic diabetes insipidus (NDI) disease, since a mutation of this by an arginine residue would impairs the interaction between AVP and V2R (Morgan et al., 2006).

Over the years, diverse scientific information has established that proteins belonging to the family of GPCRs are capable of interact as heterodimers or homodimers (Ng et al., 1996; Ward, Brown, & Harris, 1998). Experimental information identifies the V2 receptor as a homodimer (Zhu & Wess, 1998), being important for the interaction established between the monomers the N-terminal segment, specifically the transmembrane helices 1–3 (Schulz, Grosse, Schultz, Gudermann, & Schoneberg, 2000).

3.1 Post-traductional modifications

For the AVP V2 receptor, both N-glycosylation and O-glycosylation have been reported. N-glycosylation occurs in the residue of Asparagine

22 present at the N-terminus. This modification would not be fundamental for the interaction with the ligand nor the levels of expression (Innamorati, Sadeghi, & Birnbaumer, 1996). On the other hand, the presence of two bands in an electrophoresis gel under denaturant conditions and using a V2 receptor mutated in residue Q22 suggested the presence of another modification. In 1999 Sadeghi et al. carried out a study on the N-terminal end that considered the possibility of O-glycosylation on serine and threonine residues present in this extreme, determining that the change of these residues by alanine did not affect the interaction with AVP, but its research being of relevance since it was the first time that this type of modification was reported on proteins coupled to protein G (Sadeghi & Birnbaumer, 1999). Another modification presented by V2R is the palmitoylation of the residues cysteine 341 and 342. This modification is associated with the amount of receptor present in the membrane but is not associated with the interaction with the ligand, induction of the G protein, internalization and desensitization (Sadeghi, Innamorati, Dagarag, & Birnbaumer, 1997). This protein is also associated with ubiquitination processes, and the lysine residue 268 (K268) is involved. This modification is associated with the degradation process of the receptor after AVP stimulation; since the half-lives of the protein are similar when compare a wild-type and a mutated protein (K268R). This suggests the presence of two mechanisms that allow the degradation of the protein: one associated with ubiquitination and another independent of ubiquitin (Martin, Lefkowitz, & Shenoy, 2003). Later Le Gouill, Darden, Madziva, and Birnbaumer (2005) determined that K268 together with residue glutamic acid 231 (E231) were important for the activation of adenylate cyclase (Gouill et al., 2005). In 2008 Wu and colleagues described that serine 255 (S255), present in the third intracellular loop, was phosphorylated by PKA after the receptor interacted with AVP, using a set of techniques involving immunoaffinity purification, immobilized affinity chromatography in metal (IMAC), liquid chromatography/tandem mass spectrometry (Wu, Birnbaumer, & Guan, 2008). An important fact is that this type of modifications in receptors of the GPCRs family is associated to the process of internalization and degradation.

3.2 Interaction with other proteins

The interaction between V2R and α/β arrestins has been described, these proteins allow the formation of a molecular complex formed by V2R/arrestins/clathrin/Nedd4 (E3 ligase) that promotes the ubiquitination and subsequent traffic of the receptor toward lysosomal membranes that entail to the degradation of this. In 2012, Shea and colleagues determined the role of the α-arrestins ARRDC3-ARRDC4 and how, in a short period of time after AVP stimulation (5 min), the receptor was ubiquitinated (Shea, Rowell, Li, Chang, & Alvarez, 2012). One year later Shukla et al. (2013) determined the crystallographic structure of β-arrestin bound to a phosphorylated 29 amino acid peptide of V2R, allowing for the first time to establish a receptor-interacting interface on β-arrestin and how an interaction with the peptide involves conformational changes in β-arrestin, particularly the "lariat loop" implicated in maintaining the inactive state of β-arrestin1 (Shukla et al., 2013; Fig. 5).

In the process of degradation of V2R we find also associated the ALIX protein (AIP1), a protein found in the main cells of the kidney, which interacts with 29 residues present in the C-terminal end decreasing the protein levels of the receptor after AVP stimulation (Yi et al., 2007).

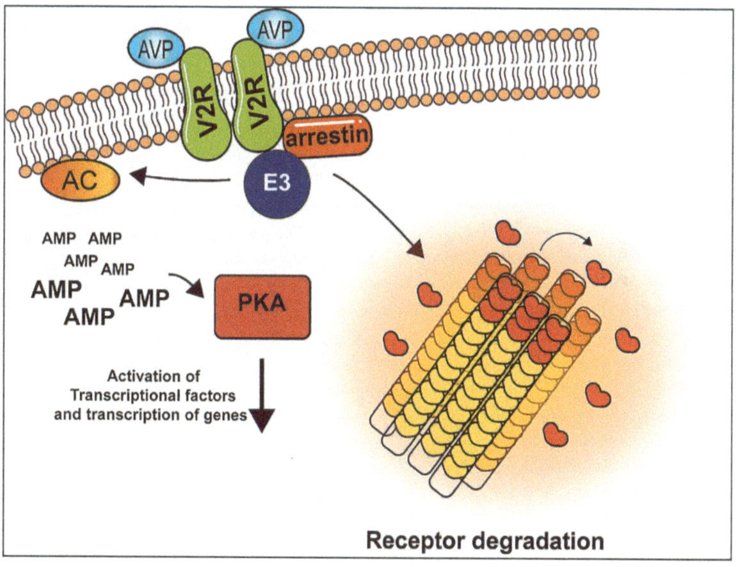

Fig. 5 Degradation process associated with β-arrestin1 activation. Experimental evidence suggests that V2R act as homodimer. After AVP interaction conformational changes induces accumulation of adenosine monophosphate (AMP) and activation of protein kinase A (PKA). This activation induces the phosphorylation of transcriptional factors and transcription of genes (molecular pathway). But, an activation by AVP promotes the ubiquitination and degradation of V2R, mediated for the formation of a complex that involucrate arrestin proteins. AVP, vasopressin; V2R, Vasopressin 2 receptor; AC, adenylate cyclase; PKA, protein kinase A; E3, Nedd4, ubiquitin E3 ligase.

Involucrate with the correct folding of the receptor in the endoplasmic reticulum, we found GC1q-R, which interacts with V2R through the arginine cluster (247RRRGRR252), all of which are relevant to the interaction process, present in the loop i3 (Granier et al., 2008).

Another protein associated with V2R that depends of the presence of intracellular calcium (Ca^{2+}) is calmodulin, a protein that can interact with receptors coupled to protein G. Nickols, Shah, Chazin, and Limbird (2004) analyzed the possibility that calmodulin interacted with V2R, since previous scientific evidence suggested this association. The scientists determined that the interaction was produced by the C-terminal end of V2R, being dependent on the presence of arginine residues (RGR) and intracellular calcium levels, representing this an independent mechanism of Gαs that increases the levels of intracellular Ca^{2+} (Nickols et al., 2004).

4. Vasopressin actions in the kidney renin angiotensin system

The cAMP-PKA and cAMP response element-binding protein (CREB) pathway is the central pathway for renin regulation in the kidney juxtaglomerular (JG) cells (Kurtz & Wagner, 1999). Activation of adenylate cyclase (AC) increased the activity of renin promoter stimulating renin gene transcription in JG cells (Klar, Sandner, Muller, & Kurtz, 2002). In the collecting duct, renin is augmented by Ang II via AT1 receptor (Gonzalez et al., 2011; Liu et al., 2012). More recently we demonstrated that AT1 receptor also increases cAMP (Gonzalez et al., 2015). Vasopressin is a potential effector of the RAS (Hogarty, Tran, & Phillips, 1994) in CD cells. The activation of the V2R stimulates the cAMP/PKA/CREB pathway and aquaporin-2 (AQP2) expression in the apical plasma membrane of principal cells in the CD (Lee et al., 2007). AT1 receptor blockade in rats treated with V2R agonist DDAVP plus low salt diet, the DDAVP-induced upregulation and phosphorylation of AQP2 via PKA is blunted (Kwon, Nielsen, Knepper, Frokiaer, & Nielsen, 2005), indicating a synergic role for Ang II in the regulation of AQP2.

In JG cells Ang II suppresses renin synthesis through PKC and Ca^{2+} in JG cells (Peti-Peterdi & Harris, 2010). In the CD, increases in intracellular Ca^{2+} and in cAMP levels are required to target AQP2 to increase osmotic water permeability (Chou et al., 2000). We have shown that Ang II-mediated renin increases can be suppressed by PKC inhibitors (Gonzalez et al., 2011). This is also observed in ex-vivo studies using isolated collecting duct cells (Gonzalez et al., 2015). Then, PKC plays seems to be important for AC/cAMP activation in the collecting duct.

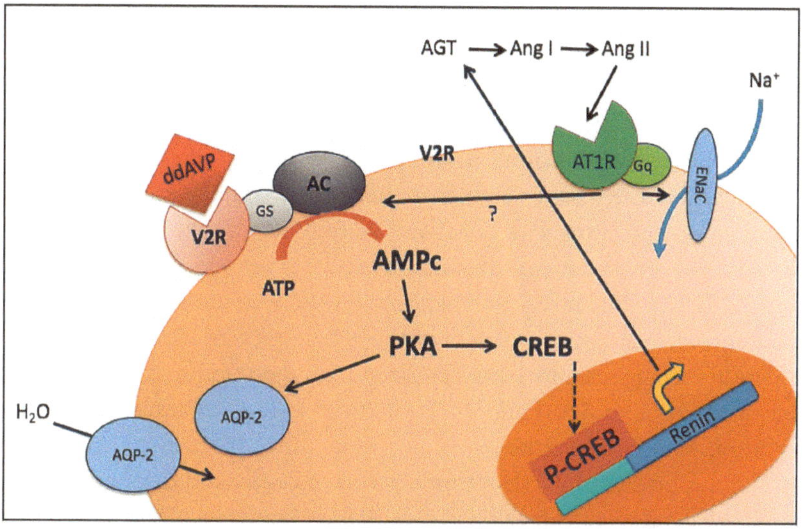

Fig. 6 Proposed mechanisms for renin regulation mediated by vasopressin type 2 receptor in the collecting duct. We have shown that desmopressin (ddAVP) is able to increase cAMP and phosphorylation of CREB, along with the stimulation of renin expression. AT1 receptor also increases renin expression through PKC and cAMP accumulation. It is expected that both signaling pathways may be involved in renin regulation.

Both, AVP or Ang II, increases renin to the same extent in collecting duct cells. Lee et al. demonstrated that Ang II enhances AQP2 targeting to the plasma membrane in collecting duct cells via AT1 receptor (Lee et al., 2007; Fig. 6). Rozengurt et al. showed that PKC activation enhanced the cAMP accumulation (Rozengurt, Murray, Zachary, & Collins, 1987), and is mediated by activation of AC6 (Roos, Strait, Raphael, Blount, & Kohan, 2012).

5. Vasopressin in hypertension

Increases in 1–2% plasma osmolarity causes increases in AVP release into the bloodstream inducing vasoconstrictor and antidiuretic effects (Bankir et al., 2017). Recent studies have shown that prolonged high salt intake promotes pathological plasticity in the circuit that controls the secretion of vasopressin. This will cause chronic vasoconstrictor activity of AVP, which is mediated by the activation of V1R receptor located on vascular smooth muscle (Kawano & Ferrario, 1990; Kawano, Matsuoka, Nishikimi, Takishita, & Omae, 1997; Nishikimi, Kawano, Saito, & Matsuoka, 1996), increasing water retention in the collecting duct and also stimulating the intratubular renin angiotensin system (Gonzalez et al., 2016).

Thus, the increase in AVP leads to water retention and vasoconstriction, contributing to an elevation of blood pressure, either by V1R or V2R activation.

In the collecting duct activation of AT1R and V2R induced increases in sodium and water reabsorption via epithelial sodium channel (ENaC) and AQP2, respectively suggesting that there is an amplification mechanism where exogenous Ang II and activation of V2R (Bankir et al., 2010). It is also likely that the exogenous Ang II and vasopressin stimulate the production of Ang II de novo in the kidney through the activation intratubular renin angiotensin system worsening the whole picture in hypertension.

Studies in preeclampsia characterized by hypertension and fetal growth restriction have shown high levels of AVP as early as the sixth week of gestation, which inherently is earlier than other biomarkers, supporting the conclusion that AVP secretion is increased early and throughout pregnancies that eventually develop preeclampsia (Sandgren, Deng, et al., 2018; Sandgren, Linggonegoro, et al., 2018; Sandgren et al., 2015; Santillan et al., 2014; Scroggins et al., 2018). Infusion of AVP throughout gestation is sufficient to initiate major preeclamptic phenotypes in mice, including elevated systolic blood pressure, proteinuria and intrauterine growth restriction (Sandgren, Deng, et al., 2018). Furthermore, the use of low doses of vasopressin and their analogues can be used to treat hypotension in patients prescribed renin angiotensin system inhibitors, demonstrating that vasopressin is enough to recover arterial blood pressure in the absence of renin angiotensin system activation.

6. Vasopressin and kidney disease

Observations regarding vasopressin (AVP) involvement in kidney disease have been made since the last three decades, these associates vasopressin levels with glomerular filtration rate, creatinine clearance or impairment of glucose metabolism, making this hormone very important in a context that goes further than just regulation of vascular tonicity or even its involvement in hypertension. Here, we summarize some highlights regarding three major pathological events where this hormone has been presumed to have a significant role.

6.1 Chronic kidney disease

Although there are early evidences pointing to a pathophysiological role of AVP in diseases like diabetes and chronic kidney disease, the research in this field was limited until remarkable advances were made, such as the cloning of three AVP receptors (V1a, V1b and V2) and the detailed information of which organs express them. Since then, the interest in studying this peptide alongside with its effect on kidney disease has gained weight. Since AVP concentrations in blood are not higher enough to be detectable by conventional methods and is not a stable molecule, it is not a suitable clinical marker. Like AVP, it comes from the propeptide preprovasopressin which is composed by a signal peptide, AVP, neurophysin II and copeptin, being the last one the carboxy-terminal peptide (Land, Schutz, Schmale, & Richter, 1982). The stability of this peptide, alongside the sensibility of the immunoabsorbent assays for this molecule and non-existing needing of pretreating samples for its measurement made copeptin a promising candidate for measuring AVP in plasma samples (Morgenthaler et al., 2007; Morgenthaler, Struck, Alonso, & Bergmann, 2006). All these findings have made possible to work with big human cohorts to find out how vasopressin can be helpful as an indicator of renal disease. Remarkably, in the last few years, several studies that associates vasopressin levels with diabetic nephropathy, chronic kidney disease, albuminuria and hypertension.

Back in 1990, Bouby and collaborators drove a series of experiments focused in studying the role of AVP in chronic kidney failure (CKF) progression by using a 5/6 nephrectomy rat model. They gave these rats water-rich food in order to decrease plasma AVP and urine concentration (Pouzet et al., 2001). Results in this study showed a reduction in hypertrophy, glomerulosclerosis and blood pressure compared to a group fed with dry food. These results were associated to a reduction in plasma AVP due to observations like the effect of the administration of AVP, desmopressin (dDAVP agonist for V2 receptor) or dibutyryl-cyclic AMP could reduce the glomerular ultrafiltration coefficient, which indicated that AVP receptors could be involved in the progression of CKF. Also, it was observed that AVP induced an Ang II-like effect in proteinuria, increasing it. Therefore, inhibiting its action or reducing its levels would have a similar effect as blocking Ang II action (Bouby, Bachmann, Bichet, & Bankir, 1990). Years later, the same group performed 5/6 nephrectomy in Brattleboro rats, which lack vasopressin, and they found that the addition of dDAVP was deleterious to renal function. Since dDAVP acts via V2 receptor, this effect was associated to that receptor in particular (Bouby, Hassler, & Bankir, 1999). Given the evidence, some researchers have hypothesized that V2 receptor is one of the implicated factor in the progression of chronic kidney disease, regarding of the etiology of this disease (Bankir, Bouby, & Ritz, 2013).

It is worthy to notice that glomerular hyperfiltration induces a cycle that leads to renal damage through a high-protein diet, causing hyperfiltration and albuminuria, leading to oxidative stress, glomerular sclerosis and thus and reducing the number of functional nephrons (Brenner, 1985). It has been suggested that AVP exerts the same effect as a high-protein diet in this cycle (Clark et al., 2016). The mechanism involving V2R is not completely understood, mostly because there are no V2R in glomerulus so the action of AVP might be through a vasopressin-dependent handling of urea which might lead to an increased concentration of flow and urea at the loops of Henle (Bankir et al., 2013) and finally a reduction in sodium levels at the macula densa which leads to a decreased tubule-glomerular feedback control of glomerular filtration rate (Clark et al., 2016).

6.2 Diabetes mellitus and diabetic nephropathy

Almost four decades ago, it was described that diabetic patients have an imbalanced quantity of plasma AVP (Zerbe, Vinicor, & Robertson, 1979). Also, it is involved in the metabolic regulation of glucose levels, stimulating hepatic gluconeogenesis and glycogenolysis (Hems, Rodrigues, & Whitton, 1978). At the beginning, there was an interest for the role of V1a receptor activation on vascular smooth muscle cells but later the interest shifted toward V1a receptor activation in pancreatic islets (Bankir et al., 2013). There is evidence that shows high AVP levels in diabetic patients (type 1 and 2) (Zerbe et al., 1979). Although the reason why this happens is not clear yet, there is a hypothesis that explains that the high levels of this peptide would be due to a contraction of the extracellular volume produced by glycosuria and from an increased sensibility of hypothalamic osmotic receptors (Zerbe et al., 1979). On the contrary, some data shows that diabetes mellitus is characterized by decreased secretion of AVP differing from nephrogenic diabetes, in which distal nephron is resistant to AVP action, causing trouble in concentrating urine (Enhorning et al., 2013; Morgenthaler et al., 2007, 2006). Also, experimental approaches using mice lacking V1 and V2 receptors has shown different effects on glucose metabolism, showing hypersensitivity to insulin, insulin resistance and impaired glucose tolerance (Koshimizu et al., 2012).

Implications of AVP in early stages of diabetes were explored by inducing diabetes through streptozotocin in vasopressin-deficient Brattleboro rats. Interestingly, when researchers measured creatinine clearance, kidney weight and albumin excretion in these rats, they found out that the lack of vasopressin was related to a failed increase in these renal function indicators (Bardoux et al., 1999). In concordance with this study, the addition of SR 121463, a V2R antagonist, to streptozotocin-treated rats for 3 months prevented the increased in albumin excretion in this model (Bardoux et al., 2003). This proved that V2R were involved in kidney disease progression in a diabetes context.

In contrast, different epidemiological studies have been made in humans, such as the Dutch ZODIAC prospective study made in 1328 patients with type 2 diabetes in which a correlation between the upper quartile of plasma copeptin and decrease glomerular filtration, additionally with an increase of the albumin-creatinine ratio was found (Boertien et al., 2012). Accordingly, Pikkemaat and collaborators described the relation between a decline in glomerular filtration rate and high copeptin in newly diagnosed diabetes in type 2 diabetes patients (Pikkemaat, Melander, & Bengtsson Bostrom, 2015). The association between copeptin and vasopressin levels and kidney disease has also been studied in type 1 diabetes, in whom there has been an association between high levels of plasma copeptin and advanced nephropathy and other mortality-associated factors, such as increased risk of coronary events. Thus, there is a clear relation between vasopressin/copeptin levels and diabetes. High vasopressin may actually be helpful in a short-term basis in regulating the amount of water necessary for the elimination of high osmolarity fluid due the presence of glucosuria but at a long-term it may cause diabetic nephropathy, aggravating the stage (Pikkemaat et al., 2015).

6.3 Autosomal polycystic kidney

Even when the role of AVP in kidney diseases is not fully understood, there is a pathology where AVP role has been elucidated. Autosomal polycystic kidney disease (APKD) is a hereditary disorder in which cysts develops within the kidney, causing major morphological changes and thus, decreased renal function. AVP can stimulate cyst growth through cyclic AMP (cAMP) stimulation via V2 receptors in human cell cultures and rodent models, specifically in collecting ducts (Belibi et al., 2004; Wang, Wu, Ward, Harris, & Torres, 2008), proved by experimental approaches that involved crossing Brattleboro rats with a model of polycystic kidney disease (PCK rats). In these studies, rats generated developed polycystic kidney disease (PKD) but had a lack of AVP, which resulted in a decreased cystogenesis compared with PCK rats at week 20, this was reverted by the addition of dDAVP (Wang et al., 2008). Also, in vitro approaches determined that addition of tolvaptan (V2 receptor inhibitor) to human APKD cells inhibited cell proliferation (Reif et al., 2011). Increased concentration of AVP in this disease is thought to be a consequence of a defect in urine-concentration mechanisms, caused by the growing of cysts in the collecting duct, disrupting their normal architecture and finally affecting the exchange of urea through the ducts (van Gastel & Torres, 2017). Human studies showed that tolvaptan efficacy and safety in management of autosomal dominant polycystic kidney disease and its outcomes. A randomized study consisting in treating 1445 patients treated with tolvaptan and following the progression of the disease up to 3 years revealed a correlation between V2 receptor inhibition by tolvaptan and a slowed progression of the disease, compared with placebo-treated patients (2.2% in the tolvaptan-treated group vs 5.5% in placebo-treated control) (Torres et al., 2017). Along with tolvaptan, different drugs have been proposed as synergistic molecules, such as statins and tetracycline antibiotics demeclocycline and doxycycline which acts as cAMP negative regulators, lowering its levels. Nonetheless, the potential of these drugs has not been effectively proved since high doses might be nephrotoxic (van Gastel & Torres, 2017).

Data obtained from a study using PCK rats, which is an orthologous for human autosomal recessive polycystic kidney disease (ARPKD), indicates that high-water intake reduces urinary vasopressin excretion by 68,3% and the expression of vasopressin receptors. Furthermore, it slowed the progression of the disease in both female and male rodents (Nagao et al., 2006). Even randomized studies has been made in patients with chronic kidney disease in advanced stages in a matter of increased water intake (Clark et al., 2016). Also, discrepancies between studies have been found (Barash, Ponda, Goldfarb, & Skolnik, 2010), probably because of the number of subjects analyzed in each study and quantity of water per kg of weight per day, indicating that there must be more randomized controlled trials to reach solid conclusions regarding the beneficial effect of high-water intake.

7. Conclusions and future directions

Despite its role in water maintenance in the body through the antidiuretic actions in the kidney through V2, vasopressin also contributes to vasoconstriction via V1. The posterior pituitary into the blood releases vasopressin formed in the hypothalamus. Hypothalamic osmotic neurons initiate the cascade for AVP actions in conditions of elevated osmolality. Chronic high salt diet is involved in continuous release of AVP leading to water retention and pressor responses. Furthermore vasopressin acting on V2 receptor stimulate the formation of cyclic AMP (cAMP) and potentiates the formation of intratubular Ang II through V2-dependent distal tubular renin formation, contributing to sodium reabsorption. Vasopressin has a role in pathological states such as the involvement of V2 dependent progression of kidney damage in diabetes and the stimulation of intracellular pathways linked to the development of polycystic kidney. Antagonists for both kind of receptor may be of importance in the management of kidney disease.

References

Bankir, L., Bichet, D. G., & Bouby, N. (2010). Vasopressin V2 receptors, ENaC, and sodium reabsorption: A risk factor for hypertension? *American Journal of Physiology. Renal Physiology*, *299*(5), F917–F928. https://doi.org/10.1152/ajprenal.00413.2010.

Bankir, L., Bichet, D. G., & Morgenthaler, N. G. (2017). Vasopressin: Physiology, assessment and osmosensation. *Journal of Internal Medicine*, *282*(4), 284–297. https://doi.org/10.1111/joim.12645.

Bankir, L., Bouby, N., & Ritz, E. (2013). Vasopressin: A novel target for the prevention and retardation of kidney disease? *Nature Reviews Nephrology*, *9*(4), 223–239. https://doi.org/10.1038/nrneph.2013.22.

Barash, I., Ponda, M. P., Goldfarb, D. S., & Skolnik, E. Y. (2010). A pilot clinical study to evaluate changes in urine osmolality and urine cAMP in response to acute and chronic water loading in autosomal dominant polycystic kidney disease. *Clinical Journal of the American Society of Nephrology*, *5*(4), 693–697. https://doi.org/10.2215/CJN.04180609.

Bardoux, P., Bichet, D. G., Martin, H., Gallois, Y., Marre, M., Arthus, M. F., et al. (2003). Vasopressin increases urinary albumin excretion in rats and humans: Involvement of V2 receptors and the renin-angiotensin system. *Nephrology, Dialysis, Transplantation*, *18*(3), 497–506.

Bardoux, P., Martin, H., Ahloulay, M., Schmitt, F., Bouby, N., Trinh-Trang-Tan, M. M., et al. (1999). Vasopressin contributes to hyperfiltration, albuminuria, and renal hypertrophy in diabetes mellitus: Study in vasopressin-deficient Brattleboro rats. *Proceedings of the National Academy of Sciences of the United States of America*, *96*(18), 10397–10402.

Belibi, F. A., Reif, G., Wallace, D. P., Yamaguchi, T., Olsen, L., Li, H., et al. (2004). Cyclic AMP promotes growth and secretion in human polycystic kidney epithelial cells. *Kidney International*, *66*(3), 964–973. https://doi.org/10.1111/j.1523-1755.2004.00843.x.

Boertien, W. E., Meijer, E., Zittema, D., van Dijk, M. A., Rabelink, T. J., Breuning, M. H., et al. (2012). Copeptin, a surrogate marker for vasopressin, is associated with kidney function decline in subjects with autosomal dominant polycystic kidney disease. *Nephrology, Dialysis, Transplantation*, *27*(11), 4131–4137. https://doi.org/10.1093/ndt/gfs070.

Bouby, N., Bachmann, S., Bichet, D., & Bankir, L. (1990). Effect of water intake on the progression of chronic renal failure in the 5/6 nephrectomized rat. *American Journal of Physiology*, *258*(4 Pt 2), F973–F979. https://doi.org/10.1152/ajprenal.1990.258.4.F973.

Bouby, N., Hassler, C., & Bankir, L. (1999). Contribution of vasopressin to progression of chronic renal failure: Study in Brattleboro rats. *Life Sciences*, *65*(10), 991–1004.

Brenner, B. M. (1985). Nephron adaptation to renal injury or ablation. *American Journal of Physiology*, *249*(3 Pt 2), F324–F337. https://doi.org/10.1152/ajprenal.1985.249.3.F324.

Burrell, L. M., Phillips, P. A., Rolls, K. A., Buxton, B. F., Johnston, C. I., & Liu, J. J. (1994). Vascular responses to vasopressin antagonists in man and rat. *Clinical Science (London, England)*, *87*(4), 389–395.

Chou, C. L., Knepper, M. A., & Layton, H. E. (1993). Urinary concentrating mechanism—The role of the inner medulla. *Seminars in Nephrology*, *13*(2), 168–181.

Chou, C. L., Yip, K. P., Michea, L., Kador, K., Ferraris, J. D., Wade, J. B., et al. (2000). Regulation of aquaporin-2 trafficking by vasopressin in the renal collecting duct—Roles of ryanodine-sensitive Ca2+ stores and calmodulin. *Journal of Biological Chemistry*, *275*(47), 36839–36846. https://doi.org/10.1074/Jbc.M005552200.

Clark, W. F., Sontrop, J. M., Huang, S. H., Moist, L., Bouby, N., & Bankir, L. (2016). Hydration and chronic kidney disease progression: A critical review of the evidence. *American Journal of Nephrology*, *43*(4), 281–292. https://doi.org/10.1159/000445959.

Enhorning, S., Bankir, L., Bouby, N., Struck, J., Hedblad, B., Persson, M., et al. (2013). Copeptin, a marker of vasopressin, in abdominal obesity, diabetes and microalbuminuria: The prospective Malmo Diet and Cancer Study cardiovascular cohort. *International Journal of Obesity*, *37*(4), 598–603. https://doi.org/10.1038/ijo.2012.88.

Fenton, R. A., Brond, L., Nielsen, S., & Praetorius, J. (2007). Cellular and subcellular distribution of the type-2 vasopressin receptor in the kidney. *American Journal of Physiology-Renal Physiology*, *293*(3), F748–F760. https://doi.org/10.1152/ajprenal.00316.2006.

Gonzalez, A. A., Cifuentes-Araneda, F., Ibaceta-Gonzalez, C., Gonzalez-Vergara, A., Zamora, L., Henriquez, R., et al. (2016). Vasopressin/V2 receptor stimulates renin synthesis in the collecting duct. *American Journal of Physiology-Renal Physiology*, *310*(4), F284–F293. https://doi.org/10.1152/ajprenal.00360.2015.

Gonzalez, A. A., Liu, L., Lara, L. S., Bourgeois, C. R., Ibaceta-Gonzalez, C., Salinas-Parra, N., et al. (2015). PKC-alpha-dependent augmentation of cAMP and CREB phosphorylation mediates the angiotensin II stimulation of renin in the collecting duct. *American Journal of Physiology. Renal Physiology*, *309*(10), F880–F888. https://doi.org/10.1152/ajprenal.00155.2015.

Gonzalez, A. A., Liu, L., Lara, L. S., Seth, D. M., Navar, L. G., & Prieto, M. C. (2011). Angiotensin II stimulates renin in inner medullary collecting duct cells via protein kinase C and independent of epithelial sodium channel and mineralocorticoid receptor activity. *Hypertension*, *57*(3), 594–599. https://doi.org/10.1161/Hypertensionaha.110.165902.

Le Gouill, C., Darden, T., Madziva, M. T., & Birnbaumer, M. (2005). A role for K268 in V2R folding. *FEBS Letters*, *579*(22), 4985–4990. https://doi.org/10.1016/j.febslet.2005.08.003.

Granier, S., Jean-Alphonse, F., Bacqueville, D., Monteil, A., Pascal, R., Poncet, J., et al. (2008). The multifunctional protein GC1q-R interacts specifically with the i3 loop arginine cluster of the vasopressin V2 receptor. *Regulatory Peptides*, *148*(1–3), 76–87. https://doi.org/10.1016/j.regpep.2008.01.007.

Hems, D. A., Rodrigues, L. M., & Whitton, P. D. (1978). Rapid stimulation by vasopressin, oxytocin and angiotensin II of glycogen degradation in hepatocyte suspensions. *Biochemical Journal*, *172*(2), 311–317.

High, S., Flint, N., & Dobberstein, B. (1991). Requirements for the membrane insertion of signal-anchor type proteins. *Journal of Cell Biology*, *113*(1), 25–34.

Hoffert, J. D., Chou, C. L., & Knepper, M. A. (2009). Aquaporin-2 in the "-omics" era. *Journal of Biological Chemistry*, *284*(22), 14683–14687. https://doi.org/10.1074/jbc.R900006200.

Hogarty, D. C., Tran, D. N., & Phillips, M. I. (1994). Involvement of angiotensin receptor subtypes in osmotically induced release of vasopressin. *Brain Research*, *637*(1–2), 126–132.

Innamorati, G., Sadeghi, H., & Birnbaumer, M. (1996). A fully active nonglycosylated V2 vasopressin receptor. *Molecular Pharmacology*, *50*(3), 467–473.

Kawano, Y., & Ferrario, C. M. (1990). Role of vasopressin in cardiovascular and neurohormonal responses to intracerebroventricular hypertonic NaCl. *Japanese Heart Journal*, *31*(2), 237–244.

Kawano, Y., Matsuoka, H., Nishikimi, T., Takishita, S., & Omae, T. (1997). The role of vasopressin in essential hypertension. Plasma levels and effects of the V1 receptor antagonist OPC-21268 during different dietary sodium intakes. *American Journal of Hypertension, 10*(11), 1240–1244.

Klar, J., Sandner, P., Muller, M. W. H., & Kurtz, A. (2002). Cyclic AMP stimulates renin gene transcription in juxtaglomerular cells. *Pflugers Archiv-European Journal of Physiology, 444*(3), 335–344. https://doi.org/10.1007/s00424-002-0818-9.

Knepper, M. A., Nielsen, S., & Chou, C. L. (2001). Physiological roles of aquaporins in the kidney. *Aqua, 51*, 121–153. https://doi.org/10.1016/S1063-5823(01)51005-4.

Knepper, M. A., Nielsen, S., Chou, C. L., & Digiovanni, S. R. (1994). Mechanism of vasopressin action in the renal collecting duct. *Seminars in Nephrology, 14*(4), 302–321.

Koshimizu, T. A., Nakamura, K., Egashira, N., Hiroyama, M., Nonoguchi, H., & Tanoue, A. (2012). Vasopressin V1a and V1b receptors: From molecules to physiological systems. *Physiological Reviews, 92*(4), 1813–1864. https://doi.org/10.1152/physrev.00035.2011.

Kurtz, A., & Wagner, C. (1999). Cellular control of renin secretion. *Journal of Experimental Biology, 202*(3), 219–225.

Kwon, T. H., Nielsen, J., Knepper, M. A., Frokiaer, J., & Nielsen, S. (2005). Angiotensin II AT1 receptor blockade decreases vasopressin-induced water reabsorption and AQP2 levels in NaCl-restricted rats. *American Journal of Physiology. Renal Physiology, 288*(4), F673–F684. https://doi.org/10.1152/ajprenal.00304.2004.

Land, H., Schutz, G., Schmale, H., & Richter, D. (1982). Nucleotide sequence of cloned cDNA encoding bovine arginine vasopressin-neurophysin II precursor. *Nature, 295*(5847), 299–303.

Lee, Y. J., Song, I. K., Jang, K. J., Nielsen, J., Frokiaer, J., Nielsen, S., et al. (2007). Increased AQP2 targeting in primary cultured IMCD cells in response to angiotensin II through AT(1) receptor. *American Journal of Physiology-Renal Physiology, 292*(1), F340–F350. https://doi.org/10.1152/ajprenal.00090.2006.

Liu, L., Lara, L. S., Gonzalez, A. A., Bourgeois, C. R., Seth, D. M., & Prieto, M. C. (2012). Angiotensin II stimulates renin synthesis and secretion via protein kinase C activation and cAMP accumulation in collecting duct M-1 cells. *Journal of Investigative Medicine, 60*(1), 456–457.

Lozic, M., Sarenac, O., Murphy, D., & Japundzic-Zigon, N. (2018). Vasopressin, central autonomic control and blood pressure regulation. *Current Hypertension Reports, 20*(2), 11. https://doi.org/10.1007/s11906-018-0811-0.

Martin, N. P., Lefkowitz, R. J., & Shenoy, S. K. (2003). Regulation of V2 vasopressin receptor degradation by agonist-promoted ubiquitination. *Journal of Biological Chemistry, 278*(46), 45954–45959. https://doi.org/10.1074/jbc.M308285200.

Morgan, T. K., Rohrwasser, A., Zhao, L., Hillas, E., Cheng, T., Ward, K. J., et al. (2006). Hypervolemia of pregnancy is not maintained in mice chronically overexpressing angiotensinogen. *American Journal of Obstetrics and Gynecology, 195*(6), 1700–1706. https://doi.org/10.1016/j.ajog.2006.04.005.

Morgenthaler, N. G., Muller, B., Struck, J., Bergmann, A., Redl, H., & Christ-Crain, M. (2007). Copeptin, a stable peptide of the arginine vasopressin precursor, is elevated in hemorrhagic and septic shock. *Shock, 28*(2), 219–226. https://doi.org/10.1097/SHK.0b013e318033e5da.

Morgenthaler, N. G., Struck, J., Alonso, C., & Bergmann, A. (2006). Assay for the measurement of copeptin, a stable peptide derived from the precursor of vasopressin. *Clinical Chemistry, 52*(1), 112–119. https://doi.org/10.1373/clinchem.2005.060038.

Mutig, K., Paliege, A., Kahl, T., Jons, T., Muller-Esterl, W., & Bachmann, S. (2007). Vasopressin V2 receptor expression along rat, mouse, and human renal epithelia with focus on TAL. *American Journal of Physiology. Renal Physiology, 293*(4), F1166–F1177. https://doi.org/10.1152/ajprenal.00196.2007.

Nagao, S., Nishii, K., Katsuyama, M., Kurahashi, H., Marunouchi, T., Takahashi, H., et al. (2006). Increased water intake decreases progression of polycystic kidney disease in the PCK rat. *Journal of the American Society of Nephrology*, *17*(8), 2220–2227. https://doi.org/10.1681/ASN.2006030251.

Ng, G. Y., O'Dowd, B. F., Lee, S. P., Chung, H. T., Brann, M. R., Seeman, P., et al. (1996). Dopamine D2 receptor dimers and receptor-blocking peptides. *Biochemical and Biophysical Research Communications*, *227*(1), 200–204. https://doi.org/10.1006/bbrc.1996.1489.

Nicco, C., Wittner, M., DiStefano, A., Jounier, S., Bankir, L., & Bouby, N. (2001). Chronic exposure to vasopressin upregulates ENaC and sodium transport in the rat renal collecting duct and lung. *Hypertension*, *38*(5), 1143–1149.

Nickols, H. H., Shah, V. N., Chazin, W. J., & Limbird, L. E. (2004). Calmodulin interacts with the V2 vasopressin receptor: Elimination of binding to the C terminus also arginine vasopressin-stimulated elevation of intracellular calcium. *Journal of Biological Chemistry*, *279*(45), 46969–46980. https://doi.org/10.1074/jbc.M407351200.

Nishikimi, T., Kawano, Y., Saito, Y., & Matsuoka, H. (1996). Effect of long-term treatment with selective vasopressin V1 and V2 receptor antagonist on the development of heart failure in rats. *Journal of Cardiovascular Pharmacology*, *27*(2), 275–282.

Peti-Peterdi, J., & Harris, R. C. (2010). Macula densa sensing and signaling mechanisms of renin release. *Journal of the American Society of Nephrology*, *21*(7), 1093–1096. https://doi.org/10.1681/ASN.2009070759.

Phillips, P. A., Abrahams, J. M., Kelly, J. M., Mooser, V., Trinder, D., & Johnston, C. I. (1990). Localization of vasopressin binding sites in rat tissues using specific V1 and V2 selective ligands. *Endocrinology*, *126*(3), 1478–1484. https://doi.org/10.1210/endo-126-3-1478.

Phillips, P. A., Abrahams, J. M., Kelly, J., Paxinos, G., Grzonka, Z., Mendelsohn, F. A., et al. (1988). Localization of vasopressin binding sites in rat brain by in vitro autoradiography using a radioiodinated V1 receptor antagonist. *Neuroscience*, *27*(3), 749–761.

Phillips, P. A., Kelly, J. M., Abrahams, J. M., Grzonka, Z., Paxinos, G., Mendelsohn, F. A., et al. (1988). Vasopressin receptors in rat brain and kidney: Studies using a radioiodinated V1 receptor antagonist. *Journal of Hypertension. Supplement*, *6*(4), S550–S553.

Pikkemaat, M., Melander, O., & Bengtsson Bostrom, K. (2015). Association between copeptin and declining glomerular filtration rate in people with newly diagnosed diabetes. The Skaraborg diabetes register. *Journal of Diabetes and its Complications*, *29*(8), 1062–1065. https://doi.org/10.1016/j.jdiacomp.2015.07.006.

Pouzet, B., Serradeil-Le Gal, C., Bouby, N., Maffrand, J. P., Le Fur, G., & Bankir, L. (2001). Selective blockade of vasopressin V2 receptors reveals significant V2-mediated water reabsorption in Brattleboro rats with diabetes insipidus. *Nephrology, Dialysis, Transplantation*, *16*(4), 725–734.

Prager-Khoutorsky, M., & Bourque, C. W. (2010). Osmosensation in vasopressin neurons: Changing actin density to optimize function. *Trends in Neurosciences*, *33*(2), 76–83. https://doi.org/10.1016/j.tins.2009.11.004.

Prager-Khoutorsky, M., Choe, K. Y., Levi, D. I., & Bourque, C. W. (2017). Role of vasopressin in rat models of salt-dependent hypertension. *Current Hypertension Reports*, *19*(5), 42. https://doi.org/10.1007/s11906-017-0741-2.

Promeneur, D., Kwon, T. H., Frokiaer, J., Knepper, M. A., & Nielsen, S. (2000). Vasopressin V(2)-receptor-dependent regulation of AQP2 expression in Brattleboro rats. *American Journal of Physiology. Renal Physiology*, *279*(2), F370–F382. https://doi.org/10.1152/ajprenal.2000.279.2.F370.

Reif, G. A., Yamaguchi, T., Nivens, E., Fujiki, H., Pinto, C. S., & Wallace, D. P. (2011). Tolvaptan inhibits ERK-dependent cell proliferation, Cl(−) secretion, and in vitro cyst growth of human ADPKD cells stimulated by vasopressin. *American Journal of Physiology. Renal Physiology*, *301*(5), F1005–F1013. https://doi.org/10.1152/ajprenal.00243.2011.

Roos, K. P., Strait, K. A., Raphael, K. L., Blount, M. A., & Kohan, D. E. (2012). Collecting duct-specific knockout of adenylyl cyclase type VI causes a urinary concentration defect in mice. *American Journal of Physiology-Renal Physiology, 302*(1), F78–F84. https://doi.org/10.1152/ajprenal.00397.2011.

Rozengurt, E., Murray, M., Zachary, I., & Collins, M. (1987). Protein-kinase-C activation enhances cAMP accumulation in Swiss 3T3 cells—Inhibition by pertussis toxin. *Proceedings of the National Academy of Sciences of the United States of America, 84*(8), 2282–2286. https://doi.org/10.1073/Pnas.84.8.2282.

Sadeghi, H., & Birnbaumer, M. (1999). O-Glycosylation of the V2 vasopressin receptor. *Glycobiology, 9*(7), 731–737.

Sadeghi, H. M., Innamorati, G., Dagarag, M., & Birnbaumer, M. (1997). Palmitoylation of the V2 vasopressin receptor. *Molecular Pharmacology, 52*(1), 21–29.

Sandgren, J. A., Deng, G., Linggonegoro, D. W., Scroggins, S. M., Perschbacher, K. J., Nair, A. R., et al. (2018). Arginine vasopressin infusion is sufficient to model clinical features of preeclampsia in mice. *JCI Insight, 3*(19). e99403https://doi.org/10.1172/jci.insight.99403.

Sandgren, J. A., Linggonegoro, D. W., Zhang, S. Y., Sapouckey, S. A., Claflin, K. E., Pearson, N. A., et al. (2018). Angiotensin AT1A receptors expressed in vasopressin-producing cells of the supraoptic nucleus contribute to osmotic control of vasopressin. *American Journal of Physiology. Regulatory, Integrative and Comparative Physiology, 314*(6), R770–R780. https://doi.org/10.1152/ajpregu.00435.2017.

Sandgren, J. A., Scroggins, S. M., Santillan, D. A., Devor, E. J., Gibson-Corley, K. N., Pierce, G. L., et al. (2015). Vasopressin: The missing link for preeclampsia? *American Journal of Physiology. Regulatory, Integrative and Comparative Physiology, 309*(9), R1062–R1064. https://doi.org/10.1152/ajpregu.00073.2015.

Sands, J. M., Nonoguchi, H., & Knepper, M. A. (1987). Vasopressin effects on urea and H2O transport in inner medullary collecting duct subsegments. *American Journal of Physiology, 253*(5 Pt 2), F823–F832. https://doi.org/10.1152/ajprenal.1987.253.5.F823.

Santillan, M. K., Santillan, D. A., Scroggins, S. M., Min, J. Y., Sandgren, J. A., Pearson, N. A., et al. (2014). Vasopressin in preeclampsia: A novel very early human pregnancy biomarker and clinically relevant mouse model. *Hypertension, 64*(4), 852–859. https://doi.org/10.1161/HYPERTENSIONAHA.114.03848.

Schulz, A., Grosse, R., Schultz, G., Gudermann, T., & Schoneberg, T. (2000). Structural implication for receptor oligomerization from functional reconstitution studies of mutant V2 vasopressin receptors. *Journal of Biological Chemistry, 275*(4), 2381–2389.

Scroggins, S. M., Santillan, D. A., Lund, J. M., Sandgren, J. A., Krotz, L. K., Hamilton, W. S., et al. (2018). Elevated vasopressin in pregnant mice induces T-helper subset alterations consistent with human preeclampsia. *Clinical Science (London, England), 132*(3), 419–436. https://doi.org/10.1042/CS20171059.

Seibold, A., Rosenthal, W., Barberis, C., & Birnbaumer, M. (1993). Cloning of the human type-2 vasopressin receptor gene. *Annals of the New York Academy of Sciences, 689*, 570–572.

Seibold, A., Rosenthal, W., Bichet, D. G., & Birnbaumer, M. (1993). The vasopressin type 2 receptor gene. Chromosomal localization and its role in nephrogenic diabetes insipidus. *Regulatory Peptides, 45*(1–2), 67–71.

Shea, F. F., Rowell, J. L., Li, Y., Chang, T. H., & Alvarez, C. E. (2012). Mammalian alpha arrestins link activated seven transmembrane receptors to Nedd4 family e3 ubiquitin ligases and interact with beta arrestins. *PLoS One,* 7(12). e50557. https://doi.org/10.1371/journal.pone.0050557.

Shukla, A. K., Manglik, A., Kruse, A. C., Xiao, K., Reis, R. I., Tseng, W. C., et al. (2013). Structure of active beta-arrestin-1 bound to a G-protein-coupled receptor phosphopeptide. *Nature, 497*(7447), 137–141. https://doi.org/10.1038/nature12120.

Slusarz, M. J., Slusarz, R., & Ciarkowski, J. (2006). Investigation of mechanism of desmopressin binding in vasopressin V2 receptor versus vasopressin V1a and oxytocin receptors: Molecular dynamics simulation of the agonist-bound state in the membrane-aqueous system. *Biopolymers*, *81*(5), 321–338. https://doi.org/10.1002/bip.20420.

Torres, V. E., Chapman, A. B., Devuyst, O., Gansevoort, R. T., Perrone, R. D., Koch, G., et al. (2017). Tolvaptan in later-stage autosomal dominant polycystic kidney disease. *The New England Journal of Medicine*, *377*(20), 1930–1942. https://doi.org/10.1056/NEJMoa1710030.

van Gastel, M. D. A., & Torres, V. E. (2017). Polycystic kidney disease and the vasopressin pathway. *Annals of Nutrition and Metabolism*, *70*(Suppl 1), 43–50. https://doi.org/10.1159/000463063.

Wang, X., Wu, Y., Ward, C. J., Harris, P. C., & Torres, V. E. (2008). Vasopressin directly regulates cyst growth in polycystic kidney disease. *Journal of the American Society of Nephrology*, *19*(1), 102–108. https://doi.org/10.1681/ASN.2007060688.

Ward, D. T., Brown, E. M., & Harris, H. W. (1998). Disulfide bonds in the extracellular calcium-polyvalent cation-sensing receptor correlate with dimer formation and its response to divalent cations in vitro. *Journal of Biological Chemistry*, *273*(23), 14476–14483.

Wu, S., Birnbaumer, M., & Guan, Z. (2008). Phosphorylation analysis of G protein-coupled receptor by mass spectrometry: Identification of a phosphorylation site in V2 vasopressin receptor. *Analytical Chemistry*, *80*(15), 6034–6037. https://doi.org/10.1021/ac8008548.

Yi, X., Bouley, R., Lin, H. Y., Bechoua, S., Sun, T. X., Del Re, E., et al. (2007). Alix (AIP1) is a vasopressin receptor (V2R)-interacting protein that increases lysosomal degradation of the V2R. *American Journal of Physiology. Renal Physiology*, *292*(5), F1303–F1313. https://doi.org/10.1152/ajprenal.00441.2005.

Zerbe, R. L., Vinicor, F., & Robertson, G. L. (1979). Plasma vasopressin in uncontrolled diabetes mellitus. *Diabetes*, *28*(5), 503–508.

Zhu, X., & Wess, J. (1998). Truncated V2 vasopressin receptors as negative regulators of wild-type V2 receptor function. *Biochemistry*, *37*(45), 15773–15784. https://doi.org/10.1021/bi981162z.

CHAPTER NINE

Vasopressin receptor subtypes and renal sodium transport

Yu. V. Natochin*, D.V. Golosova
Sechenov Institute of Evolutionary Physiology and Biochemistry of the Russian Academy of Sciences, St. Petersburg, Russia
*Corresponding author: e-mail address: natochin1@mail.ru

Contents

Abstract

In mammals, three subtypes of V-receptors have been identified in the kidney. The effects of vasopressin, a hormone synthesized in the hypothalamus, are triggered by three distinct receptor isoforms: V2, V1a, and V1b. Stimulation of V2-receptors regulates urine osmotic concentration by increasing sodium reabsorption in the thick ascending limb of the loop of Henle and enhancing osmotic permeability of the epithelium cells in the collecting duct. Stimulation of V1a-receptors inhibits renal sodium reabsorption and induces natriuresis, comparable to the effect of the diuretic furosemide, in the thick ascending limb of the loop of Henle. Stimulation of V1b-receptors induces potassium secretion in the final parts of the distal segments and initial parts of the collecting ducts. In this review, we discuss the role of vasopressin and its interaction with V-receptor subtypes in natriuresis and for stabilizing the physicochemical parameters of the internal environment and water-salt homeostasis in humans. A better understanding of these systems and their regulation is necessary to facilitate identification of additional system components and mechanisms, clarify their contribution during various normal and pathological functional states, and suggest novel strategies for the development of therapeutic interventions.

Vitamins and Hormones, Volume 113
ISSN 0083-6729
https://doi.org/10.1016/bs.vh.2019.08.013

1. Phenomenology of vasopressin-induced natriuresis

Renal Na^+ excretion (natriuresis) depends on the ratio between glomerular filtration rate, Na^+ tubular reabsorption dependent upon the participation of several transport systems, tubular intranephron redistribution of the fluid flow between the tubular segments, and secretion of the fluid containing Na^+ into the nephron tubular lumen. Numerous factors regulate Na^+ reabsorption (Schield, 2016) including vasopressin (Knepper, Kim, Fernández-Llama, & Ecelbarger, 1999), which affects the rate of Na^+ reabsorption owing to the stimulation of different subtypes of V-receptors and various Na^+ transport systems. An increase of renal Na^+ excretion is realized in terms of enhancement of Na^+ urinary loss in relation to the initial level of Na^+ excretion; however, the value of natriuresis is limited by the volume of urine produced by the kidney and urinary osmolality.

The phenomenon of vasopressin-induced natriuresis depends on arginine vasopressin (AVP) blood serum concentration or external manipulation such as through the injection of an agonist of V1a-receptors (V1aR), which consequently induces intense renal Na^+ excretion. In comparison, as the stimulation of V2-receptors (V2R) requires a lower concentration of AVP, urinary Na^+ excretion does not change whereas solute-free water reabsorption is enhanced. For example, in water-loaded rats, an injection of AVP at a dose of 0.005 nmol/100 g body weight significantly reduced solute-free water excretion, whereas the elevation of AVP dose up to 0.5 nmol/100 g body weight produced natriuresis and solute-free water reabsorption (Kutina, Golosova, Marina, Shakhmatova, & Natochin, 2016). Accordingly, when blood serum concentration of AVP is higher, both V1aR and V2R are involved in the process of renal sodium excretion. Notably, the natriuretic effect of AVP is qualitatively different from the action of loop diuretics, which also block Na^+ transport but do not affect solute-free water reabsorption.

A broad understanding of the regulatory system of sodium balance in extracellular fluid that is closely related to renal sodium transport is necessary to facilitate the development of novel strategies for therapeutic interventions.

2. Milestones of history

Investigation of the effect of neurohypophyseal hormones on renal water and solute excretion originates from the end of the 19th century, when Oliver and Schäfer (1895) demonstrated that the administration of an extract of the posterior lobe of the pituitary gland led to the rise of arterial pressure. In their experiments, increased urine flow was observed, which was subsequently explained by the enhanced diuresis consequent to blood pressure rise (Howell, 1898). In addition, von den Velden (1913) and Farini (1913) found that the introduction of neurohypophyseal extract to patients with diabetes insipidus led to an antidiuretic effect; as a result, the hormone was termed antidiuretic (ADH). For a considerable period thereafter, it was believed that vasopressor and antidiuretic effects were caused by different substances, ADH and vasopressin. Later, du Vigneaud, Gish, and Katsoyannis (1954) elucidated the chemical structure of the hormone and finally synthesized it, for which du Vigneaud received the Nobel Prize for Chemistry in 1955. It was recognized that ADH and vasopressin comprised a single chemical substance that could induce both antidiuresis and vasopressor effects.

In the latter half of the century, the effect of neurohypophyseal nonapeptides (i.e., vasopressin and oxytocin) on renal Na^+ excretion was repeatedly investigated by a number of researchers, which revealed contradictory results (Thorn, 1968). In the late 1950s, Smith, analyzing the mechanisms of regulation of water-salt homeostasis, suggested that in addition to aldosterone, which increased renal Na^+ reabsorption, and vasopressin, which increased renal solute-free water reabsorption, an additional factor likely existed that regulated kidney role in Na^+ balance (Lichardus, 2014; Smith, 1957). Years later, atrial natriuretic hormone was discovered (De Bold, 1985) and a number of natriuretic physiologically active substances were identified (Vesely, 2007), although these did not include nonapeptides of the neurohypophysis.

More recently, vasopressin was shown to be able to increase renal Na^+ reabsorption (Ecelbarger et al., 2000), which correlates with its effect to induce active Na^+ transport in osmoregulatory epithelium in amphibians (Bentley, 2002). AVP increases not only osmotic permeability but also Na^+ transport by nephron cells. In vertebrates, ADH controls the osmolality and the volume of fluid in the organism, thus participating in the regulation of urinary osmotic concentration. Moreover, at the end of the 20th century, the roles of subtypes of vasopressin receptors in the regulation of kidney functions were elucidated (Lolait, O'Carroll, & Brownstein, 1995).

3. Vasopressin and receptors in the kidney

Most mammals, including humans, express AVP, whereas lysine-vasopressin and phenypressin were discovered in pig and marsupials, respectively

(Bentley, 2002). In vertebrates AVP predecessors comprise the peptides arginine-vasotocin (AVT), hydrin 1, and hydrin 2, among others (Acher, Chauvet, & Rouillé, 1997), which function in the regulation of water-salt homeostasis. Similarly, in different species of amphibians, an injection of AVT and AVP increases water reabsorption and induces Na^+ transport in osmoregulatory organs (Bentley, 2002; Goncharevskaya, Shakhmatova, & Natochin, 1995).

Vasopressin is synthesized in the hypothalamus, released from the neurohypophysis, and influences a number of functions with the participation of different subtypes of V-receptors: V1aR, V1bR, and V2R. The regulatory effects of vasopressin are determined by the subtype of stimulated V-receptor and the function of the cell in which the receptor exhibits membrane localization. Vasopressin receptors recruit different G proteins and hence stimulate distinct signaling pathways. For example, activation of the V1aR leads to Gq recruitment, increased activity of phospholipase C, generation of inositol trisphosphate, Ca^{2+} release from intracellular stores, and protein kinase C activation (Koshimizu et al., 2012). Conversely, V2R-activation leads to Gs recruitment, activation of adenylyl cyclase, and cAMP generation, followed by increased catalytic activity of cAMP-dependent protein kinases (Juul, Bichet, Nielsen, & Nørgaard, 2014; Lolait, O'Carroll, & Brownstein, 1995).

3.1 V2R

3.1.1 Thick ascending limb (TAL)

The pioneering microdissection studies of the Morel laboratory (Imbert, Chabardès, Montegut, Clique, & Morel, 1975) demonstrated the presence of vasopressin-sensitive adenylyl cyclase in both the medullary TAL and cortical TAL of rodents and rabbits. These findings have been confirmed and extended in subsequent studies in which AVP has been demonstrated to increase the intracellular level of cAMP in microdissected TAL (Torikai, Wang, Klein, & Kurokawa, 1981). The presence of V2R in the TAL has been directly demonstrated by binding studies in microdissected rat cortical and medullary TAL segments (Ammar, Schmidt, Semmekrot, Roseau, & Butlen, 1991) and by immunohistochemistry (Nonoguchi et al., 1995). Mutig et al. (2007) localized significant V2R (*AVPR2*) mRNA synthesis to the TAL in three species, revealing axial heterogeneity with solid expression in the medullary TAL and weak signal along the cortical TAL. Similar data had already been obtained in rats at the mRNA (Lolait et al., 1992; Ostrowski, Young, Knepper, & Lolait, 1993) and immunoreactive protein

(Nonoguchi et al., 1995; Sarmiento et al., 2005) levels, although without clear assignment to the particular portions. *AVPR2* mRNA has been also detected in the TAL by in situ hybridization histochemistry (Ostrowski et al., 1992, 1993) and by reverse transcription combined with polymerase chain reaction (RT-PCR) (Firsov et al., 1994; Terada, Tomita, Nonoguchi, Yang, & Marumo, 1993). More recent evaluation of segmental and cellular distribution of V-receptors has further revealed V2R immunoreactivity in the basolateral membrane of TAL and distal convoluted tubule (DCT) epithelia (Mutig et al., 2016).

3.1.2 Collecting duct (CD)

The antidiuretic action of AVP constitutes a V2R-mediated effect in the CD. The main actions of AVP on the CD, mediated by V2R, contribute to the urinary concentrating process by influencing water permeability, urea permeability, and Na^+ transport. V2Rs are expressed in distal tubules and CD epithelia (Mutig et al., 2007; Ostrowski et al., 1993). A prominent site of action for vasopressin is the activation of the V2R in the CD, which results in increased reabsorption of water to maintain osmotic homeostasis by modulating plasma membrane levels of the water channels in CD principal cells (Kortenoeven & Fenton, 2014; Nonoguchi et al., 1995; Staruschenko, 2012). AVP increases the permeability to urea of the CD inner medulla, an effect involving the activation of preexisting urea transporters (Bankir, 2001).

3.2 V1aR

3.2.1 Glomerulus

An immunohistochemical study revealed the presence of the V1aR in the glomeruli (Tashima et al., 2001). Large signals for V1aR PCR product were also detected here (Terada et al., 1993). In contrast, Ostrowski et al. (1993) showed by in situ hybridization that V1aR is abundantly expressed in the vasa recta but is absent in glomeruli.

3.2.2 Proximal tubule

Vasopressin receptors may exist in the early proximal tubules (Jung & Endou, 1991). Moreover, small but detectable PCR signals of V1aR (*AVPR1A*) transcripts were found in proximal convoluted and straight tubules (Terada et al., 1993).

3.2.3 *Macula densa*

Macula densa cells also exhibited a basolateral signal for V1aR but were devoid of V2R (Mutig et al., 2016). AVP modulates the glomerular filtration rate and renin release through activation of V1aR (Aoyagi et al., 2008; Roald, Tenstad, & Aukland, 2004).

3.2.4 *TAL*

Detectable signals from *AVPR1A* PCR were found in TAL (Terada et al., 1993). In addition, a study using a nonquantitative RT-PCR approach reported a small quantity of *AVPR1A* mRNA in microdissected inner medullary thin limbs as well as medullary TAL (Terada et al., 1993). However, these were not detected in the TAL of rat by either quantitative RT-PCR (Firsov et al., 1994) or in situ hybridization (Ostrowski et al., 1992, 1993). An immunohistochemical study revealed the presence of V1aR in the TAL (Tashima et al., 2001). V1aR expression has been demonstrated to a lesser degree in cells of the TAL and thin ascending limb (Carmosino et al., 2007; Firsov et al., 1994; Gonzalez et al., 1997). In the study by Gonzalez et al. (1997) it was shown that TAL was stained by the antibody, suggesting the presence of the V1aR subtype in this part of the nephron.

Vasopressin at concentrations 0.1 nM can trigger a transient increase in intracellular Ca^{2+} in TAL cells (Burgess, Balment, & Beck, 1994; Jung & Endou, 1991; Morel et al., 1982; Nitschke, Fröbe, & Greger, 1991). Studies in the mouse medullary TAL revealed that AVP, but not a V2-selective agonist (desmopressin), provoked an increase in inositol-3-phosphate production (Dai & Quamme, 1994). These results indicated an ability of AVP to activate the characteristic signaling pathway in TAL cells that is associated with V1aR. The effects of AVP mediated by V1aR require a 100-fold higher hormone concentration to activate Ca^{2+} signaling in rat inner medullary CD than that required to induce an increase in cAMP accumulation (Bankir, Bichet, & Bouby, 2010).

The evidence suggests that the gene for V1aR may be expressed at low levels in TAL. V1aR simulation in TAL of Henle's loop and/or the CD may lead to increased expression of V2R in the distal tubule via hormonal or paracrine action. V1aR, mainly in the luminal membrane of the distal nephron, regulates basolateral V2R-mediated action with regard to water and ion transport through the activation of Gq/11 and phosphoinositide turnover (Inoue, Nonoguchi, & Tomita, 2001).

3.2.5 CD

Pharmacological, biochemical, and autoradiographical evidence suggests that the V1aR is present in the distal nephron and in the CD (Ammar, Roseau, & Butlen, 1992; Ando, Tabei, & Asano, 1991; Gerstberger & Fahrenholz, 1989). V1aR expression has been demonstrated in cells of the connecting tubule, and cortical and medullary portions of the CD (Firsov et al., 1994; Gonzalez et al., 1997; Tashima et al., 2001). According to the data obtained by Firsov et al. (1994), *AVPR1A* mRNA is slightly expressed in the thin ascending limb, absent in TAL, and reaches its maximum in the cortical CD. Large signals for the *AVPR1A* PCR product were detected in all parts of the initial cortical CD, cortical CD, outer medullary CD, and inner medullary CD (Terada et al., 1993). V1aR was found in principal cells in the CD (Burnatowska-Hledin & Spielman, 1989). The expression level of the V1aR is lower at deeper portions than at shallow portions of the CD and is not present in the terminal inner medullary CD (Maeda, Han, Gibson, & Knepper, 1993; Tashima et al., 2001). The V1aR in the cortical CD is distributed in the principal and intercalated cells. Carmosino et al. (2007) showed that in the CD, the V1aR is exclusively expressed in intercalated cells in the medulla, albeit in both principal and intercalated cells in the cortical CD.

3.3 V1bR

The rat VlbR constitutes a protein of 421 amino acids that has 37–50% identity with VlaR and V2R. VlbR acts via phosphotidylinositol hydrolysis and mobilization of intracellular Ca^{2+}. RNA blot analysis, reverse transcription PCR, and in situ hybridization studies revealed that *AVPR1B* mRNA is expressed in kidney (Lolait, O'Carroll, Mahan, et al., 1995). In reverse transcription-PCR analysis of the kidney, *AVPR1B* mRNA was detected only in the medulla (Saito, Tahara, Sugimoto, Abe, & Furuichi, 2000). In addition, we reported direct experimental evidence of the involvement of V1bR in selective regulation of ion transport in the kidney in the distal nephron (Kutina, Marina, & Natochin, 2014).

3.3.1 TAL

Vasopressin activates the phosphoinositide signaling pathway in TAL cells, a characteristic usually associated with V1aR and V1bR. Binding studies in isolated medullary TAL cells from rat are also consistent with the presence of a V1-like receptor (Baudouin-Legros, Bouthier, & Teulon, 1993).

3.3.2 CD

The localization and function of V1bR in the kidney, possibly in the inner medullary CD, are not well characterized. A vasopressin V1aR/V2R dual antagonist, which has no effect on V1bR, did not block the Ca^{2+} increase in the inner medullary CD cells when stimulated by desmopressin (Saito et al., 2000), which may serve as evidence of V1bR localization in the CD.

4. Renal sodium transport

Renal Na^+ transport constitutes an extremely energy-intensive activity. The main mechanisms of Na^+ transport in the kidney are represented in Fig. 1. In humans, 180 L of fluid are filtered and 178 L are usually reabsorbed daily in glomeruli. When the amount of Na^+ molecules was calculated, it appeared that 24,400 mmol of the ions entered the nephron lumen daily, 24,300 mmol was reabsorbed, and 100 mmol was excreted. The classical scheme of urine formation includes three processes: (1) glomerular filtration (C_{In}), (2) reabsorption, which is calculated as the difference between

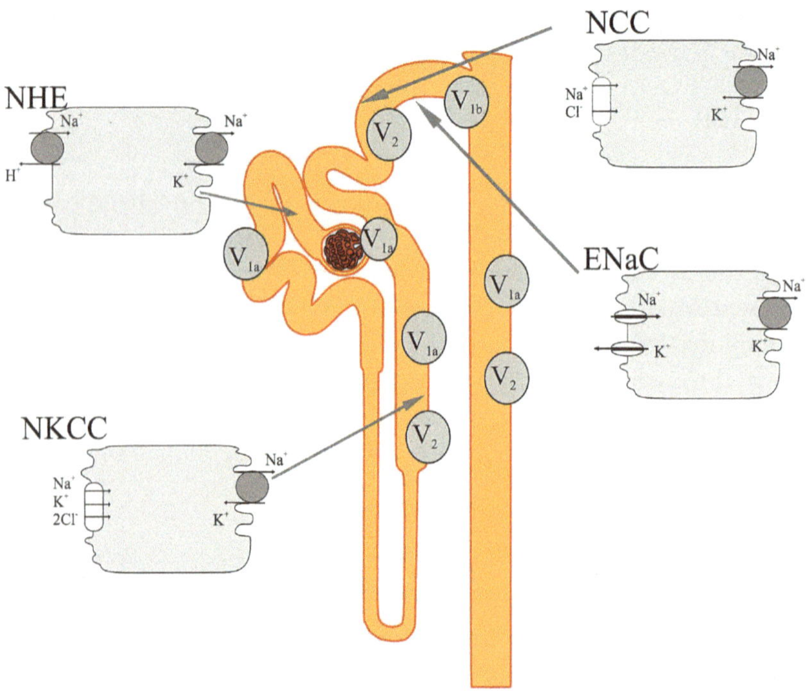

Fig. 1 Localization of V-receptors and sodium transport systems in the kidney.

$C_{In}*P_x - U_x*V$, where P_x and U_x are the concentration of the studied substance in the blood serum and in the urine, V—diuresis; and (3) secretion of the substance. Renal Na^+ reabsorption is under the control of humoral factors and the nervous system, with vasopressin playing a crucial role in these processes (Vesely, 2007). Notably, in mammals, fluid may enter the nephron lumen not only during the ultrafiltration process but also during the process of secretion, along with organic acids into the proximal tubule lumen (Grantham, Qualizza, & Irwin, 1974).

4.1 Proximal nephron segment

In mammals, approximately 2/3 of the fluid filtered in the glomeruli is reabsorbed in the proximal nephron segment, with the process being considered as obligate reabsorption. In the first half of the segment, Na^+ reabsorption is coupled with the uptake of solutes such as glucose, amino acids, phosphate, sulfate, and lactate by different cotransporter systems. In the proximal tubule, the electroneutral Na^+/H^+ exchanger links Na^+ reabsorption to that of bicarbonate (Schield, 2016). In the proximal tubule lumen, the concentration of Cl^- increases with the constant level of osmolality of the fluid. In the subsequent parts of the tubule this creates conditions for more chlorides to be driven by the high tubular concentration from the lumen to the extracellular fluid, resulting in the tubular wall becoming electropositive. Electrophysiological and isotopic estimates of the Cl^- to Na^+ permeability revealed that the superficial tubule is approximately twice as permeant to Cl^- than to Na^+ (Kawamura, Imai, Seldin, & Kukko, 1975). This indicates that a portion of the Na^+ is driven by the electrochemical gradient effected by the transport of Cl^- into the extracellular fluid by the concentration gradient.

In addition, new data indicate the possibility of redistribution of electrolyte and fluid flows between different nephron segments. In particular, cases have been identified in which obligate reabsorption in the proximal tubule is decreased and a large volume of fluid and dissolved solutes enter the distal nephron segment (Kutina et al., 2016).

4.2 Distal nephron segment

An Na-H-exchanger has also been immunolocalized to the outer medullary thin descending limb of Henle and the TAL (Rutherford et al., 1997). Vasopressin has the net effect of inhibiting Na-H-exchanger activity in the medullary TAL (Good, Watts, George, Meyer, & Shull, 2004).

The diluting mechanism is well understood as being chiefly the function of the medullary TAL of Henle's loop. As the epithelium of the TAL is water impermeable, Na^+ reabsorption in this segment results in an increased medullary interstitial osmolality and provides osmotic removal of water from the adjacent CD (Fenton & Knepper, 2007). The thin descending limb of Henle is impermeable to Na^+ ions; in contrast, the TAL contributes to the reabsorption of approximately 20% of the filtered Na^+. Two major transport mechanisms contribute to the Na^+ reabsorption in this segment, the electroneutral Na-K-2C1 cotransporter (NKCC) and the Na-H-exchanger (Schield, 2016). In the cells of TAL, the gene encoding the bumetanide-sensitive cotransporter NKCC is activated and renal Na^+ reabsorption is consequently increased (Mount et al., 1998). A particular characteristic of the TAL is the presence of a lumen positive electrical potential generated by the recycling of K^+ across the apical membrane; because of the selective cationic permeability of the tight junctions in the TAL, this lumen positive potential contributes to Na^+ reabsorption along a favorable driving force for diffusion of Na^+ through the paracellular pathway.

NKCC has several isoforms, with NKCC2 being present in the TAL (Gamba et al., 1994). Alternative splicing of the exon 4 encoding the chloride-binding protein region gives rise to the NKCC2A, NKCC2B, and NKCC2F isoforms with different ion affinities, transport kinetics, and distribution patterns along the TAL (Castrop & Schießl, 2014). The renal NKCC2 (also known as "BSC1") is expressed in the TAL, where it is localized to the apical membrane of epithelial cells (Kaplan et al., 1996; Nielsen, Maunsbach, Ecelbarger, & Knepper, 1998). NKCC2 is also expressed in the macula densa (Nielsen et al., 1998). In turn, furosemide, bumetanide, and torsemide are prototypical loop diuretics; these agents bind to the translocation pocket at the extracellular surface of NKCCs, blocking ion transport directly. Loop diuretics inhibit the NKCC2 at the apical surface of TAL cells along the loop of Henle, a transporter that reabsorbs up to 25% of filtered Na^+ and Cl^- (Ellison & Felker, 2017). NKCC2 constitutes the central element in urine dilution and generation of an osmotic gradient for urinary concentration. Physiological regulation of NKCC2 combines the effects of parathyroid hormone, calcitonin, glucagon, and catecholamines on the activating arm, counterbalanced by prostaglandins and extracellular Ca^{2+} on the inhibitory arm (Castrop & Schießl, 2014). AVP, acting via V2R, is particularly important for maintaining NKCC2 activity, as illustrated by the markedly impaired urinary concentration in transgenic rats with TAL-specific suppression of V2R (Mutig et al., 2016).

The electroneutral Na-Cl cotransporter (NCC) has been immunolocalized to the DCT (Ellison, 2013; Plotkin et al., 1996). In the DCT, Na^+ reabsorption is mediated by NCC that is specifically expressed in the apical membrane of this nephron segment. Vasopressin decreases solute-free water excretion to dilute plasma in part by promoting sodium reabsorption and consequently decreasing sodium excretion via the epithelial Na^+ channel (ENaC) activated along the distal nephron (Stockand, 2010). Further downstream, ENaC is responsible for electrogenic Na^+ reabsorption. In the late portion of the distal DCT, the expression of NCC and ENaC overlap. ENaC is predominantly expressed in the connecting tubule, and the initial and cortical CD. Further downstream its expression follows a decreasing axial gradient from the connecting tubule down to the medullary CD (Loffing & Kaissling, 2003).

4.3 CD

ENaC is distributed in the principal cells of the CD (Nonoguchi et al., 1995; Staruschenko, 2012). ENaC allows the electrogenic entry of Na^+ into the cell along a favorable electrochemical gradient. The abundances of the beta and gamma subunits of ENaC are increased by AVP. In contrast, aldosterone selectively increases the abundance of the alpha subunit protein without affecting levels of the beta and gamma subunits (Knepper, Kwon, & Nielsen, 2015).

5. Natriuretic effect of vasopressin

AVP-induced renal Na^+ transport modification is determined by (1) the region of the effect in the nephron; (2) the locus of the effect in the cell: luminal membrane, basolateral membrane, or paracellular zone; (3) the type of the macromolecule: ion channels, ion pumps, cotransporters, or claudines; (4) the subtypes of V-receptors and the mechanism of the action; and (5) the duration of the exposure. AVP action on arterial pressure depends on vasoconstrictor effect and its influence on Na^+ balance in the organism. The kidney maintains extracellular fluid volume and blood pressure by regulating urinary Na^+ excretion to precisely balance daily Na^+ intake. Thus, vasopressin takes part in regulating the ratio of the volume of intravascular fluid and the capacity of the vascular bed. The effect of vasopressin on the volume of intracellular fluid is realized consequent to removal of Na^+, the main osmotically active ion of the extracellular fluid, from the organism. Such an effect is possible owing to the natriuretic activity of vasopressin and the ability of the kidney to produce urine with osmolality exceeding that of the blood. This function is impaired in some forms of disorders in humans (i.e., chronic kidney disease) and is absent in many species of vertebrates, which are not able to perform urine osmotic concentration (Fish, Amphibians, Reptiles). The physiological role of this effect is that osmotic homeostasis requires not only the regulation of water reabsorption but an independent retention of Na^+ in the body. The regulation of Na^+ and urea transport by vasopressin is crucial for the maintenance of systemic water balance. Na^+ concentration in the renal medullary interstitium (via a countercurrent mechanism), coupled with accumulation of urea in the medullary interstitium, creates a luminal-to-interstitial osmotic gradient, allowing diffusion of water out of the connecting tubule and CD into the interstitium, from where it is subsequently returned to the general blood circulation (Fenton & Knepper, 2007).

In small doses AVP exhibits antidiuretic action, whereas it possesses severe natriuretic activity when introduced in higher doses (Perucca, Bichet, Bardoux, Bouby, & Bankir, 2008). An injection of AVT, which possesses V1aR and V2R activity, to a rat induced high natriuresis (Gao & Natochin, 2004). In experiments with unanesthetized dogs, it was shown that an infusion of AVP led to decreased diuresis and increased natriuresis. Although the V2R agonist dDAVP did not affect renal Na^+ excretion, an infusion of AVT led to natriuresis (Kompanowska-Jezierska et al., 1998). In turn, an agonist of V2R induced Na^+ reabsorption and an agonist of V1aR produced decreased Na^+ transport from the tubular lumen into the blood, consequently providing intense natriuresis (Golosova, Karavashkina, Kutina, Marina, & Natochin, 2016).

Localization of subtypes of V-receptors in the nephron makes it possible to evaluate their role in the regulation of urine formation. V2R performs a crucial role in osmoregulation: in the TAL, V2R stimulation leads to increased renal Na^+ reabsorption, whereas in the CD it leads to enhanced solute-free water reabsorption (Bankir, 2001). V1bR in the terminal parts of the distal nephron and the initial parts of the CD participates in the regulation of the urinary secretion rate of K^+. This tubular zone is located on the border of the reabsorption and secretion of K^+, with this segment determining the involvement of AVP in the regulation of renal K^+ secretion (Kutina et al., 2014).

Recent studies in vivo on rats indicated that the selective stimulation of V1aR induced urinary Na^+ excretion, whereas neither V2R nor V1bR stimulation had the same natriuretic effect (Golosova et al., 2016). However, the localization of V1aR in the kidney remains controversial. V1aR stimulation induces severe natriuresis, reaching 15–20% of filtered Na^+; as such effect is comparable to furosemide action, this may suggest TAL as representing the main site of V1aR localization. However, when mRNA was used as a criterion, only a small amount of *AVPR1A* mRNA was observed (Firsov et al., 1994). Our studies and the analysis of the existing pattern allows us provide the following explanation for this phenomenon. Unlike furosemide, which inactivates NKCC from the tubular lumen, a V1a-agonist, AVP and AVT act on V1aR on the plasma membrane and lead to the formation of second messengers, which serve as a signal for embedding or activation of NKCC from the inside of the cell luminal membrane. The obtained calculations showed that renal Na^+ excretion following V1aR stimulation is much higher than the natriuretic effect induced by furosemide. Thus, nonapeptides in molar comparison appear to be more efficient than furosemide as a natriuretic agent (Karavashkina, Kutina, Shakhmatova, & Natochin, 2011). Accordingly, the quantity of *AVPR1A* mRNA could be many times less than the number of cotransporters. If these arguments are valid, then unlike the significant amount of mRNA participating in NKCC formation, smaller amounts of *AVPR1A* would be detected. There is no need for a substantial amount of V1aR in the cell membrane of the TAL considering such a huge gain factor. In comparison, the measurement of second messengers, such as IP_3 or Ca^{2+} in toto, in experiments in either in vivo or in situ using different modern techniques is not feasible; moreover, changes in individual cells will not address the quantity of V1aR in the cells of the TAL.

The effects of V2R activation in the TAL have been studied using AVP or V2R-specific agonists, which documented an AVP-V2R-dependent rise in intracellular cAMP levels and increased NKCC2 abundance and phosphorylation, as well as stimulation of NaCl transport in TAL cells (Giménez & Forbush, 2003; Hebert & Andreoli, 1984; Morel, 1981). Vasopressin appears to exert its effects on NKCC2 trafficking, phosphorylation, and activity by adjusting a complex network of major serine/threonine kinases and phosphatases (Bachmann & Mutig, 2017). AVP is known to induce NaCl reabsorption via NKCC2 in the medullary TAL (Knepper et al., 1999). In addition, NaCl reabsorption in the medullary TAL can be increased by acute vasopressin administration (Kutina et al., 2016). This is thought to occur in part owing to increased apical membrane expression of NKCC2. Over the long-term, enhanced cAMP levels induce NKCC2 transcription and biosynthesis (Ecelbarger et al., 2000).

Renal Na^+ reabsorption in the cells of CD is based on electrically neutral and electrogenic mechanisms (Leviel et al., 2010; Tomita, Pisano, & Knepper, 1985).

The electrogenic component is caused by Na^+ transport through the plasma membrane with the participation of ENaC. The regulation of electrogenic Na^+ transport in the cortical CD can be synergistically dependent on AVP and aldosterone (Knepper et al., 2015; Staruschenko, 2012). Vasopressin acutely increases Na^+ reabsorption in the rat cortical CD by increasing apical Na^+ entry via ENaC, which is proposed to result from vasopressin-induced trafficking of ENaC-containing vesicles from intracellular stores to the apical plasma membrane (Snyder, 2005). Moreover, stimulation of V2R affects the thiazide-sensitive NCC in the luminal membrane of DCT cells (Pedersen, Hofmeister, Rosenbaek, Nielsen, & Fenton, 2010).

6. Conclusions

It is physiologically important to characterize not only the features and capabilities of each of the Na^+ transport systems in renal tubules but also their effect in conditions of different functional states of the organism. This is essential for understanding the nature of the regulation and the possibility of identifying new components of these systems, when known factors are insufficient to explain the observed effects. It can be assumed that the involvement of various subtypes of V-receptors plays a key role in stabilizing the physicochemical parameters of the internal environment and water-salt homeostasis in humans upon acute and chronic changes of these parameters. An increase in renal Na^+ excretion participates in the mechanism of osmotic homeostasis when serum concentration of Na^+ is substantially high and AVP level rises several times. Our data on dose-dependent effects of AVP on Na^+ excretion are consistent with this hypothesis. This suggests that under hypernatremia, V1aR and V2R are activated, such that in conditions of dehydration, a pattern of water and Na^+ transport regulation by AVP appears physiologically appropriate. Vasopressin is a unique physiologically active substance stimulating different subtypes of V-receptors maintains the volume and osmolality of extracellular fluid. Its clinical use might serve as a novel therapeutic agent which facilitates water-salt homeostasis.

Acknowledgment

This work was supported by the Russian Scientific Foundation (project 18-15-00358).

References

Acher, R., Chauvet, J., & Rouillé, Y. (1997). Adaptive evolution of water homeostasis regulation in amphibians: Vasotocin and hydrins. *Biology of the Cell*, *89*(5–6), 283–291. https://doi.org/10.1111/j.1768-322X.1997.tb01024.x.

Ammar, A., Roseau, S., & Butlen, D. (1992). Pharmacological characterization of V1a vasopressin receptors in the rat cortical collecting duct. *American Journal of Physiology—Renal Physiology*, *262*(4 Pt. 2), F546–F553. https://doi.org/10.1152/ajprenal.1992.262.4.F546.

Ammar, A., Schmidt, A., Semmekrot, B., Roseau, S., & Butlen, D. (1991). Receptors for neurohypophyseal hormones along the rat nephron: 125I-labelled d(CH2)5[Tyr(Me) 2, Thr4, Orn8, Tyr-NH(2)9] vasotocin binding in microdissected tubules. *Pflügers Archiv*, *418*(3), 220–227. https://doi.org/10.1007/BF00370519.

Ando, Y., Tabei, K., & Asano, Y. (1991). Luminal vasopressin modulates transport in the rabbit cortical collecting duct. *The Journal of Clinical Investigation*, *88*(3), 952–959. https://doi.org/10.1172/JCI115398.

Aoyagi, T., Izumi, Y., Hiroyama, M., Matsuzaki, T., Yasuoka, Y., Sanbe, A., et al. (2008). Vasopressin regulates the renin-angiotensin-aldosterone system via V1a receptors in macula densa cells. *American Journal of Physiology—Renal Physiology*, *295*(1), F100–F107. https://doi.org/10.1152/ajprenal.00088.2008.

Bachmann, S., & Mutig, K. (2017). Regulation of renal Na-(K)-Cl cotransporters by vasopressin. *Pflügers Archiv*, *469*(7–8), 889–897. https://doi.org/10.1007/s00424-017-2002-2.

Bankir, L. (2001). Antidiuretic action of vasopressin: Quantitative aspects and interaction between V1a and V2 receptor-mediated effects. *Cardiovascular Research*, *51*(3), 372–390. https://doi.org/10.1016/S0008-6363(01)00328-5.

Bankir, L., Bichet, D. G., & Bouby, N. (2010). Vasopressin V2 receptors, ENaC, and sodium reabsorption: A risk factor for hypertension? *American Journal of Physiology—Renal Physiology*, *299*(5), F917–F928. https://doi.org/10.1152/ajprenal.00413.2010.

Baudouin-Legros, M., Bouthier, M., & Teulon, J. (1993). [Arginine] vasopressin hydrolyses phosphoinositides in the medullary thick ascending limb of mouse nephron. *Pflügers Archiv*, *425*(5–6), 381–389. https://doi.org/10.1007/BF00374862.

Bentley, P. J. (2002). *Endocrines and osmoregulation. A comparative account in vertebrates*. [Chapter 1]. Berlin: Springer.

Burgess, W. J., Balment, R. J., & Beck, J. S. (1994). Effects of luminal vasopressin on intracellular calcium in microperfused rat medullary thick ascending limb. *Kidney and Blood Pressure Research*, *17*(1), 1–9. https://doi.org/10.1159/000173782.

Burnatowska-Hledin, M. A., & Spielman, W. S. (1989). Vasopressin V1 receptors on the principal cells of the rabbit cortical collecting tubule. Stimulation of cytosolic free calcium and inositol phosphate production via coupling to a pertussis toxin substrate. *The Journal of Clinical Investigation*, *83*(1), 84–89. https://doi.org/10.1172/JCI113888.

Carmosino, M., Brooks, H. L., Cai, Q., Davis, L. S., Opalenik, S., Hao, C., et al. (2007). Axial heterogeneity of vasopressin-receptor subtypes along the human and mouse collecting duct. *American Journal of Physiology—Renal Physiology*, *292*(1), F351–F360. https://doi.org/10.1152/ajprenal.00049.2006.

Castrop, H., & Schießl, I. M. (2014). Physiology and pathophysiology of the renal Na-K-2Cl cotransporter (NKCC2). *American Journal of Physiology—Renal Physiology*, *307*(9), F991–F1002. https://doi.org/10.1152/ajprenal.00432.2014.

Dai, L. J., & Quamme, G. A. (1994). Hormone-mediated Ca^{2+} transients in isolated renal cortical thick ascending limb cells. *Pflügers Archiv*, *427*(1–2), 1–8. https://doi.org/10.1007/BF00585935.

De Bold, A. J. (1985). Atrial natriuretic factor: A hormone produced by the heart. *Science*, *230*(4727), 767–770. https://doi.org/10.1126/science.2932797.

du Vigneaud, V., Gish, D. T., & Katsoyannis, P. G. (1954). A synthetic preparation possessing biological properties associated with arginine-vasopressin. *Journal of the American Chemical Society*, *76*(18), 4751–4752. https://doi.org/10.1021/ja01647a089.

Ecelbarger, C. A., Kim, G. H., Terris, J., Masilamani, S., Mitchell, C., Reyes, I., et al. (2000). Vasopressin-mediated regulation of epithelial sodium channel abundance in rat kidney. *American Journal of Physiology—Renal Physiology*, *279*(1), F46–F53. https://doi.org/10.1152/ajprenal.2000.279.1.F46.

Ellison, D. H. (2013). Physiology and pathophysiology of diuretic action. In R. J. Alpern, M. J. Caplan, & O. W. Moe (Eds.), *Seldin and Giebisch's the kidney physiology and pathophysiology* (5th ed., pp. 1353–1404). New York: Elsevier.

Ellison, D. H., & Felker, G. M. (2017). Diuretic treatment in heart failure. *New England Journal of Medicine*, *377*(20), 1964–1975. https://doi.org/10.1056/NEJMra1703100.

Farini, F. (1913). Diabete insipido ed opoterapia. *Gazzetta degli Ospedali e delle Cliniche*, *34*, 1135–1139.

Fenton, R. A., & Knepper, M. A. (2007). Mouse models and the urinary concentrating mechanism in the new millennium. *Physiological Reviews*, *87*(4), 1083–1112. https://doi.org/10.1152/physrev.00053.2006.

Firsov, D., Mandon, B., Morel, A., Merot, J., Le Maout, S., Bellanger, A. C., et al. (1994). Molecular analysis of vasopressin receptors in the rat nephron. Evidence for alternative splicing of the V2 receptor. *Pflügers Archiv*, *429*(1), 79–89. https://doi.org/10.1007/BF02584033.

Gamba, G., Miyanoshita, A., Lombardi, M., Lytton, J., Lee, W. S., Hediger, M. A., et al. (1994). Molecular cloning, primary structure, and characterization of two members of the mammalian electroneutral sodium-(potassium)-chloride cotransporter family expressed in kidney. *Journal of Biological Chemistry*, *269*(26), 17713–17722.

Gao, J., & Natochin, Y. V. (2004). Evolutionary advantages of participation of vasopressin instead of vasotocin in regulation of water–salt balance in mammals. *Journal of Evolutionary Biochemistry and Physiology*, *40*(2), 207–212. https://doi.org/10.1023/B:JOEY.0000033813.74233.ed.

Gerstberger, R., & Fahrenholz, F. (1989). Autoradiographic localization of V1 vasopressin binding sites in rat brain and kidney. *European Journal of Pharmacology*, *167*(1), 105–116. https://doi.org/10.1016/0014-2999(89)90752-8.

Giménez, I., & Forbush, B. (2003). Short-term stimulation of the renal Na-K-Cl cotransporter (NKCC2) by vasopressin involves phosphorylation and membrane translocation of the protein. *Journal of Biological Chemistry*, *278*(29), 26946–26951. https://doi:10.1074/jbc.M303435200.

Golosova, D. V., Karavashkina, T. A., Kutina, A. V., Marina, A. S., & Natochin, Y. V. (2016). Effects of selective agonists of V1a, V2, and V1b receptors on sodium transport in rat kidney. *Bulletin of Experimental Biology and Medicine*, *160*(6), 751–754. https://doi.org/10.1007/s10517-016-3301-x.

Goncharevskaya, O. A., Shakhmatova, E. I., & Natochin, Y. V. (1995). Modulation of the effect of arginine-vasopressin on water and ion transport in the newt early distal tubule and frog urinary bladder by V1-antagonists. *Pflügers Archiv*, *430*(6), 1004–1011.

Gonzalez, C. B., Figueroa, C. D., Reyes, C. E., Caorsi, C. E., Troncoso, S., & Menzel, D. (1997). Immunolocalization of V1 vasopressin receptors in the rat kidney using anti-receptor antibodies. *Kidney International*, *52*(5), 1206–1215. https://doi.org/10.1038/ki.1997.445.

Good, D. W., Watts, B. A., III, George, T., Meyer, J. W., & Shull, G. E. (2004). Transepithelial HCO_3^- absorption is defective in renal thick ascending limbs from Na^+/H^+ exchanger NHE1 null mutant mice. *American Journal of Physiology—Renal Physiology*, *287*(6), F1244–F1249. https://doi.org/10.1152/ajprenal.00176.2004.

Grantham, J. J., Qualizza, P. B., & Irwin, R. L. (1974). Net fluid secretion in proximal straight renal tubules in vitro: Role of PAH. *American Journal of Physiology, 226*(1), 191–197. https://doi.org/10.1152/ajplegacy.1974.226.1.191.

Hebert, S. C., & Andreoli, T. E. (1984). Effects of antidiuretic hormone on cellular conductive pathways in mouse medullary thick ascending limbs of Henle: II. Determinants of the ADH-mediated increases in transepithelial voltage and in net Cl^- absorption. *The Journal of Membrane Biology, 80*(3), 221–233. https://doi.org/10.1007/BF01868440.

Howell, W. H. (1898). The physiological effects of extracts of the hypophysis cerebri and infundibular body. *The Journal of Experimental Medicine, 3*(2), 245–258. https://doi.org/10.1084/jem.3.2.245.

Imbert, M., Chabardès, D., Montegut, M., Clique, A., & Morel, F. (1975). Vasopressin dependent adenylate cyclase in single segments of rabbit kidney tubule. *Pflügers Archiv, 357*(3), 173–186. https://doi.org/10.1007/BF00585973.

Inoue, T., Nonoguchi, H., & Tomita, K. (2001). Physiological effects of vasopressin and atrial natriuretic peptide in the collecting duct. *Cardiovascular Research, 51*(3), 470–480. https://doi.org/10.1016/S0008-6363(01)00248-6.

Jung, K. Y., & Endou, H. (1991). A novel vasopressin receptor in rat early proximal tubule. *Biochemical and Biophysical Research Communications, 180*(1), 131–137. https://doi.org/10.1016/S0006-291X(05)81265-3.

Juul, K. V., Bichet, D. G., Nielsen, S., & Nørgaard, J. P. (2014). The physiological and pathophysiological functions of renal and extrarenal vasopressin V2 receptors. *American Journal of Physiology—Renal Physiology, 306*(9), F931–F940. https://doi.org/10.1152/ajprenal.00604.2013.

Kaplan, M. R., Plotkin, M. D., Lee, W. S., Xu, Z. C., Lytton, J., & Hebert, S. C. (1996). Apical localization of the Na-K-Cl cotransporter, rBSC1, on rat thick ascending limbs. *Kidney International, 49*(1), 40–47.

Karavashkina, T. A., Kutina, A. V., Shakhmatova, E. I., & Natochin, Y. V. (2011). Mechanism of 1-deamino-arginine vasotocin induced natriuresis in rats. *General and Comparative Endocrinology, 170*(3), 460–467. https://doi.org/10.1016/j.ygcen.2010.10.020.

Kawamura, S., Imai, M., Seldin, D. W., & Kukko, J. P. (1975). Characteristics of salt and water transport in superficial and juxtamedullary straight segments of proximal tubules. *The Journal of Clinical Investigation, 55*(6), 1269–1277. https://doi.org/10.1172/JCI108046.

Knepper, M. A., Kim, G. H., Fernández-Llama, P., & Ecelbarger, C. A. (1999). Regulation of thick ascending limb transport by vasopressin. *Journal of the American Society of Nephrology, 10*(3), 628–634.

Knepper, M. A., Kwon, T. H., & Nielsen, S. (2015). Molecular physiology of water balance. *New England Journal of Medicine, 372*(14), 1349–1358. https://doi.org/10.1056/NEJMra1404726.

Kompanowska-Jezierska, E., Emmeluth, C., Grove, L., Christensen, P., Sadowski, J., & Bie, P. (1998). Mechanism of vasopressin natriuresis in the dog: Role of vasopressin receptors and prostaglandins. *American Journal of Physiology—Regulatory, Integrative and Comparative Physiology, 274*(6 Pt. 2), R1619–R1625. https://doi.org/10.1152/ajpregu.1998.274.6.R1619.

Kortenoeven, M. L., & Fenton, R. A. (2014). Renal aquaporins and water balance disorders. *Biochimica et Biophysica Acta, 1840*(5), 1533–1549. https://doi.org/10.1016/j.bbagen.2013.12.002.

Koshimizu, T. A., Nakamura, K., Egashira, N., Hiroyama, M., Nonoguchi, H., & Tanoue, A. (2012). Vasopressin V1a and V1b receptors: From molecules to physiological systems. *Physiological Reviews, 92*(4), 1813–1864. https://doi.org/10.1152/physrev.00035.2011.

Kutina, A. V., Golosova, D. V., Marina, A. S., Shakhmatova, E. I., & Natochin, Y. V. (2016). Role of vasopressin in the regulation of renal sodium excretion: Interaction with glucagon-like peptide-1. *Journal of Neuroendocrinology, 28*(4) 12367 https://doi.org/10.1111/jne.12367.

Kutina, A. V., Marina, A. S., & Natochin, Y. V. (2014). The involvement of V 1b-subtype vasopressin receptors in regulation of potassium ions excretion in the rat kidneys. *Doklady Biological Sciences, 459*(1), 338–340. https://doi.org/10.1134/S001249661406009X.

Leviel, F., Hübner, C. A., Houillier, P., Morla, L., El Moghrabi, S., Brideau, G., et al. (2010). The Na^+-dependent chloride-bicarbonate exchanger SLC4A8 mediates an electroneutral Na^+ reabsorption process in the renal cortical collecting ducts of mice. *The Journal of Clinical Investigation, 120*(5), 1627–1635. https://doi.org/10.1172/JCI40145.

Lichardus, B. (2014). Early stages of the natriuretic hormone story. *Frontiers in Endocrinology, 5*, 180. https://doi.org/10.3389/fendo.2014.00180.

Loffing, J., & Kaissling, B. (2003). Sodium and calcium transport pathways along the mammalian distal nephron: From rabbit to human. *American Journal of Physiology—Renal Physiology, 284*(4), F628–F643. https://doi.org/10.1152/ajprenal.00217.2002.

Lolait, S. J., O'Carroll, A., & Brownstein, M. J. (1995). Molecular biology of vasopressin receptors. *Annals of the New York Academy of Sciences, 771*(1), 273–292. https://doi.org/10.1111/j.1749-6632.1995.tb44688.x.

Lolait, S. J., O'Carroll, A. M., Mahan, L. C., Felder, C. C., Button, D. C., Young, W. S., 3rd, et al. (1995). Extrapituitary expression of the rat V1b vasopressin receptor gene. *Proceedings of the National Academy of Sciences of the United States of America, 92*(15), 6783–6787. https://doi.org/10.1073/pnas.92.15.6783.

Lolait, S. J., O'Carroll, A. M., McBride, O. W., Konig, M., Morel, A., & Brownstein, M. J. (1992). Cloning and characterization of a vasopressin V2 receptor and possible link to nephrogenic diabetes insipidus. *Nature, 357*(6376), 336–339. https://doi.org/10.1038/357336a0.

Maeda, Y., Han, J. S., Gibson, C. C., & Knepper, M. A. (1993). Vasopressin and oxytocin receptors coupled to Ca^{2+} mobilization in rat inner medullary collecting duct. *American Journal of Physiology—Renal Physiology, 265*(1), F15–F25. https://doi.org/10.1152/ajprenal.1993.265.1.F15.

Morel, F. (1981). Sites of hormone action in the mammalian nephron. *American Journal of Physiology, 240*, F159–F164.

Morel, F., Chabardès, D., Imbert-Teboul, M., Le Bouffant, F., Hus-Citharel, A., & Montégut, M. (1982). Multiple hormonal control of adenylate cyclase in distal segments of the rat kidney. *Kidney International. Supplement, 11*, S55–S62.

Mount, D. B., Delpire, E., Gamba, G., Hall, A. E., Poch, E., Hoover, R. S., et al. (1998). The electroneutral cation-chloride cotransporters. *Journal of Experimental Biology, 201*(14), 2091–2102.

Mutig, K., Borowski, T., Boldt, C., Borschewski, A., Paliege, A., Popova, E., et al. (2016). Demonstration of the functional impact of vasopressin signaling in the thick ascending limb by a targeted transgenic rat approach. *American Journal of Physiology—Renal Physiology, 311*(2), F411–F423. https://doi.org/10.1152/ajprenal.00126.2016.

Mutig, K., Paliege, A., Kahl, T., Jons, T., Muller-Esterl, W., & Bachmann, S. (2007). Vasopressin V2 receptor expression along rat, mouse, and human renal epithelia with focus on TAL. *American Journal of Physiology—Renal Physiology, 293*(4), F1166–F1177. https://doi.org/10.1152/ajprenal.00196.2007.

Nielsen, S., Maunsbach, A. B., Ecelbarger, C. A., & Knepper, M. A. (1998). Ultrastructural localization of Na-K-2Cl cotransporter in thick ascending limb and macula densa of rat kidney. *American Journal of Physiology—Renal Physiology, 275*(6), F885–F893. https://doi.org/10.1152/ajprenal.1998.275.6.F885.

Nitschke, R., Fröbe, U., & Greger, R. (1991). Antidiuretic hormone acts via V1 receptors on intracellular calcium in the isolated perfused rabbit cortical thick ascending limb. *Pflügers Archiv*, *417*(6), 622–632. https://doi.org/10.1007/BF00372961.

Nonoguchi, H., Owada, A., Kobayashi, N., Takayama, M., Terada, Y., Koike, J., et al. (1995). Immunohistochemical localization of V2 vasopressin receptor along the nephron and functional role of luminal V2 receptor in terminal inner medullary collecting ducts. *The Journal of Clinical Investigation*, *96*(4), 1768–1778. https://doi.org/10.1172/JCI118222.

Oliver, G., & Schäfer, E. A. (1895). The physiological effects of extracts of the suprarenal capsules. *The Journal of Physiology*, *18*(3), 230–276.

Ostrowski, N. L., Lolait, S. J., Bradley, D. J., O'Carroll, A. M., Brownstein, M. J., & Young, W. S., 3rd. (1992). Distribution of V1a and V2 vasopressin receptor messenger ribonucleic acids in rat liver, kidney, pituitary and brain. *Endocrinology*, *131*(1), 533–535. https://doi.org/10.1210/en.131.1.533.

Ostrowski, N. L., Young, W. S., 3rd, Knepper, M. A., & Lolait, S. J. (1993). Expression of vasopressin V1a and V2 receptor messenger ribonucleic acid in the liver and kidney of embryonic, developing, and adult rats. *Endocrinology*, *133*(4), 1849–1859. https://doi.org/10.1210/en.133.4.8404628.

Pedersen, N. B., Hofmeister, M. V., Rosenbaek, L. L., Nielsen, J., & Fenton, R. A. (2010). Vasopressin induces phosphorylation of the thiazide-sensitive sodium chloride cotransporter in the distal convoluted tubule. *Kidney International*, *78*(2), 160–169. https://doi.org/10.1038/ki.2010.130.

Perucca, J., Bichet, D. G., Bardoux, P., Bouby, N., & Bankir, L. (2008). Sodium excretion in response to vasopressin and selective vasopressin receptor antagonists. *Journal of the American Society of Nephrology*, *19*(9), 1721–1731. https://doi.org/10.1681/ASN.2008010021.

Plotkin, M. D., Kaplan, M. R., Verlander, J. W., Lee, W. S., Brown, D., Poch, E., et al. (1996). Localization of the thiazide sensitive Na-Cl cotransporter, rTSC1, in the rat kidney. *Kidney International*, *50*(1), 174–183. https://doi.org/10.1038/ki.1996.300.

Roald, A. B., Tenstad, O., & Aukland, K. (2004). The effect of AVP-V1 receptor stimulation on local GFR in the rat kidney. *Acta Physiologica Scandinavica*, *182*(2), 197–204. https://doi.org/10.1111/j.1365-201X.2004.01352.x.

Rutherford, P. A., Pizzonia, J. H., Biemesderfer, D., Abu-Alfa, A., Reilly, R., & Aronson, P. S. (1997). Expression of Na $(^{+})$-H^{+} exchanger isoforms NHE1 and NHE3 in kidney and blood cells of rabbit and rat. *Experimental Nephrology*, *5*(6), 490–497.

Saito, M., Tahara, A., Sugimoto, T., Abe, K., & Furuichi, K. (2000). Evidence that atypical vasopressin V2 receptor in inner medulla of kidney is V1B receptor. *European Journal of Pharmacology*, *401*(3), 289–296. https://doi.org/10.1016/S0014-2999(00)00465-9.

Sarmiento, J. M., Ehrenfeld, P., Aeazco, C. C., Reyes, C. E., Troncoso, S., Figueroa, C. D., et al. (2005). Differential distribution of the vasopressin V2 receptor along the rat nephron during renal ontogeny and maturation. *Kidney International*, *68*(2), 487–496. https://doi.org/10.1111/j.1523-1755.2005.00426.x.

Schield, L. (2016). *Sodium transport and balance: A key role for the distal nephron. In N. Turner (Ed.), Oxford textbook of clinical nephrology* (4th ed., pp. 189–196). Oxford: Oxford University Press. *https://doi.org/10.1093/med/9780199592548.003.0021.*

Smith, H. W. (1957). Salt and water volume receptors: An exercise in physiologic apologetics. *The American Journal of Medicine*, *23*(4), 623–652. https://doi.org/10.1016/0002-9343(57)90232-2.

Snyder, P. M. (2005). Minireview: Regulation of epithelial Na+ channel trafficking. *Endocrinology*, *146*(12), 5079–5085. https://doi.org/10.1210/en.2005-0894.

Staruschenko, A. (2012). Regulation of transport in the connecting tubule and cortical collecting duct. *Comprehensive Physiology, 2*, 1541–1584. https://doi.org/10.1002/cphy.c110052.

Stockand, J. D. (2010). Vasopressin regulation of renal sodium excretion. *Kidney International, 78*(9), 849–856. https://doi.org/10.1038/ki.2010.276.

Tashima, Y., Kohda, Y., Nonoguchi, H., Ikebe, M., Machida, K., Star, R. A., et al. (2001). Intranephron localization and regulation of the V1a vasopressin receptor during chronic metabolic acidosis and dehydration in rats. *Pflügers Archiv, 442*(5), 652–661. https://doi.org/10.1007/s004240100590.

Terada, Y., Tomita, K., Nonoguchi, H., Yang, T., & Marumo, F. (1993). Different localization and regulation of two types of vasopressin receptor messenger RNA in microdissected rat nephron segments using reverse transcription polymerase chain reaction. *The Journal of Clinical Investigation, 92*(5), 2339–2345. https://doi.org/10.1172/JCI116838.

Thorn, N. A. (1968). The influence of the neurohypophysial hormones and similar polypeptides on the kidneys. In B. Berde (Ed.), *Neurohypophysial hormones and similar polypeptides* (pp. 372–442). Berlin, Heidelberg: Springer.

Tomita, K., Pisano, J. J., & Knepper, M. A. (1985). Control of sodium and potassium transport in the cortical collecting duct of the rat. Effects of bradykinin, vasopressin, and deoxycorticosterone. *The Journal of Clinical Investigation, 76*(1), 132–136. https://doi.org/10.1172/JCI111935.

Torikai, S., Wang, M. S., Klein, K. L., & Kurokawa, K. (1981). Adenylate cyclase and cell cyclic AMP of rat cortical thick ascending limb of Henle. *Kidney International, 20*(5), 649–654. https://doi.org/10.1038/ki.1981.189.

Vesely, D. L. (2007). Natriuretic hormones. In R. J. Alpern & S. C. Hebert (Eds.), *Seldin and Giebisch's the kidney: Physiology and pathophysiology* (4th ed., pp. 947–977). San Diego: Elsevier/Academic Press.

von den Velden, R. (1913). Die nierenwirkung von hypophysenextrakten beim menschen. *Berliner Klinische Wochenschrift, 50*, 2083–2086.

CHAPTER TEN

Development and therapeutic potential of vasopressin synthetic analog [V^4Q^5]dDAVP as a novel anticancer agent

Juan Garona*, Marina Pifano, Giselle Ripoll, Daniel F. Alonso
Laboratory of Molecular Oncology, Science and Technology Department, National University of Quilmes, Buenos Aires, Argentina
*Corresponding author: e-mail address: juan.garona@unq.edu.ar

Contents

Vitamins and Hormones, Volume 113
ISSN 0083-6729
https://doi.org/10.1016/bs.vh.2019.08.010

Abstract

Since its discovery, arginine vasopressin (AVP) was subjected to several modifications with the aim of obtaining novel derivatives with increased potency and selectivity for biomedical use. Desmopressin (dDAVP) is a first generation synthetic analog of AVP with hemostatic and antimetastatic activity. dDAVP acts as a selective agonist of the arginine vasopressin type 2 receptor (AVPR2) present in microvascular endothelium and cancer cells. Considering its selective effects on AVPR2-expressing malignant and vascular tissue, and interesting antitumor profile, dDAVP was used as a lead compound for the development of novel peptide analogs with enhanced anticancer efficacy. After conducting different structure-activity relationship studies to determine key aminoacidic positions for its antitumor activity against AVPR2-expressing malignant cells, dDAVP was rationally modified and a wide panel of synthetic analogs with different sequence and structural modifications was assessed. As a result of this structure-based drug derivatization novel AVP analog [V^4Q^5]dDAVP (1-deamino-4-valine-5-glutamine-8-D-arginine vasopressin) was selected as the most active candidate and further developed. [V^4Q^5]dDAVP was evaluated in highly aggressive and metastatic cancer preclinical models deploying enhanced cytostatic, antimetastatic and angiostatic effects in comparison to parental peptide dDAVP. In addition, novel compound demonstrated good tolerability as evaluated in several toxicological studies, and cooperative therapeutic effects after combination with standard-of-care chemotherapy. In summary, due to its ability to inhibit growth and tumor-associated angiogenesis, as well as impairing progression of metastatic disease, AVP analogs such as novel [V^4Q^5]dDAVP are promising compounds for further development as coadjuvant agents for the management of advance or recurrent cancers.

1. AVP analogs and cancer

1.1 Introduction

Despite continuous progress in oncology research during the last decades, metastasis is still responsible for most cancer lethality. Metastases are formed when cells leave the primary tumor and establish in a distant, anatomically separate, site in the body. Nowadays, the dominant model describes the progression of systemic disease as a parallel development of primary tumor and distant metastasis, being a consequence of heterogeneous tumor subpopulations (Narod & Sopik, 2018; Pachmayr, Treese, & Stein, 2017). Excisional surgery is the primary treatment modality for solid cancers, but residual metastatic disease can remain after curative-intent surgery. In line with the idea that cancer is a complex systemic disease, increasing evidence indicates that tumor removal is capable of favoring the growth of residual cells (Coffey et al., 2003). Surgical procedure can facilitate the development of metastases through numerous mechanisms, including shedding of tumor cells into the circulation, increased levels of growth factors or decreased of antiangiogenic factors, production of neuroendocrine mediators that facilitate malignant cell survival, and suppression of antimetastatic immunity, among others. In addition, blood loss during surgery and the need for perioperative blood transfusion can also negatively influence patient outcome (Horowitz, Neeman, Sharon, & Ben-Eliyahu, 2015).

1.2 Repurposing of first generation AVP analog desmopressin for cancer therapy

Since its discovery by du Vigneaud and coworkers in 1953 (Ottenhausen, Bodhinayake, Banu, Stieg, & Schwartz, 2016), arginine vasopressin (AVP) was subjected to several modifications in order to develop analogs with enhanced pharmacological activity and improved antidiuretic/vasopressor specificity for biomedical applications (Laycock, 2010). Desmopressin (1-deamino-8-D-arginine vasopressin, also known as dDAVP) is a synthetic peptide and a first generation analog of the human antidiuretic hormone AVP (Vande Walle, Stockner, Raes, & Norgaard, 2007). Besides its traditional medical use in diabetes insipidus and enuresis, dDAVP has been used in the management of bleeding disorders and as a blood-saving agent in surgery (Mannucci, 1997; Svensson, Bergqvist, Juul, & Berntorp, 2014). In comparison to the natural AVP, dDAVP carries a deamination of cysteine in position 1, which prolongs its half-life, and substitution of L-arginine by D-arginine in position 8, which reduces the pressor effect and confers selectivity for the arginine vasopressin receptor 2 (AVPR2) (Kaufmann, Iezzi, & Vischer, 2003; Kaufmann & Vischer, 2003). This seven-transmembrane-domain G protein-coupled receptor (GPCR) is expressed in kidney collecting ducts and the endothelium of blood microvessels. Activation of AVP2R by dDAVP in endothelial cells causes a rapid cyclic adenosine monophosphate (cAMP)-induced release of von Willebrand factor (vWF), coagulation factor VIII and tissue-type plasminogen activator into the blood, thus producing a potent hemostatic effect (Kaufmann et al., 2000; Laycock, 2010).

Drug repurposing in oncology is an attractive strategy to offer options to cancer patients for the clinical management of the disease. Repurposing the large arsenal of already-approved compounds, with a nononcology primary use, that have medical safety information and robust preclinical evidence on antitumor activity has the advantage of safer and faster proof-of-concept clinical studies (Bertolini, Sukhatme, & Bouche, 2015). In the case of dDAVP, almost two decades ago it was reported that intravenous injection at high, clinically relevant doses prevented the development of blood-borne metastases in preclinical murine mammary tumor models (Alonso, Skilton, Farias, Bal de Kier Joffe, & Gomez, 1999), and also decreased axillary lymph node involvement when administered during surgical removal of the primary tumor (Giron, Tejera, Ripoll, Gomez, & Alonso, 2002). Similar results were obtained in melanoma and colorectal cancer models (Ripoll, Farina, Yoshiji, Gomez, & Alonso, 2006; Ripoll, Garona, Hermo, Gomez, & Alonso, 2010), and later experiments demonstrated that dDAVP effects were exerted during the early stages of metastasis development (Garona et al., 2014). Two different veterinary trials in dogs with locally advanced mammary cancer showed that perioperative treatment with dDAVP significantly prolonged disease-free and overall survival, with a particular survival benefit in cases with aggressive high-grade disease (Hermo et al., 2008, 2011). It seems that perioperative administration of dDAVP combats both the seeding of metastatic cells during surgical procedure and the growth of preexisting micrometastatic disease (Alonso, Ripoll, Garona, Iannucci, & Gomez, 2011; Garona & Alonso, 2016). Interestingly, vWF, the main hemostatic effector induced by dDAVP after stimulation of endothelial-AVPR2 (Kaufmann & Vischer, 2003), plays a protective role against metastatic dissemination. This large multimeric glycoprotein is able to cause resistance to metastasis by affecting survival of cancer cells early after their arrest in the target organ (Mochizuki et al., 2012; Ripoll & Alonso, 2013; Terraube, Marx, & Denis, 2007; Terraube et al., 2006).

Taking into account the well-known pharmacology profile and hemostatic effect of dDAVP, as well as its potential antimetastatic properties, a Phase II dose-escalation trial was conducted in patients with breast cancer (NCT01606072). dDAVP was administered by slow intravenous infusion before and after surgical resection of primary tumor. As expected, a reduced intraoperative bleeding was observed with increasing doses of dDAVP, and von Willebrand factor plasma levels were significantly increased. Notably, treatment was associated with a postoperative drop of circulating tumor cells, as measured by quantitative PCR of cytokeratin-19 transcripts (Weinberg et al., 2015). Another dDAVP clinical trial is ongoing in colorectal cancer patients with rectal bleeding (NCT01623206) (dDAVP preclinical and clinical hemostatic and antitumor activities are summarized in Fig. 1).

1.3 AVP and AVP-related genes in cancer cells

Besides its expression in microvascular endothelium mediating the hemostatic and antimetastatic effects of dDAVP, AVPR2 presence was also

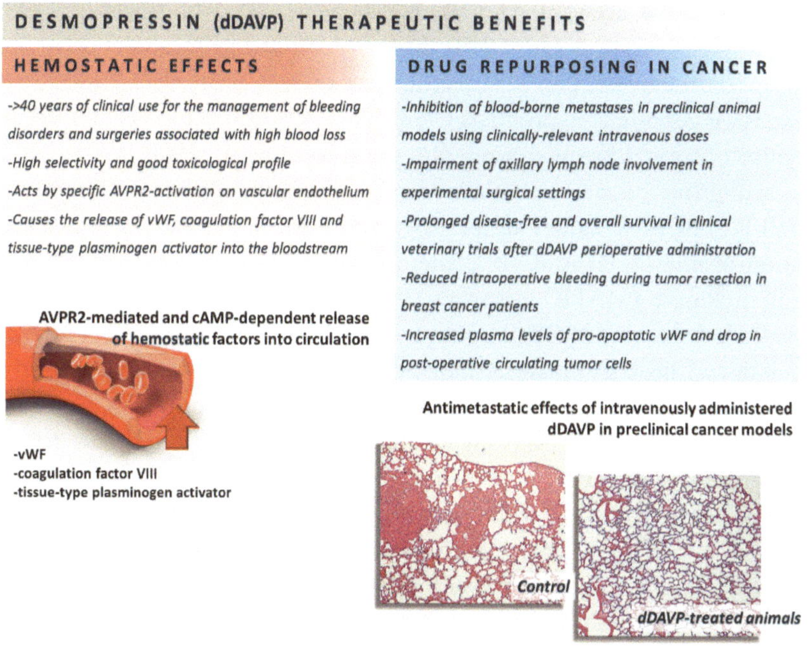

Fig. 1 First generation AVP analog and hemostatic compound desmopressin (dDAVP) as a repurposed drug in oncology. AVPR2-mediated effects on hemostatic factor release and bleeding control (left), and preclinical and clinical antitumor activity (right), are summarized. Representative microphotographs of lung tissue from control animals and mice treated with dDAVP in an experimental metastases study are shown (bottom, right). As observed in the pictures, dDAVP treatment was associated with a remarkable antimetastatic activity and a reduction in the number and size of tumor nodules, as previously reported (Alonso et al., 1999; Garona et al., 2014; Giron et al., 2002).

reported in transformed epithelial cells, as well as in a wide panel of human tumor cell lines (Garona et al., 2016; Monstein, Truedsson, Ryberg, & Ohlsson, 2008; North, Fay, & Du, 1999; Pequeux et al., 2004; Petit, Davidson, Lawrence, von Hoff, & Izbicka, 2001; Pifano et al., 2017). It is well established that tumor cell growth is stimulated by autocrine and paracrine signaling and there is accumulating evidence that supports the involvement of specific neuropeptides and peptide hormones in breast, lung, gastric, colorectal, pancreatic and prostatic cancer (Heasley, 2001; Keegan, Akerman, Pequeux, & North, 2006; Magni & Motta, 2001; Pequeux et al., 2004; Waldmann et al., 2018). In addition to its recognized endocrine action, it is known that AVP is also produced and secreted locally by many normal and pathological tissues exerting autocrine/paracrine activities

(Laycock, 2010; Mavani, DeVita, & Michelis, 2015; North et al., 1999; North, Fay, Longo, & Du, 1998). In fact, AVP and its receptors were proposed as attractive targets for breast cancer therapy more than 20 years ago after AVP gene-related products were detected by molecular techniques as a feature of several breast cancer subtypes (North et al., 1998, 1995, 2003). In malignant tissue, stimulation of arginine vasopressin receptor 1 (AVPR1) by AVP causes an increase of tumor cell growth related to activation of phospholipase C-β, protein kinase C and the mitogen-activated protein kinase (MAPK) pathway (Keegan et al., 2006; Pequeux et al., 2004). In contrast, AVPR2 agonist stimulation is associated to adenylate cyclase and protein kinase A-dependent antiproliferative cellular mechanisms and cytostatic effects (Garona et al., 2015; Iannucci et al., 2011; Manning et al., 2012; Mavani et al., 2015), showing a relevant role of AVP signaling in tumor biology and disease progression.

It should be mentioned that according to previously published data renal AVPR2 stimulation, which under normal physiological conditions mediates AVP antidiuretic action in the kidney, seems to favor abnormal cell proliferation, cyst growth and disease progression in polycystic kidney disease (PKD) through cAMP signaling and B-Raf/extracellular signal-regulated kinase (ERK) pathway activation (Yamaguchi et al., 2003). Moreover, it was recently reported that AVPR2 blockade using selective antagonists such as tolvaptan inhibits kidney cyst growth and preserves kidney function in both experimental and clinical PKD settings (Edwards et al., 2018; Rinschen, Schermer, & Benzing, 2014).

Taking all these evidence into consideration it seems to exist contrasting data regarding AVPR2-mediated effects on abnormal cell growth. Associated cytostatic or promitogenic signaling in transformed cells may depend on whether AVPR2 expression is related to a larger repertoire of molecular alterations linked to altered growth and survival (e.g., breast, colon and lung cancer, in which other AVP receptor subtypes may also be involved), or to native cell phenotype and tissue function (e.g., AVPR2 in PKD or renal cancer), respectively.

2. Design of novel AVP synthetic analogs as anticancer agents

2.1 Current limitations in cancer treatment

As previously mentioned, many cancer patients are diagnosed once the disease is at an advanced stage. In this clinical scenario therapy includes combined administration of several cytotoxic agents in addition to surgery or radiotherapy. Unfortunately, there are important limitations with its use, including tumor drug resistance caused by prior exposure, and risk of cumulative toxicities due to lack of target selectivity (Jones-Bolin, Zhao, Hunter, Klein-Szanto, & Ruggeri, 2006). Additionally, despite surgery being the mainstay of treatment for solid cancers, recent evidence suggests that surgery may also promote metastatic relapse through different deleterious mechanisms, as described above (Hirai, Matsumoto, Kubota, & Yamaguchi, 2014; Horowitz et al., 2015). Taken altogether, there is a clear need to explore novel therapeutic strategies to improve the survival of advanced or recurrent cancer patients. In this setting newly developed cytostatic agents could be used in combination with chemotherapy or during surgical removal of disease, targeting key events of cancer progression such as tumor cell growth and survival, angiogenesis, invasion and dissemination of tumor cells.

2.2 Structure-activity relationship (SAR) studies on parental AVP analog desmopressin

Our group and others previously reported the antiproliferative direct effects of AVP analog dDAVP on different experimental cancer models by agonist, AVPR2-specific interaction (Keegan et al., 2006; Ripoll, Giron, et al., 2008; Ripoll, Iannucci, et al. 2008). Peptidic compounds such as synthetic hormone derivatives have enormous potential as therapeutic agents in oncology due to their ease of rational design, target specificity, low production costs and good safety profiles (Zompra, Galanis, Werbitzky, & Albericio, 2009).

The rational design of drugs based on SAR studies constitutes an indispensable tool in the development of novel therapeutic agents (Nam, Shin, Rew, & Boger, 2007; Nicole et al., 2000; Quartara et al., 2000). SAR studies relate features of a chemical structure to a property, effect or biological activity (Macion-Dazard et al., 2006).

In this regard, due to its reported preclinical and clinical antitumor benefits and known hemostatic properties, dDAVP appears as an interesting drug acting as a lead compound for the design and development of novel peptide analogs with enhanced anticancer activity and affinity for the tumor cell AVPR2. In order to gain further knowledge about the ligand-AVP2R interaction and identify key aminoacidic positions and structural features involved in dDAVP direct antitumor activity, we synthesized and evaluated a large panel of peptide analogs to explore its SAR. First we conducted an alanine scanning (Ala-Scan) of most of its aminoacidic sequence (7 out of 9 AAs) (Pastrian et al., 2014; Pifano et al., 2018). We also assessed a series of derivatized tetrapeptides, chiral isomers and full-length nonapeptides with different rational modifications, mainly aiming at the N-terminal loop of the molecule (SAR studies on first generation AVP derivative dDAVP are summarized in Fig. 2) (Iannucci et al., 2011).

The Ala-Scan technique involves the sequential substitution of each amino acid in a lead peptide with an Alanine (Ala) residue in order to identify the amino acids that are critical for stability and biological function (Cunningham & Wells, 1989). Ala is used in this technique since it is considered to be the most neutral amino acid; it has a simple side chain, with a chemically inert methyl functional group; it is not highly hydrophobic and it has no charge. These features enable the SAR of the peptide of interest to be determined (Beck-Sickinger et al., 1994).

First generation AVP analog dDAVP was Ala-substituted using direct cytostatic effect as activity criteria on several rapidly growing and highly aggressive experimental cancer cell models, including human triple-negative breast carcinoma and SCLC. Positions 1 and 6, involved in the disulfide-bridge formation, were not Ala-substituted in order to maintain the characteristic cyclic feature of the analogs. The results for the Ala-substituted derivatives indicated that antiproliferative effect was completely abolished when amino acids 2 and 3 were replaced (Pastrian et al., 2014; Pifano et al., 2018). These results suggested that positions 2 and 3 were highly sensitive and intolerant to aminoacidic alterations given their crucial role in agonistic interaction of AVP-related ligands with tumor AVP2R, in accordance to previous studies conducted on AVP receptor/agonist complexes (Czaplewski, Kazmierkiewicz, & Ciarkowski, 1998b; Sikorska, Slusarz, & Lammek, 2006; Slusarz, Sikorska, Slusarz, & Ciarkowski, 2006). However, when amino acids in position 4 or 5 were substituted, a partial loss of cytostatic activity of about 60%, depending on tumor cell line, was observed, showing differential degrees of tolerability to sequence modifications. In addition, the biological activity of the peptide in this Ala-Scan was unaffected by the substitution of residues 7, 8 and 9 located at the C-terminal tail of the peptide, highlighting the importance of the role of the amino acids located in the peptide loop of the ligands (Pastrian et al., 2014; Pifano et al., 2018).

Taking all these data into account and with the purpose of deepening the SAR studies, we synthesized and assessed a variety of derivatized peptides, which included a panel of tetrapeptides corresponding to fragments 2–5 belonging to the conformational loop of dDAVP, chiral isomers of most

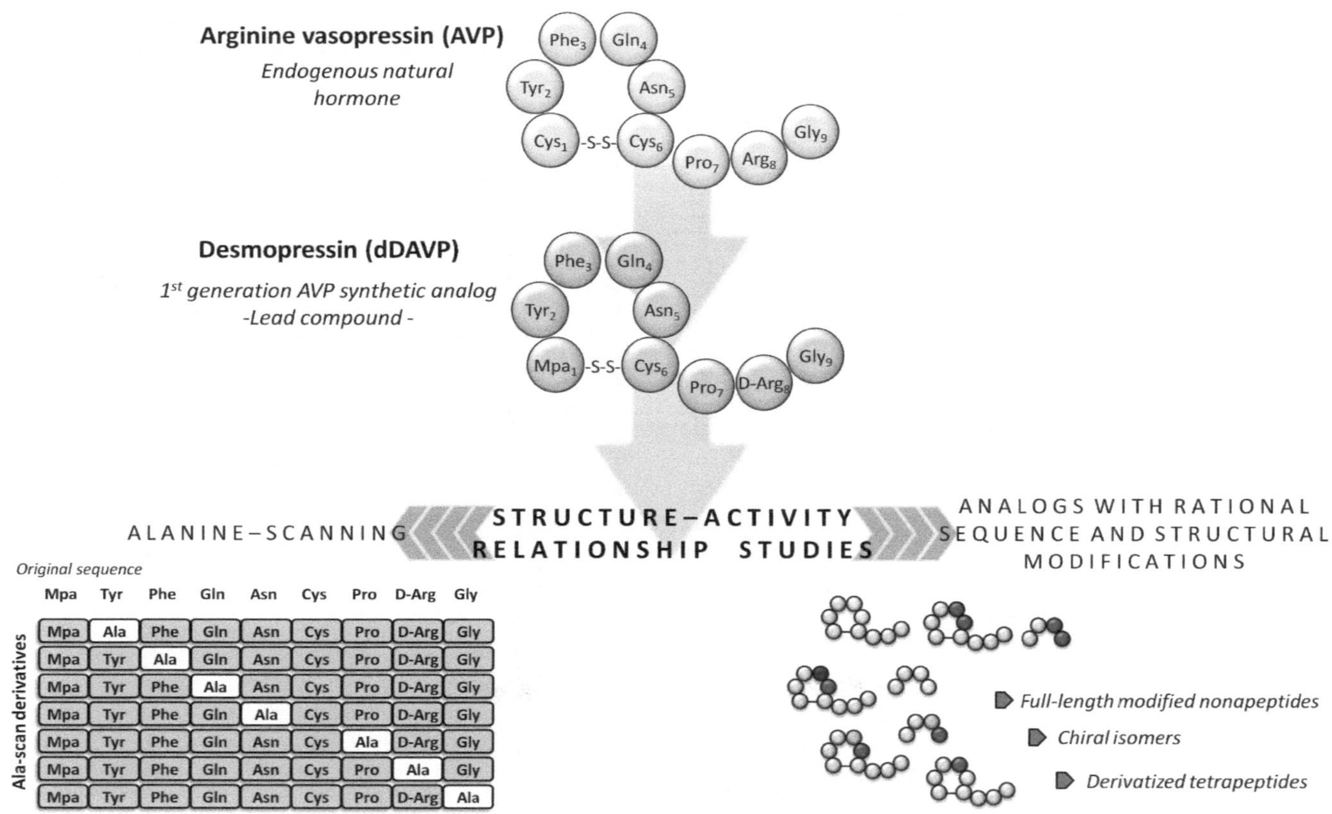

Fig. 2 Structure-activity relationship (SAR) studies conducted on first generation AVP analog desmopressin. Amino acid sequences in AVP, dDAVP and Ala-scan analogs are shown using the standard three-letter designations.

active candidates and novel full-length and cyclic variants carrying rational sequence modifications aiming at the N-terminal loop of the molecule. Once again direct cytostatic activity on log-phase growing AVPR2-expressing tumor cells was used as screening criteria (Iannucci et al., 2011).

After screening all synthetic analogs, novel nonapeptide [V^4Q^5]dDAVP (1-deamino-4-valine-5-glutamine-8-D-arginine vasopressin) resulted in the most active compound. [V^4Q^5]dDAVP preserved several first generation structural features of AVP analog desmopressin, such as deamination of cysteine in position 1 and substitution of L-arginine with its D-enantiomer in position 8, with the aim of retaining extended half-life and gain-of-AVPR2-selectivity. In addition, this novel analog added second generation sequence modifications at the N-terminal of the peptide such as replacement of glutamine by valine in position 4 and asparagine by glutamine in position 5 (sequence and structural features of [V^4Q^5]dDAVP are summarized in Fig. 3).

Several authors have postulated that enhancing hydrophobicity at position 4 may be an interesting approach to improve the interaction between AVP-related ligands and AVPR2 (Erdelyi et al., 2014; Manning et al., 2012, 2008). Using molecular modeling techniques it was reported that the N-terminal hydrophobic positions of AVP analogs, may interact

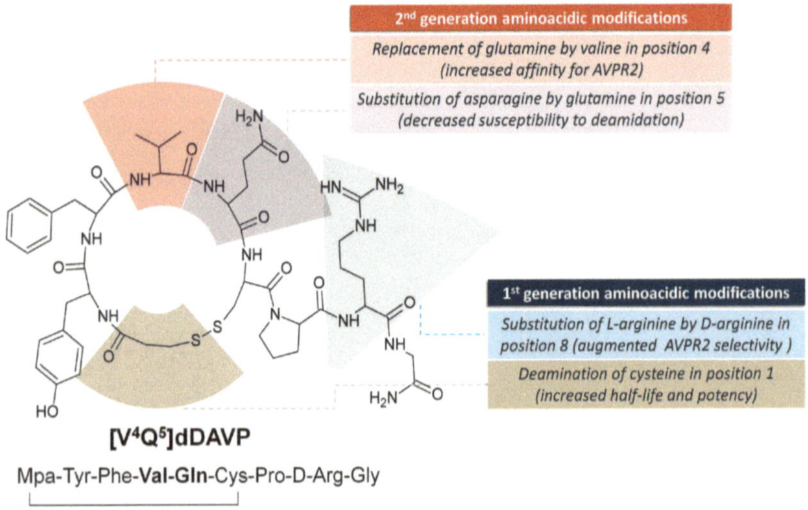

Fig. 3 Schematic view of novel AVP analog [V^4Q^5]dDAVP (1-deamino-4-valine-5-glutamine-8-D-arginine vasopressin) and rationale behind sequence modifications. Sites of first and second generation amino acid substitutions are indicated with different shaded color areas. Amino acid sequence is shown using the standard three-letter designations. Disulfide bond between positions 1 and 6 is shown with connecting line. Bold type text indicates modified amino acids in positions 4 and 5.

with hydrophobic moieties located at the bottom of extracellular AVPR2 cavity (Czaplewski, Kazmierkiewicz, & Ciarkowski, 1998a; Czaplewski et al., 1998b). Consequently, it was reported by different research groups that enhancement of hydrophobicity of this aminoacidic region in dDAVP was associated to a 10-fold increase in its AVPR2-affinity (Manning et al., 2012) and a gain of function on mutated AVPR2 in a pathological setting displaying an enhanced agonistic potency on intracellular cAMP production without cross-reacting with other AVP receptor subtypes (Erdelyi et al., 2014). Moreover, with the aim of improving the stability of the analog, we also introduced a conservative substitution at position 5, replacing asparagine with glutamine, based on its distinctive susceptibility to the deamidation process. Although both asparagine and glutamine are susceptible to deamidation, deamidation of glutamine proceeds at a much slower rate than deamidation of asparagine at peptide level (Liu, Gaza-Bulseco, & Chumsae, 2008; Robinson & Rudd, 1974). Additional SAR studies on [V^4Q^5]dDAVP also revealed that chirality in amino acids located in the loop region and 3D conformation of the nonapeptide given by full-length sequence and intrachain disulfide bonds are key properties directly related to its biological activity (Iannucci et al., 2011; Pastrian et al., 2014).

3. [V^4Q^5]dDAVP as a second generation AVP analog with enhanced biological activity

As previously mentioned, anticancer AVP analog [V^4Q^5]dDAVP was originally selected due to its increased cytostatic activity from a panel peptide derivatives with different sequence, structural and chiral modifications at the N-terminal loop of the molecule. In order to further characterize and validate its anticancer efficacy, therapeutic profile of the novel compound was evaluated using a large variety of experimental cancer cell models selected for their clinical relevance, highly characterized phenotypes and molecular profiles, including, but not limited to, small cell and non-small cell lung, pancreatic, colorectal, skin and different subtypes of breast cancer.

3.1 AVPR2-dependent cellular mechanisms triggered by [V^4Q^5] dDAVP in cancer cells

[V^4Q^5]dDAVP analog displayed significant antiproliferative effects on rapidly growing AVPR2-expressing tumor cell cultures reducing up to 40% of cell growth and viability, depending on tumor type. Growth modulating concentrations ranged from low nanomolar up to low micromolar, in accordance to previous studies conducted by our group and others on non-physiological effects of AVP-related peptides on transformed cells (Chan, Wo, Stoev, Cheng, & Manning, 2000; Iannucci et al., 2011; Juul, Bichet, Nielsen, & Norgaard, 2014; Keegan et al., 2006; North et al., 1995). In comparison to parental drug dDAVP, novel analog showed higher antitumor potency, at high (1 μM) as well as low (100 nM) concentrations.

Colony forming ability and cancel chemotaxis are hallmarks of metastatic outgrowth and key biological phenomenons in cancer biology and disease progression. In that regard, long-term 2D clonogenic and 3D multicellular tumor spheroid cultures were used as more challenging experimental systems to evaluate in vitro efficacy (Garona et al., 2015, 2016; Pifano et al., 2017). $[V^4Q^5]$dDAVP showed a much stronger effect on long-term colony formation assays, inducing a 75% reduction of clonogenic potential after 7-day drug exposure ($IC_{50}=1.13\,\mu M$) (Garona et al., 2015). Moreover, in vitro $[V^4Q^5]$dDAVP significantly decreased tumor cell chemotaxis and migration by up to 70% in aggressive lung and prostate cancer cells as evaluated by *Transwell* and wound assays (Pifano et al., 2017).

Reduction of tumor cell growth and migration by $[V^4Q^5]$dDAVP was associated with specific interaction with AVPR2 and activation of cAMP/PKA signaling axis. An increase in intracellular cAMP levels and PKA activation in cancer cells of nearly 90% and 40% vs baseline values, respectively, was observed after 60 min of incubation with $[V^4Q^5]$dDAVP. Additionally, cell cycle distribution analysis using flow cytometry showed that a 24-h treatment with low micromolar concentrations of $[V^4Q^5]$dDAVP resulted in partial arrest cells in G_0/G_1 phase (Garona et al., 2015).

cAMP, either via a PKA-dependent or PKA-independent manner, affects numerous cellular functions such as metabolism, ion channel activation, cell cycle and differentiation, gene expression and apoptosis (Antoni, 2012). It has been widely reported that increases in cAMP intracellular levels using cAMP analogs or cAMP elevating agents, such as hormones or forskolin, can trigger antimitogenic and proapoptotic responses in different cancer cell types (Carie & Sebti, 2007; Insel, Zhang, Murray, Yokouchi, & Zambon, 2012; Naviglio et al., 2009). Consequently, several authors have postulated the adenylate cyclase/cAMP/PKA axis as a growth suppressor system in cancer disease that could be targeted to block tumor formation or progression (Castoria et al., 2008; Naviglio et al., 2009). In this regard, further studies on non-canonical cAMP effectors such as exchange proteins directly activated by cAMP 1 and 2 (Epac1 and 2) and the mitogen-activated protein kinase (MAPK)/ERK axis are mandatory and are currently ongoing.

Using pharmacological blockade or gene silencing strategies we confirmed that direct antitumor effects of [V^4Q^5]dDAVP mainly result from specific AVPR2 activation. As evaluated in breast and lung cancer cells, chemical blockade using selective AVPR2 antagonist tolvaptan or small interfering RNA against AVPR2 was capable of significantly revert in vitro antitumor activity induced by [V^4Q^5]dDAVP. Furthermore, AVPR2-negative melanoma and NSCLC cells did not modify its proliferative behavior or migration ability in response to [V^4Q^5]dDAVP or other AVPR2 agonists (Pifano et al., 2017).

3.2 Acute and subacute toxicity studies of [V^4Q^5]dDAVP

Prior to conducting in vivo antitumoral efficacy studies, toxicological profile of the novel analog was evaluated on different experimental animal models. Acute toxicology studies were conducted at the Institute of Veterinary Sciences at National University of Litoral (Argentina). Wistar female rats received single intravenous doses of 1, 10 or 100 μg/kg of [V^4Q^5] dDAVP. Dosage was determined based on parental drug dDAVP clinical and preclinical therapeutic dose range. DDAVP was administered as a reference standard, showing a safety profile consistent with previous observations. A full clinical evaluation, including heart and respiratory rates, nervous system, motor activity, biochemical and hematological studies, was conducted at different time points from 1 to 72 h after drug administration. Body weight, food and water intake were monitored daily. These studies revealed that acute i.v. administration of [V^4Q^5]dDAVP at all evaluated doses had no influence on general symptoms, body weight, food and water consumption. Mild transient increases of glycemia and bilirubin were observed in treated groups in contrast to control animals administered with saline (Garona et al., 2015). All other biochemical and hematological parameters were not significantly altered. Equivalent results where obtained in subacute studies where rats were administered with [V^4Q^5]dDAVP at 0.25, 2.5 and 25 μg/kg three times per week during 4 weeks. Additional acute toxicity studies assessing single intravenous injections of [V^4Q^5] dDAVP using doses between 0.25 and 100 μg/kg were conducted on New Zealand white rabbits evaluating hemostatic parameters in addition to the above-mentioned factors. [V^4Q^5]dDAVP toxicological profile and effect on hemostasis and coagulation parameters were comparable to reference drug dDAVP (unpublished data).

3.3 [V^4Q^5]dDAVP therapeutic benefits on metastatic breast cancer

Breast cancer is the second cause of cancer-related deaths in developed countries and prognosis is strongly affected by metastatic progression of the disease (Ferlay et al., 2015). Despite advances in multimodal management of the disease, including radiotherapy, surgical techniques and adjuvant therapy based on chemotherapy and biologics, approximately 30% of patients with early-stage breast cancer will develop distant metastases (Doherty & Morris, 2015).

After confirming good toxicological profile and tolerability, [V^4Q^5]dDAVP was evaluated using metastatic hormone-independent F3II and triple-negative (cell line which lacks the estrogen receptor (ER), progesterone receptor (PR), and human epidermal growth factor receptor 2 (HER2)/neu) MDA-MB-231 breast cancer cell lines in different preclinical experimental animal models. In contrast to in vitro testing, in vivo settings represent highly challenging and complex scenarios for anticancer drug efficacy evaluation, in which other relevant parameters come into play such as drug absorption, distribution, metabolism and excretion, toxicity issues and cancer cell interactions with noncellular and cellular components of the tumor microenvironment, including fibroblasts, immune and vascular endothelial cells (Singh, Rameshwar, & Siegel, 2016).

3.3.1 Anticancer efficacy on breast cancer growth and tumor-induced angiogenesis

With the aim of studying [V^4Q^5]dDAVP activity on in vivo tumor growth and angiogenesis, human MDA-MB-231 breast cancer xenografts were generated in immunocompromised nude mice (Garona et al., 2015). MDA-MB-231 cells form rapidly growing and locally-invasive AVPR2-expressing tumors, as evaluated by immunohistochemistry (Garona et al., 2016). Treatment with clinically relevant intravenous doses of [V^4Q^5]dDAVP (0.3 μg/kg three times per week) for 8 weeks reduced final tumor load by 50% in comparison to vehicle-treated animals. Sustained [V^4Q^5]dDAVP administration was also associated to increased survival and modulation of tumor invasion and aggressiveness, evidenced by preservation of superficial layers of skin and inhibition of ulceration in subcutaneously growing xenografts.

Angiogenic induction by cancer cells represents a rate-limiting step during tumor progression. Targeting the angiogenesis phenomenon is already applied in cancer management, yet limits of antiangiogenic treatment have emerged, mainly because tumors adapt and recur after therapy. Therefore, there is a strong need to develop treatment alternatives to complement current therapeutic schemes (Langlois et al., 2014).

Histopathological studies of established breast cancer xenografts from [V^4Q^5]dDAVP-treated mice showed a decrease in tumor vascularization. Quantification of intratumoral vascular density revealed a 30% reduction in the number of blood vessels in mice administered with novel peptide. [V^4Q^5]dDAVP seems to modulate tumor angiogenesis by inducing the formation of angiostatin (ANG), a potent angiogenesis inhibitor that is generated by cancer-mediated proteolysis of plasminogen as a result of increased uPA and MMP-9 activity (Ripoll, Iannucci, et al. 2008; Ripoll et al., 2013). ANG induces apoptosis and inhibits cell growth and migration in both tumor and endothelial cells (Sakurai & Kudo, 2011; Westphal et al., 2000). Key signaling pathways related to its anticancer activity involve downregulation of mitochondrial antiapoptotic BCL-2, downregulation of c-Myc and elevated levels of another key antiangiogenic protein, thrombospondin-1, reinforcing its antitumor and antiangiogenic effects. Additionally, increased levels of tumor ANG was related to reduced recruitment and infiltration of bone marrow-derived macrophages (Lee et al., 2009). Interestingly, an enhanced production of ANG by human mammary carcinoma cells was previously reported after incubation with [V^4Q^5] dDAVP compared to first generation AVP analog dDAVP-treated cells (Ripoll, Iannucci, et al. 2008). After analyzing main angiogenic modulators using mRNA microarrays, AVPR2-activation in cancer cells was also able to modulate key proangiogenic proteins, including a dramatic reduction of vascular endothelial growth factor β (VEGF-B) and platelet-derived growth factor (PDGF), and a complete inhibition of hypoxia-inducible factor 1α (HIF-1α) expression (unpublished data).

Additionally, direct modulation of AVPR2-expressing vascular cell migration and morphogenesis was reported. Short-term incubation with [V^4Q^5] dDAVP of human lung microvascular endothelial cells resulted in a ~30% reduction of chemotaxis and capillary-like tube formation, a key early event in novel vessel development (Garona et al., 2015). Stimulation of AVPR2 in endothelial cells leads to an activation of cAMP-mediated signaling, which plays a central role in actin cytoskeletal dynamics and cell migration (Howe, 2004; Howe, Baldor, & Hogan, 2005; Kaufmann et al., 2000). Interestingly, it has been reported that PKA activation suppresses endothelial cell migration in vitro and angiogenesis in vivo (Kim, Harris, & Varner, 2000). The connection between the cAMP signaling pathway and actin structures could partially explain the direct effects of [V^4Q^5]dDAVP on microvascular cells, however, specific mechanisms responsible for [V^4Q^5]dDAVP effects on microvascular endothelial cells remain to be elucidated.

3.3.2 [V^4Q^5]dDAVP activity on breast cancer cell metastatic spread to lung

Metastatic breast cancer is still considered an incurable disease and remains a major therapeutic challenge (Morris, McArthur, & Hudis, 2009). Particularly in hormone-resistant breast cancer patients, metastatic relapse usually occurs in the lung and is strongly related to the severity of the disease and fatal outcome.

In this context, [V^4Q^5]dDAVP antimetastatic ability was assessed on spontaneous and experimental lung colonization by F3II hormone-independent breast cancer cells (Garona et al., 2015). Sustained treatment using novel AVP analog using 0.3 μg/kg i.v. doses had limited impact on primary exponentially-growing F3II disease. Nevertheless, tumors in [V^4Q^5]dDAVP-treated mice grew at a significantly lower rate compared to tumors from animals administered with saline vehicle. In contrast to control animals, which displayed 100% incidence of metastatic spread to lung, treatment with intravenous doses of [V^4Q^5]dDAVP for 6 weeks was able to completely abolish the formation of lung metastases. Reference dDAVP effects on tumor growth rate and spontaneous F3II metastases were not significant in such experimental conditions. Results were confirmed in experimental pulmonary dissemination assays in which intravenously inoculated F3II cells into syngeneic mice induced multiple macroscopic lung lesions after 3 weeks of tumor cell challenge. In this experimental setting, lung colonization was severely impaired after [V^4Q^5]dDAVP or reference dDAVP treatment. However, while dDAVP reduced the number of lung nodules by 32%, the novel analog [V^4Q^5]dDAVP displayed greater antimetastatic efficacy, reducing by 62% the formation of metastases (Garona et al., 2015).

3.3.3 Therapeutic benefits of [V^4Q^5]dDAVP addition to standard-of-care chemotherapeutic agents in breast cancer

Chemotherapy remains as a core therapeutic component for the majority of metastatic breast cancer patients. However, its effectiveness is limited by drug resistance, associated toxicity and the difficulty in finding optimal combination with biologic therapy (Lianos et al., 2014). In this context, potential combination of chemotherapy with selective cytostatic agents such as the novel [V^4Q^5]dDAVP is highly interesting.

With that purpose, [V^4Q^5]dDAVP was added to antimitotic agent paclitaxel or alkylating drug carmustine-based therapy for the treatment of MDA-MB-231 xenografts in athymic mice or metastatic F3II mammary tumors in immunocompetent host, respectively (Garona et al., 2016).

During concurrent administration of suboptimal doses of cytotoxic drugs plus [V^4Q^5]dDAVP, mammary tumor growth and aggressiveness was severely impaired compared to control or single treatment groups. In the carmustine-[V^4Q^5]dDAVP treatment regimen, mice receiving combined therapy not only showed a marked reduction of final tumor burden and skin tumor infiltration, but also a complete inhibition of spontaneous metastatic progression to the lung.

Cooperative anticancer activity was confirmed in vitro, where combination treatments caused a significant reduction in cancer cell proliferation and viability (Garona et al., 2016). Additionally, it was reported that [V^4Q^5]dDAVP plus chemotherapy treatment severely inhibits human breast cancer colony formation and multicellular tumor spheroid growth in comparison to control or individual therapeutic regimens. These results suggest that, besides vascular endothelium-related antitumor responses in host such as vWF-induced apoptosis or ANG-mediated antiangiogenic effects, direct AVPR2 stimulation in breast cancer cells using [V^4Q^5]dDAVP, in addition to standard chemotherapy, might also affect early stages of microtumor establishment and residual disease.

3.4 [V^4Q^5]dDAVP activity in lung and prostate cancer with neuroendocrine (NE) features

Neuroendocrine tumors (NETs) are a heterogeneous group of malignancies that present a wide spectrum of morphological and behavioral features. Although its low incidence, the number of registered new cases has increased in the last years. Because of the absence of specific symptoms that allow early detection, patients are generally diagnosed with advanced disease for which treatment options are limited, mainly including chemotherapy and radical tumor surgery (Kaltsas, Besser, & Grossman, 2004). Cells that undergo NE differentiation express high levels of chromogranin A (CgA) and specific neuronal enolase (NSE). These biomarkers are widely used for detecting NE differentiation either at tissue level or in the systemic circulation (Burgio et al., 2014; Diaz Perez & Curras Freixes, 2013; Isgro, Bottoni, & Scatena, 2015; Nolting et al., 2012). NETs can appear almost in any organ of the body but the most common tumors arise in lung and gastroenteropancreatic system. Small cell lung cancer (SCLC) is an infrequent and aggressive type of tumor with NE phenotype, representing 15% of bronchogenic carcinomas (Tartarone et al., 2017). The disease is characterized by high proliferation rates and early metastatic dissemination. Although early stage disease usually responds to chemotherapy and radiotherapy, eventually develops resistance to treatment and less than 5% of patients with advanced disease survive 5 years. Moreover, regardless of stage, SCLC treatment had no significant progress in the last three decades with no novel approved drugs for the specific management of such pathology (Teicher, 2014). On the other hand, prostate cancer (PC) is the second leading cause of cancer-related death in males worldwide. In early stages, androgen deprivation therapy is the choice for PC patients but invariably, disease progresses into a castration-resistant PC (CRPC) which implies poor therapy response and prognosis. Tumor progression correlates with a loss of epithelial characteristics (epithelium-mesenchymal transition; EMT), metastatic spread and development of cell foci with neuroendocrine differentiation (NED). NED is important in PC progression because androgen-deprivation therapy, chemotherapeutic agents, and radiotherapy also induce NED (Akamatsu, Inoue, Ogawa, & Gleave, 2018).

NETs growth is mainly mediated by neuropeptides, peptide hormones and bioamines. In this setting, in addition to interferon and biologic agents targeting VEGF and mTOR pathways, hormone or neuropeptide derivatives like somatostatin or vasopressin synthetic analogs are interesting therapeutic options that need to be evaluated to improve the prognosis of NET patients. Our group recently evaluated specific stimulation of AVPR2 through the use of $[V^4Q^5]$dDAVP on experimental models of human SCLC and prostate cancer. Lung NCI-H82 and prostate PC-3 cell lines are highly aggressive and express AVPR2 and high levels of NE markers NSE and CgA. Concentrations between 100 nM and 1 μM of $[V^4Q^5]$dDAVP displayed a cytostatic effect on rapidly growing cell cultures, inhibiting 40% of tumor cell growth, reducing doubling time and chemotaxis in both cell lines. Particularly in SCLC cells, $[V^4Q^5]$dDAVP caused an impressive 64% reduction in migratory ability. Antitumor profile of $[V^4Q^5]$dDAVP on PC-3 cells was in accordance to findings previously reported by Sasaki et.al, where they showed an antiproliferative effect and reduction of invasion capacity of prostate cancer cells after incubation with first generation analog dDAVP (Sasaki, Klotz, Sugar, Kiss, & Venkateswaran, 2015). Interestingly, the effect of novel analog on NCI-H82 SCLC in vivo showed that intravenous injection of clinically-relevant doses of analog $[V^4Q^5]$dDAVP impaired xenograft growth, reducing subcutaneous tumor volume compared to control group. In line with the obtained in vitro results, tumor growth rate was reduced by 35% in comparison to vehicle-treated animals (Pifano et al., 2017).

As mentioned before, NE differentiation of tumors is related with the transition from cancer cells with epithelial features to cells with invasive and metastatic phenotypes leading to unfavorable prognosis (Marcu, Radu, & Sajin, 2010). Interestingly, after treatment with AVPR2 agonists, expression of NE markers, including CgA and NSE, were reduced on lung and prostate cancer cell lines. An impaired expression of NE markers could indicate NE dedifferentiation and a loss of aggressiveness in tumor cells.

As previously described, [V^4Q^5]dDAVP ability to modulate tumor cell aggressiveness was associated to activation of the AVPR2/cAMP/PKA axis. The cAMP/PKA pathway has been reported to inhibit cell growth, cell cycle progression and differentiation in many normal or malignant cell types while stimulating them in others (Antoni, 2012; Insel et al., 2012; Lucchi et al., 2011). It was previously reported that, besides inhibiting their proliferation, cAMP-elevating agents, such as forskolin or pituitary adenylate cyclase-activating polypeptide (PACAP) promote neuronal differentiation of PC12 NET cells of the adrenal medulla through different canonical and non-canonical effectors. This process involves expression of differentiation-specific genes, such as sodium-channel proteins, production of specific neurotransmitters and, interestingly, NSE (Gerdin & Eiden, 2007; Vaudry et al., 2002; Vinores, Marangos, Parma, & Guroff, 1981). As a result, in order to gain deeper insight into the specific molecular mechanisms ruling activity on NET cell growth and differentiation further studies must be conducted on cAMP signaling pathways triggered by [V^4Q^5]dDAVP treatment, including noncanonical effectors such as Ras/Raf/MEK/ERK or EPAC1-2/Rap1 axis (Schmidt, Dekker, & Maarsingh, 2013), as well as on additional differentiation markers.

Taken all together, experimental studies conducted on cancer models with NE features showed encouraging findings to continue our research and gain more knowledge about the mechanisms of action and potential therapeutic windows of application of [V^4Q^5]dDAVP that would benefit patients with NETs.

3.5 Antiangiogenic and antimetastatic effects of [V^4Q^5]dDAVP in colorectal cancer

Colorectal cancer (CRC) represents a public health concern worldwide with an annual incidence of nearly 1 million cases (Ferlay et al., 2015). Metastatic disease in liver and lung is associated to high mortality rates with only 10% of patients surviving more than 5 years. With the aim of developing novel therapeutic strategies to improve the survival of advanced or recurrent CRC patients, drug development and efficacy validation should focus in key pathological processes related to disease progression such as metastatic spread to liver and lung, cellular growth and angiogenesis induction. Studies using [V^4Q^5]dDAVP were conducted on the CT-26 metastatic KRAS-mutated murine colon carcinoma model (Castle et al., 2014) and the Colo-205 human colon adenocarcinoma cell line bearing the BRAF (V600E) mutation (Khan et al., 2014). These CRC cells represent highly challenging and aggressive experimental models characterized by molecular profiles associated to limited therapeutic options and resistance to biologic drugs.

As previously mentioned, liver and lung metastases are the leading cause of cancer-related deaths in CRC (Kim et al., 2015; Riihimaki, Hemminki, Sundquist, & Hemminki, 2016). Taking this clinical concern into account [V^4Q^5]dDAVP impact on metastatic spread to liver in animals surgically implanted in the spleen with was assayed (Garona et al., 2019). Perioperative plus sustained administration of [V^4Q^5]dDAVP using 0.3 μg/kg i.v. doses caused a remarkable decrease in liver micro and macrometastatic burden. Total liver metastases in [V^4Q^5]dDAVP-treated animals was three times lower than control group. Moreover, the number of macrometastatic hepatic lesions dropped from 20 nodules per mice, as found in the control group, to 2.5 after [V^4Q^5]dDAVP treatment. The effect of [V^4Q^5] dDAVP administration on experimental metastatic spread to lung of CRC cells was also assessed. After analyzing colonized lung weights as a gross measure of metastatic tumor burden, [V^4Q^5]dDAVP i.v. treatment given before and after tumor cell injection inhibited pulmonary metastasis formation by 36% in comparison to saline vehicle-treated animals.

It is well known that high vascular density and overexpression of proangiogenic biomarkers in CRC correlate with disease progression (Rmali, Puntis, & Jiang, 2007). The capacity of [V^4Q^5]dDAVP administration to modify in vivo angiogenesis triggered by human AVPR2-expressing CRC cells was evaluated in incipient tumor implants using basement membrane extract plugs in nude mice. 2 Weeks of i.v. treatment using [V^4Q^5] dDAVP was capable of reducing Colo-205 cell-induced angiogenic response by nearly 60% in contrast with saline vehicle-treated animals (Garona et al., 2019).

Unfortunately, as occurs in many types of cancers, CRC patients are diagnosed once the disease is at an advanced stage. In this clinical scenario patients receive multimodality therapy based on several cytotoxic agents addition to surgery or radiotherapy (Marques, Araujo, & de Mello, 2013). Potential therapeutic benefits of [V^4Q^5]dDAVP addition to 5-fluorouracil (5-FU), a standard chemotherapy agent widely used in the treatment of malignancies in the gastrointestinal tract, were also explored in vitro. Combined treatments on log-phase growing CRC cell cultures resulted in a synergistic inhibition of CRC cell growth in comparison to cytotoxic or [V^4Q^5]dDAVP single agent therapies (Garona et al., 2019).

Summary of drug information and [V^4Q^5]dDAVP preclinical data are included in Table 1.

Table 1 [V^4Q^5]dDAVP compound information and summarized preclinical data.

Characteristic	Description	References
Full name	[V^4Q^5]dDAVP (1-deamino-4-valine-5-glutamine-8-D-arginine vasopressin)	Garona et al. (2015), Iannucci et al. (2011)
Drug type	Synthetic peptide, hormone analog	Pifano et al. (2018)
Aminoacid sequence	Mpa-Tyr-Phe-Val-Gln-Cys-Pro-D-Arg-Gly	Iannucci et al. (2011)
Molecular weight (g/mol)[a]	1053.53	Iannucci et al. (2011)
Chemical synthesis[b,c]	Solid phase, using Nα-Fmoc protection and the tea-bag strategy	Iannucci et al. (2011), Pastrian et al. (2014)
Molecular target, type, cellular location	Arginine vasopressin receptor type 2 (AVPR2), G-protein coupled receptor, plasma membrane	Garona et al. (2019), Ripoll et al. (2010), Ripoll et al. (2013), Weinberg et al. (2015)
Target tissue distribution	Microvascular endothelium. Kidney collecting ducts. Tumor cells	Garona et al. (2016), Juul et al. (2014), North et al. (1999), Pifano et al. (2017)
Mechanisms of action	Selective agonist activation of AVPR2 in tumor cells. Cystostatic and antiangiogenic. Impairs metastatic progression and combats residual disease. Others less characterized: hemostatic and endothelial AVPR2-mediated effectors such as proapoptotic vWF	Garona et al. (2015), Garona et al. (2016), Garona et al. (2019), Iannucci et al. (2011), Mochizuki et al. (2012) Pifano et al. (2017), Ripoll & Alonso, (2013), Terraube et al. (2007)
Cancer types	Breast cancer (hormone independent and TNBC), SCLC, prostate cancer, NET and colorectal cancer	Garona et al. (2015), Garona et al. (2019), Iannucci et al. (2011), Pifano et al. (2017), Pifano et al. (2018)

Continued

Table 1 $[V^4Q^5]$dDAVP compound information and summarized preclinical data—cont'd

Characteristic	Description	References
Dosage	0.3–1 μg/kg i.v.	Garona et al. (2015), Garona et al. (2019), Pifano et al. (2017)
Toxicological profile	Good tolerability in acute (1–100 μg/kg i.v.) and subacute (0.25–25 μg/kg i.v.) studies in rats and rabbits	Garona et al. (2015)
Potential clinical therapeutic scenarios	Surgical adjuvant during tumor resection (perioperative use). Coadjuvant agent in combination with standard-of-care chemotherapy (sustained administration in addition to taxanes, antimetabolites and/or alkylating agents)	Garona et al. (2015), Garona et al. (2019), Hirai et al. (2014), Horowitz et al. (2015), Pifano et al. (2017), Weinberg et al. (2015)

[a]Theoretical molecular weight: 1053.26 (calculated with Compute pI/MW ExPASy server).
[b]Peptide was purified by reversed-phase high-performance liquid chromatography (RP-HPLC).
[c]Peptide was identified by electrospray ionization mass spectrometry (ESI-MS).
Three-letter code for amino acid sequence description was used. Special amino acids; *Mpa*, β-mercaptopropionic acid (deamino-cysteine); *D-Arg*, D-arginine; *vWF*, von willebrand factor; *TNBC*, triple-negative breast cancer; *SCLC*, small cell lung cancer; *NET*, neuroendocrine tumors; *i.v.*, intravenously.

4. Conclusions and future perspectives

Despite major advances in detection, surgical techniques and adjuvant therapy, metastatic disease is still associated with a poor prognosis and alarmingly high mortality rates. In patients with resectable cancer, surgical excision of the primary tumor is commonly the first step toward long-term control of the disease. However, recent findings suggest that cancer surgery may also favor the spread and progression of residual cells, as well as the outgrowth of pre-existing subclinical metastasis through different deleterious mechanisms, such as tissue damage-related inflammation and enhanced production of growth factors, decreased production of endogenous anti-angiogenic factors, malignant cell dissemination due to surgical manipulation and immunosuppression, among others (Hirai et al., 2014; Horowitz et al., 2015). In this regard, the perioperative and early postsurgical timeframe represent an underexploited window of opportunity that could be targeted in order to modulate tumor-host interactions and minimize the risk of both local and metastatic recurrence.

$[V^4Q^5]$dDAVP is second generation AVP analog obtained as a result of a series of structure-activity relationship studies aimed at obtaining new peptide derivatives with enhanced anticancer activity. Interestingly, this novel nonapeptide has selective agonist activity on AVPR2 receptors present in both endothelium and tumor cells. $[V^4Q^5]$dDAVP activates AVPR2-dependent cytostatic signaling pathways involving the cAMP/PKA axis leading to cell cycle arrest, inhibition of chemotaxis and impairment of colony formation. Preclinical results also indicate that $[V^4Q^5]$dDAVP impairs angiogenesis by acting on AVPR2-expressing cells triggering different mechanisms including increased production of endogenous angiostatic molecules and reduction of the expression of different promoters of endothelial growth. Moreover, as evaluated in multiple cancer animal models with clinical relevance $[V^4Q^5]$dDAVP administration was associated with control of metastatic disease.

In this regard, perioperative use of $[V^4Q^5]$dDAVP could provide multiple therapeutic benefits aiding surgical management of resectable cancers by helping to destroy minimal residual disease in resection site and impair tumor cell dissemination by blocking cell growth and induction of angiogenesis. Additionally, as shown in combination with taxanes, antimetabolites and alkylating agents, $[V^4Q^5]$dDAVP could complement standard-of-care chemotherapy, suppressing primary tumor progression as well as spread and development of distant metastasis (Fig. 4).

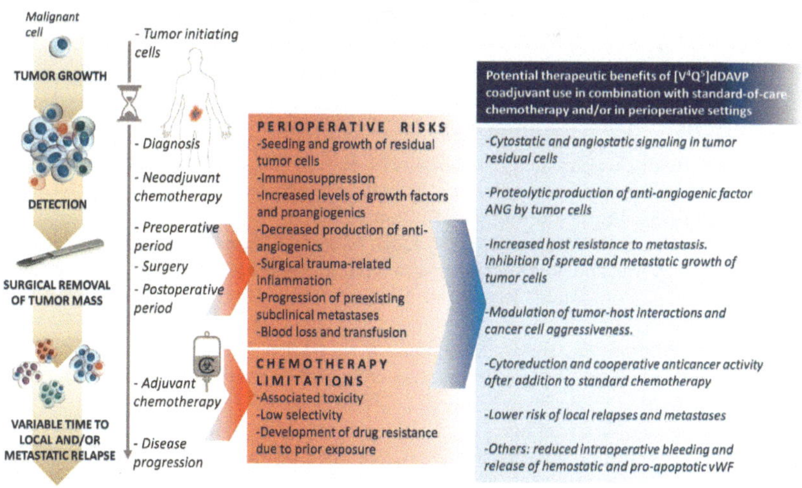

Fig. 4 Potential antitumor benefits of $[V^4Q^5]$dDAVP administration as a surgical coadjuvant and/or in combination with cytotoxic therapy. Cancer perioperative risks and therapeutic limitations associated with standard-of-care chemotherapy are also described.

Despite its high selectivity and biological activity, ease of production and good toxicological profile, integration of novel and promising anticancer agents such as [V^4Q^5]dDAVP into current cancer management is highly challenging. Preclinical validation efforts associated to [V^4Q^5]dDAVP development still continue, including studies on cell signaling and mechanisms of action, additional clinically relevant tumor types and unexplored biological processes such as modulation of immune microenvironment by ANG and vWF effector molecules. Additionally, nanoencapsulation of AVP-related peptides using biodegradable copolymeric nanocarriers is being evaluated by our group as a very promising approach for increasing drug delivery to disease site and to allow controlled drug release.

Taking into account its capacity to modulate tumor cell behavior, tumor-associated angiogenesis and cancer cell dissemination, selective vasopressin analogs like the novel compound [V^4Q^5]dDAVP are interesting candidates for further development as coadjuvant agents for cancer treatment in combination to standard chemotherapy or used in a perioperative setting.

Acknowledgments

The authors would like to acknowledge the support of the National University of Quilmes (UNQ, Argentina), the National Agency for the Promotion of Science and Technology (ANPCYT, Argentina), the National Institute of Cancer (INC, Argentina) and Chemo-Romikin. J. Garona, G. Ripoll and D.F. Alonso are members of the National Scientific and Technical Research Council (CONICET, Argentina).

References

Akamatsu, S., Inoue, T., Ogawa, O., & Gleave, M. E. (2018). Clinical and molecular features of treatment-related neuroendocrine prostate cancer. *International Journal of Urology*, *25*(4), 345–351. https://doi.org/10.1111/iju.13526.

Alonso, D. F., Ripoll, G. V., Garona, J., Iannucci, N. B., & Gomez, D. E. (2011). Metastasis: Recent discoveries and novel perioperative treatment strategies with particular interest in the hemostatic compound desmopressin. *Current Pharmaceutical Biotechnology*, *12*(11), 1974–1980.

Alonso, D. F., Skilton, G., Farias, E. F., Bal de Kier Joffe, E., & Gomez, D. E. (1999). Antimetastatic effect of desmopressin in a mouse mammary tumor model. *Breast Cancer Research and Treatment*, *57*(3), 271–275.

Antoni, F. A. (2012). New paradigms in cAMP signalling. *Molecular and Cellular Endocrinology*, *353*(1–2), 3–9. https://doi.org/10.1016/j.mce.2011.10.034.

Beck-Sickinger, A. G., Wieland, H. A., Wittneben, H., Willim, K. D., Rudolf, K., & Jung, G. (1994). Complete L-alanine scan of neuropeptide Y reveals ligands binding to Y1 and Y2 receptors with distinguished conformations. *European Journal of Biochemistry*, *225*(3), 947–958.

Bertolini, F., Sukhatme, V. P., & Bouche, G. (2015). Drug repurposing in oncology—Patient and health systems opportunities. *Nature Reviews. Clinical Oncology*, *12*(12), 732–742. https://doi.org/10.1038/nrclinonc.2015.169.

Burgio, S. L., Conteduca, V., Menna, C., Carretta, E., Rossi, L., Bianchi, E., et al. (2014). Chromogranin A predicts outcome in prostate cancer patients treated with abiraterone. *Endocrine-Related Cancer*, *21*(3), 487–493. https://doi.org/10.1530/ERC-14-0071.

Carie, A. E., & Sebti, S. M. (2007). A chemical biology approach identifies a beta-2 adrenergic receptor agonist that causes human tumor regression by blocking the Raf-1/Mek-1/Erk1/2 pathway. *Oncogene*, *26*(26), 3777–3788. https://doi.org/10.1038/sj.onc.1210172.

Castle, J. C., Loewer, M., Boegel, S., de Graaf, J., Bender, C., Tadmor, A. D., et al. (2014). Immunomic, genomic and transcriptomic characterization of CT26 colorectal carcinoma. *BMC Genomics*, *15*, 190. https://doi.org/10.1186/1471-2164-15-190.

Castoria, G., Migliaccio, A., D'Amato, L., Di Stasio, R., Ciociola, A., Lombardi, M., et al. (2008). Integrating signals between cAMP and MAPK pathways in breast cancer. *Frontiers in Bioscience*, *13*, 1318–1327.

Chan, W. Y., Wo, N. C., Stoev, S. T., Cheng, L. L., & Manning, M. (2000). Discovery and design of novel and selective vasopressin and oxytocin agonists and antagonists: The role of bioassays. *Experimental Physiology*, *85*, 7S–18S. Spec No.

Coffey, J. C., Wang, J. H., Smith, M. J., Bouchier-Hayes, D., Cotter, T. G., & Redmond, H. P. (2003). Excisional surgery for cancer cure: Therapy at a cost. *The Lancet Oncology*, *4*(12), 760–768.

Cunningham, B. C., & Wells, J. A. (1989). High-resolution epitope mapping of hGH-receptor interactions by alanine-scanning mutagenesis. *Science*, *244*(4908), 1081–1085.

Czaplewski, C., Kazmierkiewicz, R., & Ciarkowski, J. (1998a). Molecular modeling of the human vasopressin V2 receptor/agonist complex. *Journal of Computer-Aided Molecular Design*, *12*(3), 275–287.

Czaplewski, C., Kazmierkiewicz, R., & Ciarkowski, J. (1998b). Molecular modelling of the vasopressin V2 receptor/antagonist interactions. *Acta Biochimica Polonica*, *45*(1), 19–26.

Diaz Perez, J. A., & Curras Freixes, M. (2013). Chromogranin A and neuroendocrine tumors. *Endocrinología y Nutrición*, *60*(7), 386–395. https://doi.org/10.1016/j.endonu.2012.10.003.

Doherty, M. K., & Morris, P. G. (2015). Eribulin for the treatment of metastatic breast cancer: An update on its safety and efficacy. *International Journal of Women's Health*, 7, 47–58. https://doi.org/10.2147/IJWH.S74462.

Edwards, M. E., Chebib, F. T., Irazabal, M. V., Ofstie, T. G., Bungum, L. A., Metzger, A. J., et al. (2018). Long-term administration of tolvaptan in autosomal dominant polycystic kidney disease. *Clinical Journal of the American Society of Nephrology*, *13*, 1153–1161. https://doi.org/10.2215/CJN.01520218.

Erdelyi, L. S., Balla, A., Patocs, A., Toth, M., Varnai, P., & Hunyady, L. (2014). Altered agonist sensitivity of a mutant v2 receptor suggests a novel therapeutic strategy for nephrogenic diabetes insipidus. *Molecular Endocrinology*, *28*(5), 634–643. https://doi.org/10.1210/me.2013-1424.

Ferlay, J., Soerjomataram, I., Dikshit, R., Eser, S., Mathers, C., Rebelo, M., et al. (2015). Cancer incidence and mortality worldwide: Sources, methods and major patterns in GLOBOCAN 2012. *International Journal of Cancer*, *136*(5), E359–E386. https://doi.org/10.1002/ijc.29210.

Garona, J., & Alonso, D. F. (2016). Perioperative biology in primary breast cancer: Selective targeting of vasopressin type 2 receptor using desmopressin as a novel therapeutic approach. *Breast Cancer Research and Treatment*, *158*(3), 597–599. https://doi.org/10.1007/s10549-016-3899-4.

Garona, J., Pifano, M., Orlando, U. D., Pastrian, M. B., Iannucci, N. B., Ortega, H. H., et al. (2015). The novel desmopressin analogue [V4Q5]dDAVP inhibits angiogenesis, tumour growth and metastases in vasopressin type 2 receptor-expressing breast cancer models. *International Journal of Oncology*, *46*(6), 2335–2345. https://doi.org/10.3892/ijo.2015.2952.

Garona, J., Pifano, M., Pastrian, M. B., Gomez, D. E., Ripoll, G. V., & Alonso, D. F. (2016). Addition of vasopressin synthetic analogue [V(4)Q(5)]dDAVP to standard chemotherapy enhances tumour growth inhibition and impairs metastatic spread in aggressive breast tumour models. *Clinical & Experimental Metastasis*, *33*(6), 589–600. https://doi.org/10.1007/s10585-016-9799-5.

Garona, J., Pifano, M., Scursoni, A. M., Gomez, D. E., Alonso, D. F., & Ripoll, G. V. (2014). Insight into the effect of the vasopressin analog desmopressin on lung colonization by mammary carcinoma cells in BALB/c mice. *Anticancer Research*, *34*(9), 4761–4765.

Garona, J., Sobol, N. T., Pifano, M., Segatori, V. I., Gomez, D. E., Ripoll, G. V., et al. (2019). Preclinical efficacy of [V4Q5]dDAVP, a second generation vasopressin analog, on metastatic spread and tumor-associated angiogenesis in colorectal cancer. *Cancer Research and Treatment*, *51*, 438–450. https://doi.org/10.4143/crt.2018.040.

Gerdin, M. J., & Eiden, L. E. (2007). Regulation of PC12 cell differentiation by cAMP signaling to ERK independent of PKA: Do all the connections add up? *Science's STKE*, *2007*(382), pe15. https://doi.org/10.1126/stke.3822007pe15.

Giron, S., Tejera, A. M., Ripoll, G. V., Gomez, D. E., & Alonso, D. F. (2002). Desmopressin inhibits lung and lymph node metastasis in a mouse mammary carcinoma model of surgical manipulation. *Journal of Surgical Oncology*, *81*(1), 38–44. https://doi.org/10.1002/jso.10141.

Heasley, L. E. (2001). Autocrine and paracrine signaling through neuropeptide receptors in human cancer. *Oncogene*, *20*(13), 1563–1569. https://doi.org/10.1038/sj.onc.1204183.

Hermo, G. A., Torres, P., Ripoll, G. V., Scursoni, A. M., Gomez, D. E., Alonso, D. F., et al. (2008). Perioperative desmopressin prolongs survival in surgically treated bitches with mammary gland tumours: A pilot study. *Veterinary Journal*, *178*(1), 103–108. https://doi.org/10.1016/j.tvjl.2007.06.015.

Hermo, G. A., Turic, E., Angelico, D., Scursoni, A. M., Gomez, D. E., Gobello, C., et al. (2011). Effect of adjuvant perioperative desmopressin in locally advanced canine mammary carcinoma and its relation to histologic grade. *Journal of the American Animal Hospital Association*, *47*(1), 21–27. https://doi.org/10.5326/JAAHA-MS-5509.

Hirai, T., Matsumoto, H., Kubota, H., & Yamaguchi, Y. (2014). Regulating surgical oncotaxis to improve the outcomes in cancer patients. *Surgery Today*, *44*(5), 804–811. https://doi.org/10.1007/s00595-013-0627-0.

Horowitz, M., Neeman, E., Sharon, E., & Ben-Eliyahu, S. (2015). Exploiting the critical perioperative period to improve long-term cancer outcomes. *Nature Reviews. Clinical Oncology*, *12*(4), 213–226. https://doi.org/10.1038/nrclinonc.2014.224.

Howe, A. K. (2004). Regulation of actin-based cell migration by cAMP/PKA. *Biochimica et Biophysica Acta*, *1692*(2–3), 159–174. https://doi.org/10.1016/j.bbamcr.2004.03.005.

Howe, A. K., Baldor, L. C., & Hogan, B. P. (2005). Spatial regulation of the cAMP-dependent protein kinase during chemotactic cell migration. *Proceedings of the National Academy of Sciences of the United States of America*, *102*(40), 14320–14325. https://doi.org/10.1073/pnas.0507072102.

Iannucci, N. B., Ripoll, G. V., Garona, J., Cascone, O., Ciccia, G. N., Gomez, D. E., et al. (2011). Antiproliferative effect of 1-deamino-8-D-arginine vasopressin analogs on human breast cancer cells. *Future Medicinal Chemistry*, *3*(16), 1987–1993. https://doi.org/10.4155/fmc.11.152.

Insel, P. A., Zhang, L., Murray, F., Yokouchi, H., & Zambon, A. C. (2012). Cyclic AMP is both a pro-apoptotic and anti-apoptotic second messenger. *Acta Physiologica (Oxford, England)*, *204*(2), 277–287. https://doi.org/10.1111/j.1748-1716.2011.02273.x.

Isgro, M. A., Bottoni, P., & Scatena, R. (2015). Neuron-specific enolase as a biomarker: Biochemical and clinical aspects. *Advances in Experimental Medicine and Biology*, *867*, 125–143. https://doi.org/10.1007/978-94-017-7215-0_9.

Jones-Bolin, S., Zhao, H., Hunter, K., Klein-Szanto, A., & Ruggeri, B. (2006). The effects of the oral, pan-VEGF-R kinase inhibitor CEP-7055 and chemotherapy in orthotopic models of glioblastoma and colon carcinoma in mice. *Molecular Cancer Therapeutics*, *5*(7), 1744–1753. https://doi.org/10.1158/1535-7163.MCT-05-0327.

Juul, K. V., Bichet, D. G., Nielsen, S., & Norgaard, J. P. (2014). The physiological and pathophysiological functions of renal and extrarenal vasopressin V2 receptors. *American Journal of Physiology. Renal Physiology*, *306*(9), F931–F940. https://doi.org/10.1152/ajprenal.00604.2013.

Kaltsas, G. A., Besser, G. M., & Grossman, A. B. (2004). The diagnosis and medical management of advanced neuroendocrine tumors. *Endocrine Reviews*, *25*(3), 458–511. https://doi.org/10.1210/er.2003-0014.

Kaufmann, J. E., Iezzi, M., & Vischer, U. M. (2003). Desmopressin (DDAVP) induces NO production in human endothelial cells via V2 receptor- and cAMP-mediated signaling. *Journal of Thrombosis and Haemostasis*, *1*(4), 821–828.

Kaufmann, J. E., Oksche, A., Wollheim, C. B., Gunther, G., Rosenthal, W., & Vischer, U. M. (2000). Vasopressin-induced von Willebrand factor secretion from endothelial cells involves V2 receptors and cAMP. *The Journal of Clinical Investigation*, *106*(1), 107–116. https://doi.org/10.1172/JCI9516.

Kaufmann, J. E., & Vischer, U. M. (2003). Cellular mechanisms of the hemostatic effects of desmopressin (DDAVP). *Journal of Thrombosis and Haemostasis*, *1*(4), 682–689.

Keegan, B. P., Akerman, B. L., Pequeux, C., & North, W. G. (2006). Provasopressin expression by breast cancer cells: Implications for growth and novel treatment strategies. *Breast Cancer Research and Treatment*, *95*(3), 265–277. https://doi.org/10.1007/s10549-005-9024-8.

Khan, S., Cameron, S., Blaschke, M., Moriconi, F., Naz, N., Amanzada, A., et al. (2014). Differential gene expression of chemokines in KRAS and BRAF mutated colorectal cell lines: Role of cytokines. *World Journal of Gastroenterology*, *20*(11), 2979–2994. https://doi.org/10.3748/wjg.v20.i11.2979.

Kim, S., Harris, M., & Varner, J. A. (2000). Regulation of integrin alpha vbeta 3-mediated endothelial cell migration and angiogenesis by integrin alpha5beta1 and protein kinase A. *The Journal of Biological Chemistry*, *275*(43), 33920–33928. https://doi.org/10.1074/jbc.M003668200.

Kim, K. Y., Kim, N. K., Cha, I. H., Ahn, J. B., Choi, J. S., Choi, G. H., et al. (2015). Novel methods for clinical risk stratification in patients with colorectal liver metastases. *Cancer Research and Treatment*, *47*(2), 242–250. https://doi.org/10.4143/crt.2014.066.

Langlois, B., Saupe, F., Rupp, T., Arnold, C., van der Heyden, M., Orend, G., et al. (2014). AngioMatrix, a signature of the tumor angiogenic switch-specific matrisome, correlates with poor prognosis for glioma and colorectal cancer patients. *Oncotarget*, *5*(21), 10529–10545. https://doi.org/10.18632/oncotarget.2470.

Laycock, J. F. (2010). *Perspectives on vasopressin*. London Hackensack, NJ: Imperial College Press; Distributed by World Scientific.

Lee, T. Y., Muschal, S., Pravda, E. A., Folkman, J., Abdollahi, A., & Javaherian, K. (2009). Angiostatin regulates the expression of antiangiogenic and proapoptotic pathways via targeted inhibition of mitochondrial proteins. *Blood*, *114*(9), 1987–1998. https://doi.org/10.1182/blood-2008-12-197236.

Lianos, G. D., Vlachos, K., Zoras, O., Katsios, C., Cho, W. C., & Roukos, D. H. (2014). Potential of antibody-drug conjugates and novel therapeutics in breast cancer management. *OncoTargets and Therapy*, *7*, 491–500. https://doi.org/10.2147/OTT.S34235.

Liu, H., Gaza-Bulseco, G., & Chumsae, C. (2008). Glutamine deamidation of a recombinant monoclonal antibody. *Rapid Communications in Mass Spectrometry*, *22*(24), 4081–4088. https://doi.org/10.1002/rcm.3831.

Lucchi, S., Calebiro, D., de Filippis, T., Grassi, E. S., Borghi, M. O., & Persani, L. (2011). 8-Chloro-cyclic AMP and protein kinase A I-selective cyclic AMP analogs inhibit cancer cell growth through different mechanisms. *PLoS One*, *6*(6). e20785https://doi.org/10.1371/journal.pone.0020785.

Macion-Dazard, R., Callahan, N., Xu, Z., Wu, N., Thibonnier, M., & Shoham, M. (2006). Mapping the binding site of six nonpeptide antagonists to the human V2-renal vasopressin receptor. *The Journal of Pharmacology and Experimental Therapeutics*, *316*(2), 564–571. https://doi.org/10.1124/jpet.105.095554.

Magni, P., & Motta, M. (2001). Expression of neuropeptide Y receptors in human prostate cancer cells. *Annals of Oncology*, *12*(Suppl. 2), S27–S29.

Manning, M., Misicka, A., Olma, A., Bankowski, K., Stoev, S., Chini, B., et al. (2012). Oxytocin and vasopressin agonists and antagonists as research tools and potential therapeutics. *Journal of Neuroendocrinology*, *24*(4), 609–628. https://doi.org/10.1111/j.1365-2826.2012.02303.x.

Manning, M., Stoev, S., Chini, B., Durroux, T., Mouillac, B., & Guillon, G. (2008). Peptide and non-peptide agonists and antagonists for the vasopressin and oxytocin V1a, V1b, V2 and OT receptors: Research tools and potential therapeutic agents. *Progress in Brain Research*, *170*, 473–512. https://doi.org/10.1016/S0079-6123(08)00437-8.

Mannucci, P. M. (1997). Desmopressin (DDAVP) in the treatment of bleeding disorders: The first 20 years. *Blood*, *90*(7), 2515–2521.

Marcu, M., Radu, E., & Sajin, M. (2010). Neuroendocrine transdifferentiation of prostate carcinoma cells and its prognostic significance. *Romanian Journal of Morphology and Embryology*, *51*(1), 7–12.

Marques, I., Araujo, A., & de Mello, R. A. (2013). Anti-angiogenic therapies for metastatic colorectal cancer: Current and future perspectives. *World Journal of Gastroenterology*, *19*(44), 7955–7971. https://doi.org/10.3748/wjg.v19.i44.7955.

Mavani, G. P., DeVita, M. V., & Michelis, M. F. (2015). A review of the nonpressor and nonantidiuretic actions of the hormone vasopressin. *Frontiers in Medicine*, *2*, 19. https://doi.org/10.3389/fmed.2015.00019.

Mochizuki, S., Soejima, K., Shimoda, M., Abe, H., Sasaki, A., Okano, H. J., et al. (2012). Effect of ADAM28 on carcinoma cell metastasis by cleavage of von Willebrand factor. *Journal of the National Cancer Institute*, *104*(12), 906–922. https://doi.org/10.1093/jnci/djs232.

Monstein, H. J., Truedsson, M., Ryberg, A., & Ohlsson, B. (2008). Vasopressin receptor mRNA expression in the human gastrointestinal tract. *European Surgical Research*, *40*(1), 34–40. https://doi.org/10.1159/000108655.

Morris, P. G., McArthur, H. L., & Hudis, C. A. (2009). Therapeutic options for metastatic breast cancer. *Expert Opinion on Pharmacotherapy*, *10*(6), 967–981. https://doi.org/10.1517/14656560902834961.

Nam, J., Shin, D., Rew, Y., & Boger, D. L. (2007). Alanine scan of [L-Dap(2)]ramoplanin A2 aglycon: Assessment of the importance of each residue. *Journal of the American Chemical Society*, *129*(28), 8747–8755. https://doi.org/10.1021/ja068573k.

Narod, S. A., & Sopik, V. (2018). Is invasion a necessary step for metastases in breast cancer? *Breast Cancer Research and Treatment*, *169*(1), 9–23. https://doi.org/10.1007/s10549-017-4644-3.

Naviglio, S., Di Gesto, D., Romano, M., Sorrentino, A., Illiano, F., Sorvillo, L., et al. (2009). Leptin enhances growth inhibition by cAMP elevating agents through apoptosis of MDA-MB-231 breast cancer cells. *Cancer Biology & Therapy*, *8*(12), 1183–1190.

Nicole, P., Lins, L., Rouyer-Fessard, C., Drouot, C., Fulcrand, P., Thomas, A., et al. (2000). Identification of key residues for interaction of vasoactive intestinal peptide with human VPAC1 and VPAC2 receptors and development of a highly selective VPAC1 receptor agonist. Alanine scanning and molecular modeling of the peptide. *The Journal of Biological Chemistry*, *275*(31), 24003–24012. https://doi.org/10.1074/jbc.M002325200.

Nolting, S., Kuttner, A., Lauseker, M., Vogeser, M., Haug, A., Herrmann, K. A., et al. (2012). Chromogranin a as serum marker for gastroenteropancreatic neuroendocrine tumors: A single center experience and literature review. *Cancers (Basel)*, *4*(1), 141–155. https://doi.org/10.3390/cancers4010141.

North, W. G., Fay, M. J., & Du, J. (1999). MCF-7 breast cancer cells express normal forms of all vasopressin receptors plus an abnormal V2R. *Peptides*, *20*(7), 837–842.

North, W. G., Fay, M. J., Longo, K. A., & Du, J. (1998). Expression of all known vasopressin receptor subtypes by small cell tumors implies a multifaceted role for this neuropeptide. *Cancer Research*, *58*(9), 1866–1871.

North, W. G., Pai, S., Friedmann, A., Yu, X., Fay, M., & Memoli, V. (1995). Vasopressin gene related products are markers of human breast cancer. *Breast Cancer Research and Treatment*, *34*(3), 229–235.

North, W. G., Wells, W., Fay, M. J., Mathew, R. S., Donnelly, E. M., & Memoli, V. A. (2003). Immunohistochemical evaluation of vasopressin expression in breast fibrocystic disease and ductal carcinoma in situ (DCIS). *Endocrine Pathology*, *14*(3), 257–262.

Ottenhausen, M., Bodhinayake, I., Banu, M. A., Stieg, P. E., & Schwartz, T. H. (2016). Vincent du Vigneaud: Following the sulfur trail to the discovery of the hormones of the posterior pituitary gland at Cornell Medical College. *Journal of Neurosurgery*, *124*(5), 1538–1542. https://doi.org/10.3171/2015.5.JNS141952.

Pachmayr, E., Treese, C., & Stein, U. (2017). Underlying mechanisms for distant metastasis—Molecular biology. *Visceral Medicine*, *33*(1), 11–20. https://doi.org/10.1159/000454696.

Pastrian, M. B., Guzman, F., Garona, J., Pifano, M., Ripoll, G. V., Cascone, O., et al. (2014). Structure-activity relationship of 1-desamino-8-D-arginine vasopressin as an antiproliferative agent on human vasopressin V2 receptor-expressing cancer cells. *Molecular Medicine Reports*, *9*(6), 2568–2572. https://doi.org/10.3892/mmr.2014.2140.

Pequeux, C., Keegan, B. P., Hagelstein, M. T., Geenen, V., Legros, J. J., & North, W. G. (2004). Oxytocin- and vasopressin-induced growth of human small-cell lung cancer is mediated by the mitogen-activated protein kinase pathway. *Endocrine-Related Cancer*, *11*(4), 871–885. https://doi.org/10.1677/erc.1.00803.

Petit, T., Davidson, K. K., Lawrence, R. A., von Hoff, D. D., & Izbicka, E. (2001). Neuropeptide receptor status in human tumor cell lines. *Anti-Cancer Drugs*, *12*(2), 133–136.

Pifano, M., Garona, J., Capobianco, C. S., Gonzalez, N., Alonso, D. F., & Ripoll, G. V. (2017). Peptide agonists of vasopressin V2 receptor reduce expression of neuroendocrine markers and tumor growth in human lung and prostate tumor cells. *Frontiers in Oncology*, 7, 11. https://doi.org/10.3389/fonc.2017.00011.

Pifano, M., Garona, J., Sobol, N. T., Alberto, M., Alonso, D. F., & Ripoll, G. V. (2018). Search of vasopressin analogs with antiproliferative activity on small-cell lung cancer: Drug design based on two different approaches. *Future Medicinal Chemistry*, *10*(8), 879–894. https://doi.org/10.4155/fmc-2017-0178.

Quartara, L., Ricci, R., Meini, S., Patacchini, R., Giolitti, A., Amadesi, S., et al. (2000). Ala scan analogues of HOE 140. Synthesis and biological activities. *European Journal of Medicinal Chemistry*, *35*(11), 1001–1010.

Riihimaki, M., Hemminki, A., Sundquist, J., & Hemminki, K. (2016). Patterns of metastasis in colon and rectal cancer. *Scientific Reports*, *6*, 29765. https://doi.org/10.1038/srep29765.

Rinschen, M. M., Schermer, B., & Benzing, T. (2014). Vasopressin-2 receptor signaling and autosomal dominant polycystic kidney disease: From bench to bedside and back again. *Journal of the American Society of Nephrology*, *25*(6), 1140–1147. https://doi.org/10.1681/ASN.2013101037.

Ripoll, G. V., & Alonso, D. F. (2013). Implication of von Willebrand factor as a regulator of tumor cell metastasis: Potential perioperative use of desmopressin and novel peptide analogs. *Acta Haematologica*, *129*(4), 223–224. https://doi.org/10.1159/000346071.

Ripoll, G. V., Farina, H. G., Yoshiji, H., Gomez, D. E., & Alonso, D. F. (2006). Desmopressin reduces melanoma lung metastasis in transgenic mice overexpressing tissue inhibitor of metalloproteinases-1. *In Vivo*, *20*(6B), 881–885.

Ripoll, G. V., Garona, J., Hermo, G. A., Gomez, D. E., & Alonso, D. F. (2010). Effects of the synthetic vasopressin analog desmopressin in a mouse model of colon cancer. *Anticancer Research*, *30*(12), 5049–5054.

Ripoll, G. V., Garona, J., Pifano, M., Farina, H. G., Gomez, D. E., & Alonso, D. F. (2013). Reduction of tumor angiogenesis induced by desmopressin in a breast cancer model. *Breast Cancer Research and Treatment*, *142*(1), 9–18. https://doi.org/10.1007/s10549-013-2724-6.

Ripoll, G. V., Giron, S., Krzymuski, M. J., Hermo, G. A., Gomez, D. E., & Alonso, D. F. (2008). Antitumor effects of desmopressin in combination with chemotherapeutic agents in a mouse model of breast cancer. *Anticancer Research*, *28*(5A), 2607–2611.

Ripoll, G., Iannucci, N., Giron, S., Cascone, O., Gomez, D., & Alonso, D. (2008). Angiostatic activity of 1-deamino-8-D-arginine vasopressin and novel peptide analogues in breast cancer cells. In *AACR meeting abstracts* (p. 295).

Rmali, K. A., Puntis, M. C., & Jiang, W. G. (2007). Tumour-associated angiogenesis in human colorectal cancer. *Colorectal Disease*, *9*(1), 3–14. https://doi.org/10.1111/j.1463-1318.2006.01089.x.

Robinson, A. B., & Rudd, C. J. (1974). Deamidation of glutaminyl and asparaginyl residues in peptides and proteins. *Current Topics in Cellular Regulation*, *8*, 247–295.

Sakurai, T., & Kudo, M. (2011). Signaling pathways governing tumor angiogenesis. *Oncology*, *81*(Suppl. 1), 24–29. https://doi.org/10.1159/000333256.

Sasaki, H., Klotz, L. H., Sugar, L. M., Kiss, A., & Venkateswaran, V. (2015). A combination of desmopressin and docetaxel inhibit cell proliferation and invasion mediated by urokinase-type plasminogen activator (uPA) in human prostate cancer cells. *Biochemical and Biophysical Research Communications*, *464*(3), 848–854. https://doi.org/10.1016/j.bbrc.2015.07.050.

Schmidt, M., Dekker, F. J., & Maarsingh, H. (2013). Exchange protein directly activated by cAMP (epac): A multidomain cAMP mediator in the regulation of diverse biological functions. *Pharmacological Reviews*, *65*(2), 670–709. https://doi.org/10.1124/pr.110.003707.

Sikorska, E., Slusarz, M. J., & Lammek, B. (2006). Conformational studies of vasopressin analogues modified with N-methylphenylalanine enantiomers in dimethyl sulfoxide solution. *Biopolymers*, *82*(6), 603–614. https://doi.org/10.1002/bip.20514.

Singh, S. R., Rameshwar, P., & Siegel, P. (2016). Targeting tumor microenvironment in cancer therapy. *Cancer Letters*, *380*(1), 203–204. https://doi.org/10.1016/j.canlet.2016.04.009.

Slusarz, M. J., Sikorska, E., Slusarz, R., & Ciarkowski, J. (2006). Molecular docking-based study of vasopressin analogues modified at positions 2 and 3 with N-methylphenylalanine: Influence on receptor-bound conformations and interactions with vasopressin and oxytocin receptors. *Journal of Medicinal Chemistry*, *49*(8), 2463–2469. https://doi.org/10.1021/jm051075m.

Svensson, P. J., Bergqvist, P. B., Juul, K. V., & Berntorp, E. (2014). Desmopressin in treatment of haematological disorders and in prevention of surgical bleeding. *Blood Reviews*, *28*(3), 95–102. https://doi.org/10.1016/j.blre.2014.03.001.

Tartarone, A., Giordano, P., Lerose, R., Rodriquenz, M. G., Conca, R., & Aieta, M. (2017). Progress and challenges in the treatment of small cell lung cancer. *Medical Oncology*, *34*(6), 110. https://doi.org/10.1007/s12032-017-0966-6.

Teicher, B. A. (2014). Targets in small cell lung cancer. *Biochemical Pharmacology*, *87*(2), 211–219. https://doi.org/10.1016/j.bcp.2013.09.014.

Terraube, V., Marx, I., & Denis, C. V. (2007). Role of von Willebrand factor in tumor metastasis. *Thrombosis Research*, *120*(Suppl. 2), S64–S70. https://doi.org/10.1016/S0049-3848(07)70132-9.

Terraube, V., Pendu, R., Baruch, D., Gebbink, M. F., Meyer, D., Lenting, P. J., et al. (2006). Increased metastatic potential of tumor cells in von Willebrand factor-deficient mice. *Journal of Thrombosis and Haemostasis*, *4*(3), 519–526. https://doi.org/10.1111/j.1538-7836.2005.01770.x.

Vande Walle, J., Stockner, M., Raes, A., & Norgaard, J. P. (2007). Desmopressin 30 years in clinical use: A safety review. *Current Drug Safety*, *2*(3), 232–238.

Vaudry, D., Chen, Y., Ravni, A., Hamelink, C., Elkahloun, A. G., & Eiden, L. E. (2002). Analysis of the PC12 cell transcriptome after differentiation with pituitary adenylate cyclase-activating polypeptide (PACAP). *Journal of Neurochemistry*, *83*(6), 1272–1284.

Vinores, S. A., Marangos, P. J., Parma, A. M., & Guroff, G. (1981). Increased levels of neuron-specific enolase in PC12 pheochromocytoma cells as a result of nerve growth factor treatment. *Journal of Neurochemistry*, *37*(3), 597–600.

Waldmann, J., Fendrich, V., Reichert, M., Hecker, A., Bartsch, D. K., Padberg, W., et al. (2018). Expression of neuropeptide Y and its receptors Y1 and Y2 in pancreatic intraepithelial neoplasia and invasive pancreatic cancer in a transgenic mouse model and human samples of pancreatic cancer. *The Journal of Surgical Research*, *223*, 230–236. https://doi.org/10.1016/j.jss.2017.11.010.

Weinberg, R. S., Grecco, M. O., Ferro, G. S., Seigelshifer, D. J., Perroni, N. V., Terrier, F. J., et al. (2015). A phase II dose-escalation trial of perioperative desmopressin (1-desamino-8-d-arginine vasopressin) in breast cancer patients. *Springerplus*, *4*, 428. https://doi.org/10.1186/s40064-015-1217-y.

Westphal, J. R., Van't Hullenaar, R., Geurts-Moespot, A., Sweep, F. C., Verheijen, J. H., Bussemakers, M. M., et al. (2000). Angiostatin generation by human tumor cell lines: Involvement of plasminogen activators. *International Journal of Cancer*, *86*(6), 760–767.

Yamaguchi, T., Nagao, S., Wallace, D. P., Belibi, F. A., Cowley, B. D., Pelling, J. C., et al. (2003). Cyclic AMP activates B-Raf and ERK in cyst epithelial cells from autosomal-dominant polycystic kidneys. *Kidney International*, *63*(6), 1983–1994. https://doi.org/10.1046/j.1523-1755.2003.00023.x.

Zompra, A. A., Galanis, A. S., Werbitzky, O., & Albericio, F. (2009). Manufacturing peptides as active pharmaceutical ingredients. *Future Medicinal Chemistry*, *1*(2), 361–377. https://doi.org/10.4155/fmc.09.23.

CHAPTER ELEVEN

Vasopressin and vasopressin receptors in brain edema

Emil Zeynalov[a,*], Susan M. Jones[a], J. Paul Elliott[b]
[a]Swedish Medical Center, Englewood, CO, United States
[b]Colorado Brain and Spine Institute, Denver, CO, United States
*Corresponding author: e-mail address: zeynalov@healthonecares.com

Contents

Abstract

Vasopressin is a peptide hormone produced in the hypothalamus and released from the posterior pituitary. Secretion of vasopressin is followed by activation of its receptors V1a, V1b, and V2 throughout the body. Each receptor type is responsible for a specific function of vasopressin. For example, V1a receptor activation triggers vasoconstriction, V1b receptor is responsible for modulation of mood and behavior, and V2 receptor induces water reabsorption in the kidney. Vasopressin is known to regulate blood pressure, blood osmolality, and blood volume. The effects of V1a and V2 receptors can be amplified when vasopressin is secreted in excessive amounts, and this condition may be experienced by patients undergoing a disease or stress. In pathological conditions such as stroke, traumatic brain injury, subarachnoid hemorrhage, liver disease, and other diseases, vasopressin can exacerbate brain edema. Oversecretion of vasopressin unleashes deleterious pathways leading to hyponatremia and brain edema. This book chapter describes important mechanisms and pathways linking vasopressin and brain edema triggered by various conditions.

Vitamins and Hormones, Volume 113
ISSN 0083-6729
https://doi.org/10.1016/bs.vh.2019.08.015

1. Vasopressin and vasopressin receptors

Vasopressin, also known as arginine-vasopressin (AVP) or antidiuretic hormone (ADH), was discovered by Oliver and Schafer (1895). The discovery was prompted by an observation of remarkable elevation of blood pressure in individuals who received intravenous injection of pituitary extract (Oliver & Schafer, 1895). Today, vasopressin is known to be a cyclic nonapeptide hormone with physiological significance reaching far beyond regulation of arterial blood pressure (Zhang, Hense, Riegger, & Schunkert, 1999), blood volume (Robertson & Athar, 1976), plasma osmolality (Robertson & Athar, 1976), cerebrospinal fluid (CSF) production (Faraci, Mayhan, & Heistad, 1990), and renal tubular water reabsorption (Hollander Jr., Williams, Fordham 3rd, & Welt, 1957). Vasopressin is synthesized in the magnocellular and parvocellular neurons of supraoptic and paraventricular nuclei of the hypothalamus and transported along their axons to the posterior pituitary gland where is it secreted (Evers & Wellmann, 2016). Elevated blood osmolality (Inaba, Umeda, Yamane, Urakami, & Inada, 1988; Robertson & Athar, 1976) or non-osmotic stimuli such as trauma or stress (Schrier & Goldberg, 1980) can trigger release of vasopressin along with catecholamine and corticosteroid hormones (Axelrod & Reisine, 1984; Schrier & Goldberg, 1980). This hypothalamic–pituitary response to stress produces vasoconstriction (Kihara, Shioyama, Okuda, & Takahashi, 2002; Sasaki & Low, 1996) and changes in plasma sodium concentration and osmolality (Robertson & Athar, 1976). Because of the association of vasopressin with cerebrovascular physiology (Hockel, Scholler, Trabold, Nussberger, & Plesnila, 2012) and water homeostasis (Schrier, 2008) it became a focus of studies related to stroke-evoked brain edema (Zeynalov & Jones, 2016). Studies linking vasopressin and traumatic brain injury (TBI) (Filippidis et al., 2014) or subarachnoid hemorrhage (SAH)-triggered brain edema (Hockel et al., 2012) have also contributed greatly to current understanding of this hormone.

Vasopressin has multiple sites of action due to its activation of receptors V1a, V1b, and V2 (Fig. 1) (Koshimizu et al., 2012; Miller, Sandoval, Pisitkun, Knepper, & Hoffert, 2013). Vasopressin receptors are known to be G protein-coupled receptor (Jans & Hemmings, 1991; Orcel et al., 2009; Strakova, Kumar, Watson, & Soloff, 1997), and they are localized on the plasma membrane (Evers & Wellmann, 2016) in many different organs. V1a receptors are found in neurons (Yamazaki, Chen, Schreiber, & Brinton, 1997) and glia

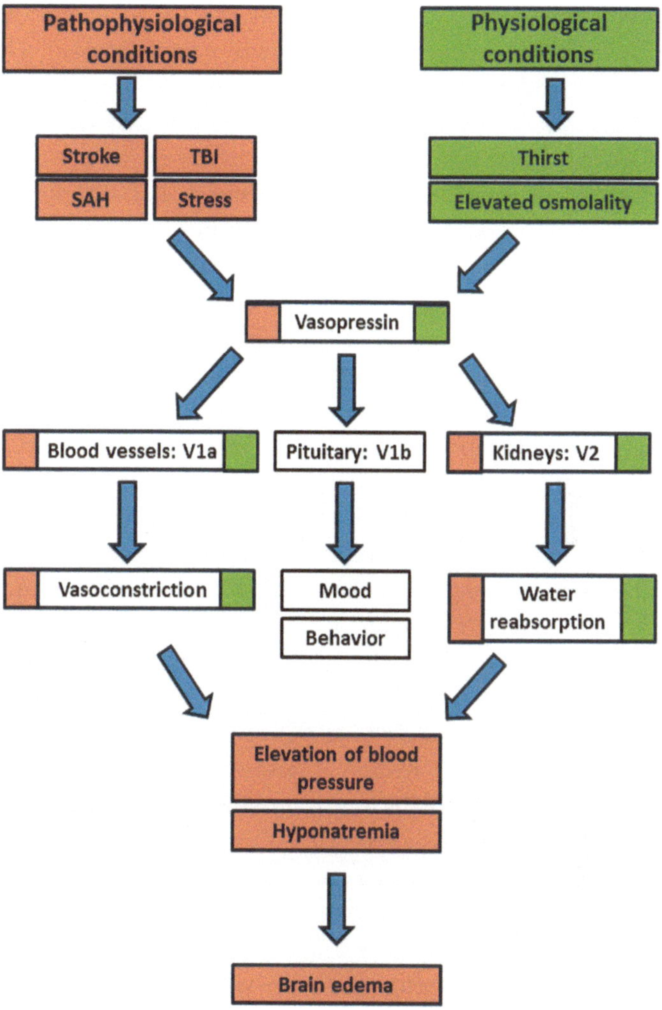

Fig. 1 Vasopressin and its receptors play important roles in physiological (green colored box) and pathophysiological (red colored box) conditions. Vasopressin is released from the posterior pituitary in response to thirst or changes in plasma osmolality activating V1a (blood vessels), V1b (pituitary gland) and V2 receptors (kidneys). A disease and/or trauma may cause an excessive release of vasopressin activating its receptors. In pathophysiological settings, vasopressin can trigger development or exacerbation of brain edema.

(Yamazaki et al., 1997) in various regions of the brain (Kleindienst, Dunbar, Glisson, & Marmarou, 2013; Milik, Szczepanska-Sadowska, Dobruch, Cudnoch-Jedrzejewska, & Maslinski, 2014; Szmydynger-Chodobska et al., 2004). V1a receptors are also localized on blood vessels (Thibonnier, Coles, Thibonnier, & Shoham, 2002), platelets (Launay et al., 1987), liver (Lolait et al., 1995), kidney (Lolait et al., 1995), uterus (Akerlund, Bossmar, Brouard, & Steinwall, 1998), adrenal cortex (Mavani, DeVita, & Michelis, 2015), and adipose tissue (Mavani et al., 2015). On blood vessels, V1a receptors facilitate vasoconstriction (Aoyagi, Koshimizu, & Tanoue, 2009), and V1a receptor agonist drugs are effective in elevating blood pressure to physiological levels after septic shock (Rehberg et al., 2011). In pathophysiological processes, activation of V1a receptors in the brain can exacerbate brain edema triggering elevation of intracranial hypertension (Jia, Liu, Wang, & Wang, 2016; Krieg, Sonanini, Plesnila, & Trabold, 2015; Taya et al., 2008). Chronic stress and myocardial infarction have also been shown to increase V1a receptor protein expression in the brain and the kidney (Milik et al., 2014). Localization of V1a receptors on platelets could be related to the ability of vasopressin to induce platelet aggregation (Brouard et al., 1998) helping blood clot formation. Another class of vasopressin receptors, V1b, are found in the anterior pituitary gland on specialized cells (corticotrophs) where they promote the release of adrenocorticotropic hormone (ACTH) and corticotropin-releasing hormone (CRH) (Evers & Wellmann, 2016; Thibonnier et al., 2002). The action of V1b receptor on the anterior pituitary gland is associated with regulation of behavioral and mood changes (Stemmelin, Lukovic, Salome, & Griebel, 2005; Wersinger, Ginns, O'Carroll, Lolait, & Young 3rd., 2002). Both V1a and V1b receptors activate phosphatidylinositol hydrolysis pathway and mobilize intracellular Ca^{2+} (Lolait et al., 1995) which explains the vasoconstrictive effects of vasopressin at V1a receptors.

The V2 receptors are localized in the kidney (Schliebe et al., 2008), alveolar epithelial cells in the lung (Mavani et al., 2015), and many other organs suggesting that it may have a wide range of functions. In the renal collecting tubules, V2 receptors facilitate water reabsorption decreasing urine formation (Hollander Jr. et al., 1957) with the help of aquaporin-2 (AQP-2) water channels (Evers & Wellmann, 2016). In pathological conditions, such as stroke or TBI, excessive vasopressin secretion activates V2 receptors in the collecting tubules of the kidney which results in increased water reabsorption, dilutional hyponatremia (Takagi, Otake, Nakao, Takashashi, & Hirooka, 2003), and accumulation of water in the brain and in the body overall (Walcott, Kahle, & Simard, 2012). This association of V2 receptors with renal physiology has inspired drug development studies to explore the potential for a new family of diuretics (aquaretics) (Peters, Robben, Deen, & Wetzels, 2007; Vishram-Nielsen & Gustafsson, 2017). One of the representative drugs from this class of diuretics, conivaptan, is currently used to treat hyponatremia induced by the syndrome of inappropriate release of anti-diuretic hormone (SIADH) Conivaptan: YM 087, 2004). SIADH is commonly diagnosed in hospitalized patients with TBI (Rajagopal, Swaminathan, Nair, & Joseph, 2017) or stroke (Hannon & Thompson, 2014; Lim & Yap, 2001). Conivaptan blocks both V1a and V2 receptors which may have a potential to reduce stroke or TBI-induced brain edema (Dhar & Murphy-Human, 2011; Hedna et al., 2014; Zeynalov, Jones, Seo, Snell, & Elliott, 2015).

2. Brain edema

Brain edema is a pathophysiological process manifested by accumulation of water in brain cells (Nase, Helm, Enger, & Ottersen, 2008) and/or in the extracellular space (Unterberg, Stover, Kress, & Kiening, 2004). Classification of brain edema into the cytotoxic and vasogenic types was proposed by Klatzo over 3 decades ago (Klatzo, 1987), and it is still being used to study and describe this phenomenon. Other brain edema subtypes such as osmotic and hydrostatic are also useful in its characterization and were added to the literature later (Go, 1997). Cytotoxic brain edema is defined as accumulation of water within the brain cells causing swelling. Cytotoxic brain edema often precedes the vasogenic type, which is a result of blood-brain barrier (BBB) breakdown manifested by accumulation of water in the extracellular space (Klatzo, 1987). Both cytotoxic and vasogenic brain edema can be triggered by neurological disorders such as stroke (Kleindienst et al., 2013), TBI (Unterberg et al., 2004), hydrocephalus (Leinonen, Vanninen, & Rauramaa, 2017), and seizure (Cipolla, 2009). Some non-neurological diseases and conditions such as SIADH (Kleindienst, Hannon, Buchfelder, & Verbalis, 2016), liver disease (Rama Rao, Jayakumar, & Norenberg, 2014), inflammation (Stamatovic, Dimitrijevic, Keep, & Andjelkovic, 2006), acute hypertension (Cipolla, 2009), and many others can also trigger development of brain edema. Regardless of the cause and the type, brain edema is detrimental to the brain, and, because of the rigidity of the skull, intracranial pressure (ICP) can be dramatically increased (Kotwica, Persson, & Thuomas, 1989). High ICP, in turn, decreases blood flow to the brain creating perfusion deficit (Westermaier, Stetter, Raslan, Vince, & Ernestus, 2012). If not resolved, brain edema is followed by a secondary brain injury (Tucker et al., 2017; Wang et al., 2014) which gradually spreads to the healthy regions of the brain causing more tissue damage.

3. Water diffusion and vasopressin in brain edema

Water constitutes the most essential environment in the brain and the body overall. Therefore, the importance of balanced water distribution between cells, extracellular space, and intravascular compartment should not be overlooked. Water is a solvent for all organic and inorganic molecules (Go, 1997) which constantly diffuse across the plasma membrane. Water serves as a medium and a substrate for biochemical reactions, and it helps eliminate waste products from the cell (Martinek, Levashov, Khmelnitsky, Klyachko, & Berezin, 1982). The amount of water in the brain cells and the extracellular space is controlled by the hypothalamic homeostasis mechanism where vasopressin plays an important role. Brain insult may disturb this function of the hypothalamus and result in water accumulation and edema. Understanding these mechanisms of water movement between compartments of the brain, and learning how to manipulate them will ultimately help develop more efficient treatment against brain edema.

On the molecular level brain edema owes its development, as well as resolution, in part to the most abundant water channel in the brain, aquaporin-4 (AQP-4) (Fukuda et al., 2012; Wang et al., 2011). AQP-4 is embedded in the plasma membrane (Jung et al., 1994; Salman et al., 2017), and permits diffusion of water molecules in a direction determined by the osmotic gradient between compartments (Wambo, Rodriguez, & Chen, 2017). However, osmolality constantly fluctuates because Na^+ and K^+ ion channels and pumps allow these osmolytes to cross the plasma membrane (Jacob, Chappell, & Becker, 2016). The movement of water through AQPs is bidirectional, one molecule at a time, and it occurs from the lower to the higher solute-concentrated environment (Fig. 2) (Agre et al., 1993; Jung et al., 1994). Additionally, vasopressin regulates the diffusion of water across the plasma membrane by changing AQP-4 protein expression after a brain injury or stroke (Kleindienst et al., 2013; Liu, Nakayama, Amiry-Moghaddam, Ottersen, & Bhardwaj, 2010). To study the diffusion of water in the brain all these factors must be considered (Fig. 2).

In the CNS, water is exchanged between the following compartments: (1) the cell and the extracellular space (Krizaj, Rice, Wardle, & Nicholson, 1996); (2) the extracellular space and the vascular compartment, or brain–blood interface (Hawkins & Davis, 2005); and (3) the choroid plexus and ependymal cells in the ventricles or pia mater on the brain surface and cerebrospinal fluid (CSF), or brain–CSF interface (Go, 1997; Johanson et al., 2008).

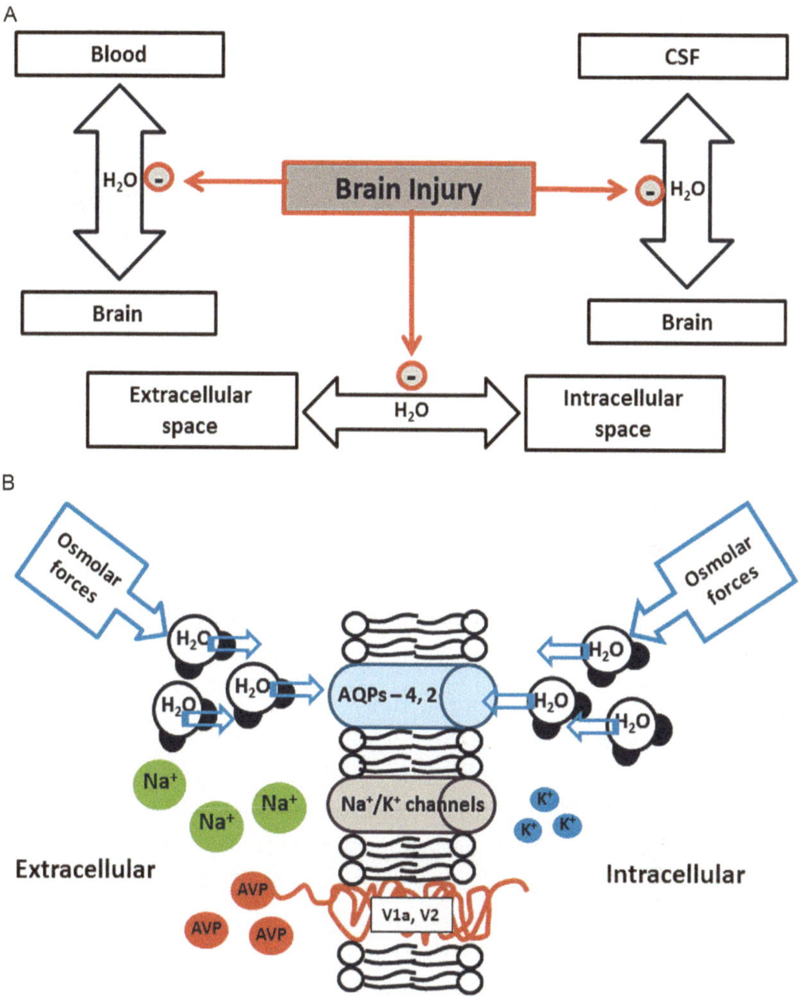

Fig. 2 Diffusion of water across the plasma membrane separating compartments of the three physiologically important sites of the brain: the blood-brain interface, extracellular-intracellular space, and brain–CSF interface (A). Brain injury can disable water diffusion at all three sites resulting in disturbances of water homeostasis. A hypothetical small fragment of plasma membrane with imbedded aquaporin-4, or 2 (AQP-4, 2) water channels, Na^+/K^+ channels, and vasopressin receptors V1a, or V2 (B). All three proteins are shown next to each other for simplicity. Osmolar gradient is a driving force of water diffusion through AQP-4,2 across plasma membrane. Hydrostatic force of water diffusion is not shown.

All these compartments are separated by the plasma membrane (lipid bilayer) which contains functionally linked vasopressin receptors, AQP-4, and Na^+/K^+ ion channels (Fig. 2) (Niermann, Amiry-Moghaddam, Holthoff, Witte, & Ottersen, 2001).

An insult to the brain such as TBI or stroke may disable water diffusion in cells adjacent to the injury site (Fig. 2) causing brain edema formation. For example, interruption of water movement from the intracellular to the extracellular space may contribute to cytotoxic brain edema (Kleindienst et al., 2013). Injury-triggered BBB breakdown drives water by hydrostatic forces from the capillaries into the extracellular space of brain, described above as vasogenic brain edema (Nagashima, Horwitz, & Rapoport, 1990). Increased intraventricular pressure can cause the ependymal cells of the choroid plexus at the brain–CSF interface to produce and secrete vasopressin locally (Johanson et al., 2008). This autocrine action of vasopressin is seen in the hydrocephalus condition with elevated ICP and intraventricular pressure. Increased production of local vasopressin at the choroid plexus activates V1a receptors and results in decreased CSF production (Johanson et al., 2008), and possible accumulation of water within the cells and intracellular space.

4. Injury-induced brain edema and vasopressin

A large number of patients with brain injury develop brain edema where increased vasopressin production worsens their condition because of hyponatremia as a result of SIADH (Cuesta, Hannon, & Thompson, 2016; Kleindienst et al., 2016; Sherlock et al., 2009). Different brain injury types are classified and diagnosed based on their etiology. For example, ischemic stroke is a result of occlusion of a cerebral artery. In clinical or experimental settings, ischemic stroke has been shown to cause cellular deprivation of oxygen and glucose, energy depletion, disruption of BBB, inflammation, apoptosis and necrosis, and SIADH (Ameli et al., 2014; Jia et al., 2016; Vakili, Kataoka, & Plesnila, 2005; Zeynalov et al., 2015). Hemorrhagic stroke is a result of rupture of an aneurysm which leads to complications including SIADH and accumulation of blood in the extracellular space of the brain parenchyma (Beez, Steiger, & Etminan, 2017; Martin & Rymer, 2011; Saleem, Yousuf, Gul, Gupta, & Verma, 2014). Though hemorrhagic and ischemic events in the brain may seem to be completely different injury stimuli, they do have similarities such as disruption of brain perfusion causing tissue deprivation of oxygen and glucose. Additionally, hemorrhagic transformation is often seen in patients with ischemic stroke which can be a consequence of BBB disruption exacerbated by administration of tissue plasminogen activator (tPA) (Castro, Azevedo, Serrador, Rocha, & Sorond, 2017; Hom et al., 2011; Wang et al., 2009, 2004).

TBI is another example of a brain injury, and similar to ischemic and hemorrhagic stroke may be associated with excessive vasopressin production, brain edema, and secondary brain tissue damage (Pascale, Szmydynger-Chodobska, Sarri, & Chodobski, 2006; Szmydynger-Chodobska et al., 2004; Trabold, Krieg, Scholler, & Plesnila, 2008). Impact to the head due to traumatic accidents such as falls and sports related injuries, or combat injuries such as gunshot or blast injuries can produce extensive brain tissue damage. The range and severity of pathophysiological complications after TBI is a reflection of complex pathological events which include but are not limited to contusion and hemorrhage also leading to edema, BBB disruption, inflammation, excessive vasopressin secretion, and perfusion deficit (Dekker et al., 2016; Pearn et al., 2017; Rauen, Trabold, Brem, Terpolilli, & Plesnila, 2013; Salehi, Zhang, & Obenaus, 2017; Szmydynger-Chodobska, Fox, Lynch, Zink, & Chodobski, 2010).

Extensive research has been done on the subject of stroke or TBI-induced brain edema, and it has become evident that different etiologies of brain injury may cause similar pathophysiological sequelae with some variations. Therefore, it is not a surprise to observe the same molecular mechanisms being involved in development of brain edema triggered by different injurious stimuli to the brain. We and others have demonstrated an association of vasopressin with brain edema formation after ischemic stroke. Preclinical and clinical studies have found that simultaneous blocking of V1a and V2 vasopressin receptors halts development of brain edema and BBB disruption after stroke (Decker et al., 2016; Hedna et al., 2014; Zeynalov, Jones, & Elliott, 2016, 2017; Zeynalov et al., 2015). Other studies have shown that TBI-induced vasopressin secretion can influence brain edema formation (Huang et al., 2008; Krieg et al., 2015; Rauen et al., 2013), BBB disruption (Pascale et al., 2006), astrocytic swelling, water flux through changes in AQP-4 (Marmarou et al., 2014; Taya et al., 2008), and changes in extracellular sodium (Filippidis et al., 2014). The relevance of these findings to injury-induced brain edema continues to spark new research ideas with V1a and V2 receptors, AQP-4, and Na^+-K^+ channels being the main focus of the studies.

In normal conditions properly distributed and functional AQP-4 and Na^+-K^+ channels maintain water and electrolyte homeostasis in the brain tissue (Ayus, Achinger, & Arieff, 2008). Balanced "housekeeping" secretion of vasopressin (Bevilacqua et al., 1985) with activation of vasopressin receptors (Ribeiro, Panizza Hdo, Santos, Ferreira-Neto, & Antunes, 2015) is a physiological response of hypothalamus to fluctuations in blood osmolality.

Brain injury with substantial tissue damage can trigger vasopressin secretion from the posterior pituitary. It has been shown that injury-induced vasopressin production can affect AQP-4 and Na^+/K^+ ion channels which are involved in water accumulation in the brain after injury (Filippidis, Carozza, & Rekate, 2016; Filippidis et al., 2014; Hertz et al., 2014; Kleindienst et al., 2013; Liu et al., 2010; Manaenko et al., 2011; Marmarou et al., 2014; Moeller, Fenton, Zeuthen, & Macaulay, 2009; O'Donnell et al., 2013; Okuno et al., 2008; Rutkowsky, Wallace, Wise, & O'Donnell, 2011; Taya et al., 2008; Vakili et al., 2005). Activation of V1a receptors in the brain following injury affects function of Na^+/K^+ channels associated with both triggering and resolving brain edema (Filippidis et al., 2014). In their study, Filippidis et al. described V1a receptor-dependent mechanisms of brain edema development. Extracellular Na^+ content was decreased by almost 40% after TBI, presumably due to opening of electrolyte channels and the movement of Na^+ into the cells and K^+-in the opposite direction (Filippidis et al., 2014). This shift in Na^+ balance triggered by TBI likely contributes to cytotoxic brain edema. It is possible that because of the energy depletion after the injury, $Na^{+/}K^+$/ATPase and other transporters are not able to restore the electrolyte balance between the extracellular and intracellular compartments which leads to sustained edema. Treatment with a selective V1a receptor blocker accelerated the recovery of the extracellular Na^+ concentration which prevented increase of intracranial pressure (ICP) (Filippidis et al., 2014). Because TBI induces increased AQP-4 expression resulting in cytotoxic brain edema (Kleindienst et al., 2013; Manaenko et al., 2011; Marmarou et al., 2014; Taya et al., 2008), a reduction in AQP-4 protein expression after the shift of Na^+ would be desirable to prevent water from following Na^+ inside the cell after the injury. The discovery of a new V1a receptor-dependent mechanism of Na^+ restoration in the extracellular space may lead to an effective therapeutic strategy to restore electrolyte balance and minimize edema after an insult to the brain.

5. Subarachnoid hemorrhage (SAH)-induced brain edema and vasopressin

SAH-induced brain edema may be linked to vasopressin secretion and V1a receptor-dependent vasoconstrictive responses of cerebral arteries (Hockel et al., 2012), perfusion deficit in the brain, and secondary hypoxia/ischemia (Westermaier et al., 2012). After SAH, increased ICP can trigger vasopressin-induced vasoconstriction with systemic hypertension (Cushing syndrome) which dramatically increases the rebleeding rate (Hockel et al., 2012). In addition, vasopressin-triggered vasoconstriction of the cerebral blood vessels may cause significant reduction of regional blood flow. In the setting of SAH-caused vasopressin release and vasoconstriction-induced perfusion deficit, water diffusion from the extracellular into the vascular compartment may be impaired producing accumulation of water in the extracellular space of the brain. SAH-caused disruption of BBB can cause the capillary hydrostatic pressure to force blood from the vascular compartment into the extracellular space with minimal resistance. Under conditions of SAH-induced high vasopressin release, sustained generalized vasoconstriction can further increase hydrostatic pressure in cerebral microvessels accelerating extravasation of blood contents and further increasing vasogenic brain edema. V1a antagonists reduce the Cushing response, decreasing ICP and rebleeding (Hockel et al., 2012). Here, the uncontrolled vasopressin production can greatly damage the brain tissue if treatment is not initiated.

6. Kidneys, brain edema and vasopressin

Kidney function can influence water balance in the brain especially when the body undergoes a severe trauma or disease. Kidneys facilitate changes in body water content by allowing secretion or reabsorption of water, and these processes are controlled by vasopressin and V2 receptors on the plasma membrane of the renal collecting tubules. Brain injury-triggered release of abnormally large amounts of vasopressin stimulates the pathological condition known as SIADH (Hannon & Thompson, 2014). Activated V2 receptors increase renal tubular AQP-2 protein expression and its insertion into the plasma membrane. Overly abundant AQP-2 protein at the tubular plasma membrane facilitates excessive reabsorption water (Robertson, 2006) and development of dilutional hyponatremia (Hannon & Thompson, 2014). Consequently, hyponatremia creates a hypo-osmolar environment in the blood when plasma sodium levels fall below 125 mEq/dL. This produces an osmotic gradient between the blood and the extracellular/cellular compartments which forces water to diffuse from the capillary into the brain (Hannon & Thompson, 2014). Currently, vasopressin receptor V1a and V2 antagonists are successfully used to treat SIADH, and potentially, brain stroke or TBI-induced brain edema (Hedna et al., 2014; Zeynalov et al., 2015). But in spite of availability, the use of vasopressin receptor antagonists against brain edema is not yet fully implemented into the clinical practice. Future clinical trials and pre-clinical studies will be necessary to demonstrate potential benefits of blocking V1a and V2 receptors to prevent injury-induced brain edema and secondary brain injury. Promotion of increased water secretion in the kidney (Conivaptan: YM 087, 2004) by antagonism of V2 receptors may be helpful in alleviating brain edema after an injury. Blocking both V1a and V2 receptors by conivaptan has been shown to have a positive effect on recovery from brain edema compared to the V2 receptor blocker tolvaptan (Zeynalov et al., 2015).

7. Inflammation in brain edema and vasopressin

Recently, it has become evident that vasopressin could be an important player in the pathophysiology of stroke or TBI-induced inflammation associated with brain edema (Decker et al., 2016; Szmydynger-Chodobska et al., 2010). Numerous studies demonstrate that vasopressin is a facilitator in activation of downstream pathways of inflammation and edema. Inflammation and vasopressin are bound by a complex relationship. Inflammation can trigger increased secretion of vasopressin which, in turn, affects the inflammatory response. It has been shown that in the presence of vasopressin, the inflammatory response of brain cells is augmented, and V1a receptors act in a synergistic fashion with TNFα and inflammatory chemokines (Szmydynger-Chodobska et al., 2010). Inflammatory cytokines interleukin 1 (IL-1) and interleukin 6 (IL-6) can also stimulate production of vasopressin (Baker, Richards, Dayan, & Jessop, 2003; Chikanza & Grossman, 1998) which often results in hyponatremia (Park & Shin, 2013). In another study by Boyd et al., vasopressin has been shown to downregulate production of the inflammatory cytokine IL-6 which demonstrates that vasopressin has an immunomodulatory property as well. This reduction of IL-6 expression in the lung after sepsis was V2 receptor dependent (Boyd, Holmes, Wang, Roberts, & Walley, 2008). In a study by Baker et al., vasopressin and its receptors were found in rat lymphocytes B, thymus, and human peripheral blood mononuclear cells (PBMC) which suggests that vasopressin is produced not only in the brain but peripherally as well (Baker et al., 2003). It is still unclear why vasopressin has such a diverse influence on inflammation, and more research is needed to demystify this phenomenon.

8. Liver disease-caused brain edema and vasopressin

A liver disease may progress into hepatic encephalopathy and brain edema, and this condition is accompanied by acute liver failure and associated with poor survival rate (Bernal, Lee, Wendon, Larsen, & Williams, 2015; Wijdicks, 2016). Pathophysiology of acute liver failure-induced brain edema is multifactorial. Liver disease-triggered apoptosis and necrosis of hepatic cells results in release of cellular contents into the bloodstream which produces generalized inflammation, cardiovascular instability, and vasodilatory septic shock (Bernal et al., 2015). As a result, reduction of systemic vascular resistance and blood pressure causes dramatic decrease in cerebral perfusion pressure and failure of autoregulation of cerebral blood flow (Bernal et al., 2015). Furthermore, acute liver failure-induced cerebrovascular insufficiency results in exacerbation of brain edema and BBB breakdown (Bernal et al., 2015). On the biochemical level, liver failure coincides with disabled urea synthesis which produces and releases into the blood stream large amounts of ammonia (hyperammonemia) (Bernal et al., 2015). Ammonia is known to be a potent neurotoxin because it easily crosses the BBB (Rama Rao et al., 2014). Ammonia is used by astrocytic glutamine synthase for conversion of glutamate to glutamine—the detoxification process (Bernal et al., 2015). Excessive production and accumulation of glutamine in the astrocytes results in their swelling because glutamine is an osmolyte which attracts water into the intracellular compartment (Bernal et al., 2015). In addition, acute liver failure-caused cytotoxic brain edema with astrocytic swelling (Rama Rao & Norenberg, 2007) can stimulate uncontrolled release of vasopressin leading to SIADH and hyponatremia (Shchekochikhin, Tkachenko, & Schrier, 2013). One molecular player identified to be associated with hepatic encephalopathy is alpha-1 antichymotrypsin (Park et al., 2017). In their study, alpha-1 antichymotrypsin (ACT) was released following hepatic encephalopathy. In vitro studies revealed that ACT, in concert with vasopressin, triggered astrocytic swelling through increased expression of AQP-4 which was reduced by V1a antagonism (Park et al., 2017).

9. Summary

Vasopressin is implicated in brain edema formation induced by neurological or non-neurological causes. Vasopressin and its receptors can exacerbate brain edema triggered by TBI, stroke, SAH, liver disease, SIADH, and many other conditions. On the cellular and molecular levels, vasopressin receptors are associated with AQP-4 in the brain and AQP-2 in the kidney, which facilitate water transport in corresponding organs leading to brain edema formation. Molecular players such as Na^+/K^+ ion channels, inflammatory cytokines, and chemokines also assist vasopressin receptors in triggering brain edema. Vasopressin is of biological significance in a variety of responses of the body to stress, ranging from vasoconstriction and edema to inflammation and pain. Knowledge of deleterious molecular pathways triggered by vasopressin and its receptors will assist in the development of drugs against disease-induced complications such as brain edema.

Acknowledgments

We acknowledge and thank the Swedish Medical Center Englewood Colorado for ongoing research support.

References

Agre, P., Preston, G. M., Smith, B. L., Jung, J. S., Raina, S., Moon, C., et al. (1993). Aquaporin CHIP: The archetypal molecular water channel. *The American Journal of Physiology*, *265*(4 Pt. 2), F463–F476. https://doi.org/10.1152/ajprenal.1993.265.4.F463.

Akerlund, M., Bossmar, T., Brouard, R., & Steinwall, M. (1998). Evidence for an involvement of vasopressin in mechanism of primary dysmenorrhea and effect of the non-peptide vasopressin V1a receptor antagonist, SR 49059, on the uterus of non-pregnant women. *Advances in Experimental Medicine and Biology*, *449*, 467–472.

Ameli, P. A., Ameli, N. J., Gubernick, D. M., Ansari, S., Mohan, S., Satriotomo, I., et al. (2014). Role of vasopressin and its antagonism in stroke related edema. *Journal of Neuroscience Research*, *92*(9), 1091–1099. https://doi.org/10.1002/jnr.23407.

Aoyagi, T., Koshimizu, T. A., & Tanoue, A. (2009). Vasopressin regulation of blood pressure and volume: Findings from V1a receptor-deficient mice. *Kidney International*, *76*(10), 1035–1039. https://doi.org/10.1038/ki.2009.319.

Axelrod, J., & Reisine, T. D. (1984). Stress hormones: Their interaction and regulation. *Science*, *224*(4648), 452–459.

Ayus, J. C., Achinger, S. G., & Arieff, A. (2008). Brain cell volume regulation in hyponatremia: Role of sex, age, vasopressin, and hypoxia. *American Journal of Physiology. Renal Physiology*, *295*(3), F619–F624. https://doi.org/10.1152/ajprenal.00502.2007.

Baker, C., Richards, L. J., Dayan, C. M., & Jessop, D. S. (2003). Corticotropin-releasing hormone immunoreactivity in human T and B cells and macrophages: Colocalization with arginine vasopressin. *Journal of Neuroendocrinology*, *15*(11), 1070–1074.

Beez, T., Steiger, H. J., & Etminan, N. (2017). Pharmacological targeting of secondary brain damage following ischemic or hemorrhagic stroke, traumatic brain injury, and bacterial meningitis—A systematic review and meta-analysis. *BMC Neurology*, *17*(1), 209. https://doi.org/10.1186/s12883-017-0994-z.

Bernal, W., Lee, W. M., Wendon, J., Larsen, F. S., & Williams, R. (2015). Acute liver failure: A curable disease by 2024? *Journal of Hepatology*, *62*(Suppl. 1), S112–S120. https://doi.org/10.1016/j.jhep.2014.12.016.

Bevilacqua, M., Meroni, R., Dagani, R., Renesto, E., Baruto, C., & Norbiato, G. (1985). Role of blood osmolality in the regulation of vasopressin secretion in man: Application of a new radioimmunoassay method for vasopressin. *Journal of Endocrinological Investigation, 8*(2), 97–101. https://doi.org/10.1007/BF03350655.

Boyd, J. H., Holmes, C. L., Wang, Y., Roberts, H., & Walley, K. R. (2008). Vasopressin decreases sepsis-induced pulmonary inflammation through the V2R. *Resuscitation, 79*(2), 325–331. https://doi.org/10.1016/j.resuscitation.2008.07.006.

Brouard, R., Laporte, V., Serradeil Le Gal, C., Pignol, R., Jang, H., Donat, F., et al. (1998). Safety, tolerability, and pharmacokinetics of SR 49059, a V1a vasopressin receptor antagonist, after repeated oral administration in healthy volunteers. *Advances in Experimental Medicine and Biology, 449*, 455–465.

Castro, P., Azevedo, E., Serrador, J., Rocha, I., & Sorond, F. (2017). Hemorrhagic transformation and cerebral edema in acute ischemic stroke: Link to cerebral autoregulation. *Journal of the Neurological Sciences, 372*, 256–261. https://doi.org/10.1016/j.jns.2016.11.065.

Chikanza, I. C., & Grossman, A. S. (1998). Hypothalamic-pituitary-mediated immunomodulation: Arginine vasopressin is a neuroendocrine immune mediator. *British Journal of Rheumatology, 37*(2), 131–136.

Cipolla, M. J. (2009). *The cerebral circulation.* San Rafael, CA: Morgan & Claypool Life Sciences.

Conivaptan: YM 087. (2004). *Drugs in R&D, 5*(2), 94–97.

Cuesta, M., Hannon, M. J., & Thompson, C. J. (2016). Diagnosis and treatment of hyponatraemia in neurosurgical patients. *Endocrinología y Nutrición, 63*(5), 230–238. https://doi.org/10.1016/j.endonu.2015.12.007.

Decker, D., Collier, L., Lau, T., Olivera, R., Roma, G., Leonardo, C., et al. (2016). The effects of clinically relevant hypertonic saline and conivaptan administration on ischemic stroke. *Acta Neurochirurgica. Supplement, 121*, 243–250. https://doi.org/10.1007/978-3-319-18497-5_43.

Dekker, S. E., Duvekot, A., de Vries, H. M., Geeraedts, L. M., Jr., Peerdeman, S. M., de Waard, M. C., et al. (2016). Relationship between tissue perfusion and coagulopathy in traumatic brain injury. *The Journal of Surgical Research, 205*(1), 147–154. https://doi.org/10.1016/j.jss.2016.06.023.

Dhar, R., & Murphy-Human, T. (2011). A bolus of conivaptan lowers intracranial pressure in a patient with hyponatremia after traumatic brain injury. *Neurocritical Care, 14*(1), 97–102. https://doi.org/10.1007/s12028-010-9366-x.

Evers, K. S., & Wellmann, S. (2016). Arginine vasopressin and copeptin in perinatology. *Frontiers in Pediatrics, 4*, 75. https://doi.org/10.3389/fped.2016.00075.

Faraci, F. M., Mayhan, W. G., & Heistad, D. D. (1990). Effect of vasopressin on production of cerebrospinal fluid: Possible role of vasopressin (V1)-receptors. *The American Journal of Physiology, 258*(1 Pt. 2), R94–R98. https://doi.org/10.1152/ajpregu.1990.258.1.R94.

Filippidis, A. S., Carozza, R. B., & Rekate, H. L. (2016). Aquaporins in brain edema and neuropathological conditions. *International Journal of Molecular Sciences, 18*(1). https://doi.org/10.3390/ijms18010055.

Filippidis, A. S., Liang, X., Wang, W., Parveen, S., Baumgarten, C. M., & Marmarou, C. R. (2014). Real-time monitoring of changes in brain extracellular sodium and potassium concentrations and intracranial pressure after selective vasopressin-1a receptor inhibition following focal traumatic brain injury in rats. *Journal of Neurotrauma, 31*(14), 1258–1267. https://doi.org/10.1089/neu.2013.3063.

Fukuda, A. M., Pop, V., Spagnoli, D., Ashwal, S., Obenaus, A., & Badaut, J. (2012). Delayed increase of astrocytic aquaporin 4 after juvenile traumatic brain injury: Possible role in edema resolution? *Neuroscience, 222*, 366–378. https://doi.org/10.1016/j.neuroscience.2012.06.033.

Go, K. G. (1997). The normal and pathological physiology of brain water. *Advances and Technical Standards in Neurosurgery*, *23*, 47–142.

Hannon, M. J., & Thompson, C. J. (2014). Neurosurgical hyponatremia. *Journal of Clinical Medicine*, *3*(4), 1084–1104. https://doi.org/10.3390/jcm3041084.

Hawkins, B. T., & Davis, T. P. (2005). The blood-brain barrier/neurovascular unit in health and disease. *Pharmacological Reviews*, *57*(2), 173–185. https://doi.org/10.1124/pr.57.2.4.

Hedna, V. S., Bidari, S., Gubernick, D., Ansari, S., Satriotomo, I., Khan, A. A., et al. (2014). Treatment of stroke related refractory brain edema using mixed vasopressin antagonism: A case report and review of the literature. *BMC Neurology*, *14*, 213. https://doi.org/10.1186/s12883-014-0213-0.

Hertz, L., Xu, J., Chen, Y., Gibbs, M. E., Du, T., Hertz, L., et al. (2014). Antagonists of the vasopressin V1 receptor and of the beta(1)-adrenoceptor inhibit cytotoxic brain edema in stroke by effects on astrocytes—But the mechanisms differ. *Current Neuropharmacology*, *12*(4), 308–323. https://doi.org/10.2174/1570159X12666140828222723.

Hockel, K., Scholler, K., Trabold, R., Nussberger, J., & Plesnila, N. (2012). Vasopressin V(1a) receptors mediate posthemorrhagic systemic hypertension thereby determining rebleeding rate and outcome after experimental subarachnoid hemorrhage. *Stroke*, *43*(1), 227–232. https://doi.org/10.1161/STROKEAHA.111.626168.

Hollander, W., Jr., Williams, T. F., Fordham, C. C., 3rd, & Welt, L. G. (1957). A study of the quantitative relationship between antidiuretic hormone (vasopressin) and the renal tubular reabsorption of water. *The Journal of Clinical Investigation*, *36*(7), 1059–1071. https://doi.org/10.1172/JCI103501.

Hom, J., Dankbaar, J. W., Soares, B. P., Schneider, T., Cheng, S. C., Bredno, J., et al. (2011). Blood-brain barrier permeability assessed by perfusion CT predicts symptomatic hemorrhagic transformation and malignant edema in acute ischemic stroke. *AJNR. American Journal of Neuroradiology*, *32*(1), 41–48. https://doi.org/10.3174/ajnr.A2244.

Huang, W. D., Pan, J., Xu, M., Su, W., Lu, Y. Q., Chen, Z. J., et al. (2008). Changes and effects of plasma arginine vasopressin in traumatic brain injury. *Journal of Endocrinological Investigation*, *31*(11), 996–1000. https://doi.org/10.1007/BF03345638.

Inaba, K., Umeda, Y., Yamane, Y., Urakami, M., & Inada, M. (1988). Characterization of human platelet vasopressin receptor and the relation between vasopressin-induced platelet aggregation and vasopressin binding to platelets. *Clinical Endocrinology*, *29*(4), 377–386.

Jacob, M., Chappell, D., & Becker, B. F. (2016). Regulation of blood flow and volume exchange across the microcirculation. *Critical Care*, *20*(1), 319. https://doi.org/10.1186/s13054-016-1485-0.

Jans, D. A., & Hemmings, B. A. (1991). cAMP-dependent protein kinase activation affects vasopressin V2-receptor number and internalization in LLC-PK1 renal epithelial cells. *FEBS Letters*, *281*(1–2), 267–271.

Jia, S. W., Liu, X. Y., Wang, S. C., & Wang, Y. F. (2016). Vasopressin hypersecretion-associated brain edema formation in ischemic stroke: Underlying mechanisms. *Journal of Stroke and Cerebrovascular Diseases*, *25*(6), 1289–1300. https://doi.org/10.1016/j.jstrokecerebrovasdis.2016.02.002.

Johanson, C. E., Duncan, J. A., 3rd, Klinge, P. M., Brinker, T., Stopa, E. G., & Silverberg, G. D. (2008). Multiplicity of cerebrospinal fluid functions: New challenges in health and disease. *Cerebrospinal Fluid Research*, *5*, 10. https://doi.org/10.1186/1743-8454-5-10.

Jung, J. S., Bhat, R. V., Preston, G. M., Guggino, W. B., Baraban, J. M., & Agre, P. (1994). Molecular characterization of an aquaporin cDNA from brain: Candidate osmoreceptor and regulator of water balance. *Proceedings of the National Academy of Sciences of the United States of America*, *91*(26), 13052–13056.

Kihara, M., Shioyama, M., Okuda, K., & Takahashi, M. (2002). The impact of aging on vasa nervorum, nerve blood flow and vasopressin responsiveness. *The Canadian Journal of Neurological Sciences*, *29*(2), 164–168.

Klatzo, I. (1987). Pathophysiological aspects of brain edema. *Acta Neuropathologica*, *72*(3), 236–239.

Kleindienst, A., Dunbar, J. G., Glisson, R., & Marmarou, A. (2013). The role of vasopressin V1A receptors in cytotoxic brain edema formation following brain injury. *Acta Neurochirurgica*, *155*(1), 151–164. https://doi.org/10.1007/s00701-012-1558-z.

Kleindienst, A., Hannon, M. J., Buchfelder, M., & Verbalis, J. G. (2016). Hyponatremia in neurotrauma: The role of vasopressin. *Journal of Neurotrauma*, *33*(7), 615–624. https://doi.org/10.1089/neu.2015.3981.

Koshimizu, T. A., Nakamura, K., Egashira, N., Hiroyama, M., Nonoguchi, H., & Tanoue, A. (2012). Vasopressin V1a and V1b receptors: From molecules to physiological systems. *Physiological Reviews*, *92*(4), 1813–1864. https://doi.org/10.1152/physrev.00035.2011.

Kotwica, Z., Persson, L., & Thuomas, K. A. (1989). The effects of brain edema on intracranial pressure in focal cerebral ischemia. An experimental study in a rat using magnetic resonance imaging. *Zentralblatt für Neurochirurgie*, *50*(2), 68–71.

Krieg, S. M., Sonanini, S., Plesnila, N., & Trabold, R. (2015). Effect of small molecule vasopressin V1a and V2 receptor antagonists on brain edema formation and secondary brain damage following traumatic brain injury in mice. *Journal of Neurotrauma*, *32*(4), 221–227. https://doi.org/10.1089/neu.2013.3274.

Krizaj, D., Rice, M. E., Wardle, R. A., & Nicholson, C. (1996). Water compartmentalization and extracellular tortuosity after osmotic changes in cerebellum of Trachemys scripta. *The Journal of Physiology*, *492*(Pt. 3), 887–896.

Launay, J. M., Vittet, D., Vidaud, M., Rondot, A., Mathieu, M. N., Lalau-Keraly, C., et al. (1987). V1a-vasopressin specific receptors on human platelets: Potentiation by ADP and epinephrine and evidence for homologous down-regulation. *Thrombosis Research*, *45*(4), 323–331.

Leinonen, V., Vanninen, R., & Rauramaa, T. (2017). Raised intracranial pressure and brain edema. *Handbook of Clinical Neurology*, *145*, 25–37. https://doi.org/10.1016/B978-0-12-802395-2.00004-3.

Lim, J. K., & Yap, K. B. (2001). Hyponatraemia in hospitalised elderly patients. *The Medical Journal of Malaysia*, *56*(2), 232–235.

Liu, X., Nakayama, S., Amiry-Moghaddam, M., Ottersen, O. P., & Bhardwaj, A. (2010). Arginine-vasopressin V1 but not V2 receptor antagonism modulates infarct volume, brain water content, and aquaporin-4 expression following experimental stroke. *Neurocritical Care*, *12*(1), 124–131. https://doi.org/10.1007/s12028-009-9277-x.

Lolait, S. J., O'Carroll, A. M., Mahan, L. C., Felder, C. C., Button, D. C., Young, W. S., 3rd, et al. (1995). Extrapituitary expression of the rat V1b vasopressin receptor gene. *Proceedings of the National Academy of Sciences of the United States of America*, *92*(15), 6783–6787.

Manaenko, A., Fathali, N., Khatibi, N. H., Lekic, T., Hasegawa, Y., Martin, R., et al. (2011). Arginine-vasopressin V1a receptor inhibition improves neurologic outcomes following an intracerebral hemorrhagic brain injury. *Neurochemistry International*, *58*(4), 542–548. https://doi.org/10.1016/j.neuint.2011.01.018.

Marmarou, C. R., Liang, X., Abidi, N. H., Parveen, S., Taya, K., Henderson, S. C., et al. (2014). Selective vasopressin-1a receptor antagonist prevents brain edema, reduces astrocytic cell swelling and GFAP, V1aR and AQP4 expression after focal traumatic brain injury. *Brain Research*, *1581*, 89–102. https://doi.org/10.1016/j.brainres.2014.06.005.

Martin, C. O., & Rymer, M. M. (2011). Hemorrhagic stroke: Aneurysmal subarachnoid hemorrhage. *Missouri Medicine*, *108*(2), 124–127.

Martinek, K., Levashov, A. V., Khmelnitsky, Y. L., Klyachko, N. L., & Berezin, I. V. (1982). Colloidal solution of water in organic solvents: A microheterogeneous medium for enzymatic reactions. *Science*, *218*(4575), 889–891.

Mavani, G. P., DeVita, M. V., & Michelis, M. F. (2015). A review of the nonpressor and nonantidiuretic actions of the hormone vasopressin. *Frontiers in Medicine (Lausanne)*, *2*, 19. https://doi.org/10.3389/fmed.2015.00019.

Milik, E., Szczepanska-Sadowska, E., Dobruch, J., Cudnoch-Jedrzejewska, A., & Maslinski, W. (2014). Altered expression of V1a receptors mRNA in the brain and kidney after myocardial infarction and chronic stress. *Neuropeptides*, *48*(5), 257–266. https://doi.org/10.1016/j.npep.2014.07.004.

Miller, R. L., Sandoval, P. C., Pisitkun, T., Knepper, M. A., & Hoffert, J. D. (2013). Vasopressin inhibits apoptosis in renal collecting duct cells. *American Journal of Physiology. Renal Physiology*, *304*(2), F177–F188. https://doi.org/10.1152/ajprenal.00431.2012.

Moeller, H. B., Fenton, R. A., Zeuthen, T., & Macaulay, N. (2009). Vasopressin-dependent short-term regulation of aquaporin 4 expressed in Xenopus oocytes. *Neuroscience*, *164*(4), 1674–1684. https://doi.org/10.1016/j.neuroscience.2009.09.072.

Nagashima, T., Horwitz, B., & Rapoport, S. I. (1990). A mathematical model for vasogenic brain edema. *Advances in Neurology*, *52*, 317–326.

Nase, G., Helm, P. J., Enger, R., & Ottersen, O. P. (2008). Water entry into astrocytes during brain edema formation. *Glia*, *56*(8), 895–902. https://doi.org/10.1002/glia.20664.

Niermann, H., Amiry-Moghaddam, M., Holthoff, K., Witte, O. W., & Ottersen, O. P. (2001). A novel role of vasopressin in the brain: Modulation of activity-dependent water flux in the neocortex. *The Journal of Neuroscience*, *21*(9), 3045–3051.

O'Donnell, M. E., Chen, Y. J., Lam, T. I., Taylor, K. C., Walton, J. H., & Anderson, S. E. (2013). Intravenous HOE-642 reduces brain edema and Na uptake in the rat permanent middle cerebral artery occlusion model of stroke: Evidence for participation of the blood-brain barrier Na/H exchanger. *Journal of Cerebral Blood Flow and Metabolism*, *33*(2), 225–234. https://doi.org/10.1038/jcbfm.2012.160.

Okuno, K., Taya, K., Marmarou, C. R., Ozisik, P., Fazzina, G., Kleindienst, A., et al. (2008). The modulation of aquaporin-4 by using PKC-activator (phorbol myristate acetate) and V1a receptor antagonist (SR49059) following middle cerebral artery occlusion/reperfusion in the rat. *Acta Neurochirurgica. Supplement*, *102*, 431–436.

Oliver, G., & Schafer, E. A. (1895). On the Physiological Action of Extracts of Pituitary Body and certain other Glandular Organs: Preliminary Communication. *The Journal of Physiology*, *18*(3), 277–279.

Orcel, H., Albizu, L., Perkovska, S., Durroux, T., Mendre, C., Ansanay, H., et al. (2009). Differential coupling of the vasopressin V1b receptor through compartmentalization within the plasma membrane. *Molecular Pharmacology*, *75*(3), 637–647. https://doi.org/10.1124/mol.108.049031.

Park, J., Masaki, T., Mezaki, Y., Yokoyama, H., Nakamura, M., Maehashi, H., et al. (2017). Alpha-1 antichymotrypsin is involved in astrocyte injury in concert with arginine-vasopressin during the development of acute hepatic encephalopathy. *PLoS One*, *12*(12), e0189346. https://doi.org/10.1371/journal.pone.0189346.

Park, S. J., & Shin, J. I. (2013). Inflammation and hyponatremia: An underrecognized condition? *Korean Journal of Pediatrics*, *56*(12), 519–522. https://doi.org/10.3345/kjp.2013.56.12.519.

Pascale, C. L., Szmydynger-Chodobska, J., Sarri, J. E., & Chodobski, A. (2006). Traumatic brain injury results in a concomitant increase in neocortical expression of vasopressin and its V1a receptor. *Journal of Physiology and Pharmacology*, *57*(Suppl. 11), 161–167.

Pearn, M. L., Niesman, I. R., Egawa, J., Sawada, A., Almenar-Queralt, A., Shah, S. B., et al. (2017). Pathophysiology associated with traumatic brain injury: Current treatments and potential novel therapeutics. *Cellular and Molecular Neurobiology*, *37*(4), 571–585. https://doi.org/10.1007/s10571-016-0400-1.

Peters, H. P., Robben, J. H., Deen, P. M., & Wetzels, J. F. (2007). Water in health and disease: New aspects of disturbances in water metabolism. *The Netherlands Journal of Medicine*, *65*(9), 325–332.

Rajagopal, R., Swaminathan, G., Nair, S., & Joseph, M. (2017). Hyponatremia in traumatic brain injury: A practical management protocol. *World Neurosurgery*, *108*, 529–533. https://doi.org/10.1016/j.wneu.2017.09.013.

Rama Rao, K. V., Jayakumar, A. R., & Norenberg, M. D. (2014). Brain edema in acute liver failure: Mechanisms and concepts. *Metabolic Brain Disease*, *29*(4), 927–936. https://doi.org/10.1007/s11011-014-9502-y.

Rama Rao, K. V., & Norenberg, M. D. (2007). Aquaporin-4 in hepatic encephalopathy. *Metabolic Brain Disease*, *22*(3-4), 265–275. https://doi.org/10.1007/s11011-007-9063-4.

Rauen, K., Trabold, R., Brem, C., Terpolilli, N. A., & Plesnila, N. (2013). Arginine vasopressin V1a receptor-deficient mice have reduced brain edema and secondary brain damage following traumatic brain injury. *Journal of Neurotrauma*, *30*(16), 1442–1448. https://doi.org/10.1089/neu.2012.2807.

Rehberg, S., Ertmer, C., Vincent, J. L., Morelli, A., Schneider, M., Lange, M., et al. (2011). Role of selective V1a receptor agonism in ovine septic shock. *Critical Care Medicine*, *39*(1), 119–125. https://doi.org/10.1097/CCM.0b013e3181fa3898.

Ribeiro, N., Panizza Hdo, N., Santos, K. M., Ferreira-Neto, H. C., & Antunes, V. R. (2015). Salt-induced sympathoexcitation involves vasopressin V1a receptor activation in the paraventricular nucleus of the hypothalamus. *American Journal of Physiology. Regulatory, Integrative and Comparative Physiology*, *309*(11), R1369–R1379. https://doi.org/10.1152/ajpregu.00312.2015.

Robertson, G. L. (2006). Regulation of arginine vasopressin in the syndrome of inappropriate antidiuresis. *The American Journal of Medicine*, *119*(7 Suppl. 1), S36–S42. https://doi.org/10.1016/j.amjmed.2006.05.006.

Robertson, G. L., & Athar, S. (1976). The interaction of blood osmolality and blood volume in regulating plasma vasopressin in man. *The Journal of Clinical Endocrinology and Metabolism*, *42*(4), 613–620. https://doi.org/10.1210/jcem-42-4-613.

Rutkowsky, J. M., Wallace, B. K., Wise, P. M., & O'Donnell, M. E. (2011). Effects of estradiol on ischemic factor-induced astrocyte swelling and AQP4 protein abundance. *American Journal of Physiology. Cell Physiology*, *301*(1), C204–C212. https://doi.org/10.1152/ajpcell.00399.2010.

Saleem, S., Yousuf, I., Gul, A., Gupta, S., & Verma, S. (2014). Hyponatremia in stroke. *Annals of Indian Academy of Neurology*, *17*(1), 55–57. https://doi.org/10.4103/0972-2327.128554.

Salehi, A., Zhang, J. H., & Obenaus, A. (2017). Response of the cerebral vasculature following traumatic brain injury. *Journal of Cerebral Blood Flow and Metabolism*, *37*(7), 2320–2339. https://doi.org/10.1177/0271678X17701460.

Salman, M. M., Kitchen, P., Woodroofe, M. N., Brown, J. E., Bill, R. M., Conner, A. C., et al. (2017). Hypothermia increases aquaporin 4 (AQP4) plasma membrane abundance in human primary cortical astrocytes via a calcium/transient receptor potential vanilloid 4 (TRPV4)- and calmodulin-mediated mechanism. *The European Journal of Neuroscience*, *46*(9), 2542–2547. https://doi.org/10.1111/ejn.13723.

Sasaki, H., & Low, P. A. (1996). Extreme vasoreactivity of rat epineurial arterioles to vasopressin. *The American Journal of Physiology*, *271*(4 Pt. 2), H1307–H1313. https://doi.org/10.1152/ajpheart.1996.271.4.H1307.

Schliebe, N., Strotmann, R., Busse, K., Mitschke, D., Biebermann, H., Schomburg, L., et al. (2008). V2 vasopressin receptor deficiency causes changes in expression and function of renal and hypothalamic components involved in electrolyte and water homeostasis. *American Journal of Physiology. Renal Physiology, 295*(4), F1177–F1190. https://doi.org/10.1152/ajprenal.00465.2007.

Schrier, R. W. (2008). Vasopressin and aquaporin 2 in clinical disorders of water homeostasis. *Seminars in Nephrology, 28*(3), 289–296. https://doi.org/10.1016/j.semnephrol.2008.03.009.

Schrier, R. W., & Goldberg, J. P. (1980). The physiology of vasopressin release and the pathogenesis of impaired water excretion in adrenal, thyroid, and edematous disorders. *The Yale Journal of Biology and Medicine, 53*(6), 525–541.

Shchekochikhin, D., Tkachenko, O., & Schrier, R. W. (2013). Hyponatremia: An update on current pharmacotherapy. *Expert Opinion on Pharmacotherapy, 14*(6), 747–755. https://doi.org/10.1517/14656566.2013.781584.

Sherlock, M., O'Sullivan, E., Agha, A., Behan, L. A., Owens, D., Finucane, F., et al. (2009). Incidence and pathophysiology of severe hyponatraemia in neurosurgical patients. *Postgraduate Medical Journal, 85*(1002), 171–175. https://doi.org/10.1136/pgmj.2008.072819.

Stamatovic, S. M., Dimitrijevic, O. B., Keep, R. F., & Andjelkovic, A. V. (2006). Inflammation and brain edema: New insights into the role of chemokines and their receptors. *Acta Neurochirurgica. Supplement, 96*, 444–450.

Stemmelin, J., Lukovic, L., Salome, N., & Griebel, G. (2005). Evidence that the lateral septum is involved in the antidepressant-like effects of the vasopressin V1b receptor antagonist, SSR149415. *Neuropsychopharmacology, 30*(1), 35–42. https://doi.org/10.1038/sj.npp.1300562.

Strakova, Z., Kumar, A., Watson, A. J., & Soloff, M. S. (1997). A new linear V1A vasopressin antagonist and its use in characterizing receptor/G protein interactions. *Molecular Pharmacology, 51*(2), 217–224.

Szmydynger-Chodobska, J., Chung, I., Kozniewska, E., Tran, B., Harrington, F. J., Duncan, J. A., et al. (2004). Increased expression of vasopressin v1a receptors after traumatic brain injury. *Journal of Neurotrauma, 21*(8), 1090–1102. https://doi.org/10.1089/0897715041651033.

Szmydynger-Chodobska, J., Fox, L. M., Lynch, K. M., Zink, B. J., & Chodobski, A. (2010). Vasopressin amplifies the production of proinflammatory mediators in traumatic brain injury. *Journal of Neurotrauma, 27*(8), 1449–1461. https://doi.org/10.1089/neu.2010.1331.

Takagi, J., Otake, K., Nakao, N., Takashashi, M., & Hirooka, Y. (2003). Urinary excretion of aquaporin-2 and inappropriate secretion of vasopressin in hyponatremic patients after cerebral infarction. *Hormone and Metabolic Research, 35*(1), 62–66. https://doi.org/10.1055/s-2003-38393.

Taya, K., Gulsen, S., Okuno, K., Prieto, R., Marmarou, C. R., & Marmarou, A. (2008). Modulation of AQP4 expression by the selective V1a receptor antagonist, SR49059, decreases trauma-induced brain edema. *Acta Neurochirurgica. Supplement, 102*, 425–429.

Thibonnier, M., Coles, P., Thibonnier, A., & Shoham, M. (2002). Molecular pharmacology and modeling of vasopressin receptors. *Progress in Brain Research, 139*, 179–196.

Trabold, R., Krieg, S., Scholler, K., & Plesnila, N. (2008). Role of vasopressin V(1a) and V2 receptors for the development of secondary brain damage after traumatic brain injury in mice. *Journal of Neurotrauma, 25*(12), 1459–1465. https://doi.org/10.1089/neu.2008.0597.

Tucker, B., Aston, J., Dines, M., Caraman, E., Yacyshyn, M., McCarthy, M., et al. (2017). Early brain edema is a predictor of in-hospital mortality in traumatic brain injury. *The Journal of Emergency Medicine, 53*(1), 18–29. https://doi.org/10.1016/j.jemermed.2017.02.010.

Unterberg, A. W., Stover, J., Kress, B., & Kiening, K. L. (2004). Edema and brain trauma. *Neuroscience*, *129*(4), 1021–1029. https://doi.org/10.1016/j.neuroscience.2004.06.046.

Vakili, A., Kataoka, H., & Plesnila, N. (2005). Role of arginine vasopressin V1 and V2 receptors for brain damage after transient focal cerebral ischemia. *Journal of Cerebral Blood Flow and Metabolism*, *25*(8), 1012–1019. https://doi.org/10.1038/sj.jcbfm.9600097.

Vishram-Nielsen, J. K., & Gustafsson, F. (2017). Vasopressin and vasopressin antagonists in heart failure. *Handbook of Experimental Pharmacology*, *243*, 307–328. https://doi.org/10.1007/164_2017_28.

Walcott, B. P., Kahle, K. T., & Simard, J. M. (2012). Novel treatment targets for cerebral edema. *Neurotherapeutics*, *9*(1), 65–72. https://doi.org/10.1007/s13311-011-0087-4.

Wambo, T. O., Rodriguez, R. A., & Chen, L. Y. (2017). Computing osmotic permeabilities of aquaporins AQP4, AQP5, and GlpF from near-equilibrium simulations. *Biochimica et Biophysica Acta*, *1859*(8), 1310–1316. https://doi.org/10.1016/j.bbamem.2017.04.022.

Wang, G., Guo, Q., Hossain, M., Fazio, V., Zeynalov, E., Janigro, D., et al. (2009). Bone marrow-derived cells are the major source of MMP-9 contributing to blood-brain barrier dysfunction and infarct formation after ischemic stroke in mice. *Brain Research*, *1294*, 183–192. https://doi.org/10.1016/j.brainres.2009.07.070.

Wang, P., Ni, R. Y., Chen, M. N., Mou, K. J., Mao, Q., & Liu, Y. H. (2011). Expression of aquaporin-4 in human supratentorial meningiomas with peritumoral brain edema and correlation of VEGF with edema formation. *Genetics and Molecular Research*, *10*(3), 2165–2171. https://doi.org/10.4238/vol10-3gmr1212.

Wang, H. C., Sun, C. F., Chen, H., Chen, M. S., Shen, G., Ma, Y. B., et al. (2014). Where are we in the modelling of traumatic brain injury? Models complicated by secondary brain insults. *Brain Injury*, *28*(12), 1491–1503. https://doi.org/10.3109/02699052.2014.943288.

Wang, X., Tsuji, K., Lee, S. R., Ning, M., Furie, K. L., Buchan, A. M., et al. (2004). Mechanisms of hemorrhagic transformation after tissue plasminogen activator reperfusion therapy for ischemic stroke. *Stroke*, *35*(11 Suppl. 1), 2726–2730. https://doi.org/10.1161/01.STR.0000143219.16695.af.

Wersinger, S. R., Ginns, E. I., O'Carroll, A. M., Lolait, S. J., & Young, W. S., 3rd. (2002). Vasopressin V1b receptor knockout reduces aggressive behavior in male mice. *Molecular Psychiatry*, 7(9), 975–984. https://doi.org/10.1038/sj.mp.4001195.

Westermaier, T., Stetter, C., Raslan, F., Vince, G. H., & Ernestus, R. I. (2012). Brain edema formation correlates with perfusion deficit during the first six hours after experimental subarachnoid hemorrhage in rats. *Experimental and Translational Stroke Medicine*, *4*(1), 8. https://doi.org/10.1186/2040-7378-4-8.

Wijdicks, E. F. (2016). Hepatic encephalopathy. *The New England Journal of Medicine*, *375*(17), 1660–1670. https://doi.org/10.1056/NEJMra1600561.

Yamazaki, R. S., Chen, Q., Schreiber, S. S., & Brinton, R. D. (1997). Localization of V1a vasopressin receptor mRNA expression in cultured neurons, astroglia, and oligodendroglia of rat cerebral cortex. *Brain Research. Molecular Brain Research*, *45*(1), 138–140.

Zeynalov, E., & Jones, S. M. (2016). Recent advances and future directions in preclinical research of arginine-vasopressin (AVP) receptor blocker conivaptan in the context of stroke. *Neural Regeneration Research*, *11*(3), 392–393. https://doi.org/10.4103/1673-5374.179038.

Zeynalov, E., Jones, S. M., & Elliott, J. P. (2016). Continuous IV infusion is the choice treatment route for arginine-vasopressin receptor blocker conivaptan in mice to study stroke-evoked brain edema. *Journal of Visualized Experiments*, *115*. https://doi.org/10.3791/54170.

Zeynalov, E., Jones, S. M., & Elliott, J. P. (2017). Therapeutic time window for conivaptan treatment against stroke-evoked brain edema and blood-brain barrier disruption in mice. *PLoS One*, *12*(8), e0183985. https://doi.org/10.1371/journal.pone.0183985.
Zeynalov, E., Jones, S. M., Seo, J. W., Snell, L. D., & Elliott, J. P. (2015). Arginine-vasopressin receptor blocker conivaptan reduces brain edema and blood-brain barrier disruption after experimental stroke in mice. *PLoS One*, *10*(8), e0136121. https://doi.org/10.1371/journal.pone.0136121.
Zhang, X., Hense, H. W., Riegger, G. A., & Schunkert, H. (1999). Association of arginine vasopressin and arterial blood pressure in a population-based sample. *Journal of Hypertension*, *17*(3), 319–324.

FSC
www.fsc.org
MIX
Papier | Fördert gute Waldnutzung
FSC® C083411

Kolibri 360 GmbH
Ferdinand-Jühlke-Straße 7
99095 Erfurt, Deutschland
produktsicherheit@kolibri360.de